GELS HANDBOOK

Volume 1

GELS HANDBOOK

Volume 1
The Fundamentals

Editors-in-Chief

Yoshihito Osada and Kanji Kajiwara

Associate Editors

Takao Fushimi, Okihiko Hirasa,
Yoshitsugu Hirokawa, Tsutomu Matsunaga,
Tadao Shimomura, and Lin Wang

Translated by

Hatsuo Ishida

ACADEMIC PRESS

A Harcourt Science and Technology Company

San Diego   San Francisco   New York   Boston
London   Sydney   Tokyo

Copyright © 2001 by Academic Press

ACADEMIC PRESS
A Harcourt Science and Technology Company
525 B Street, Suite 1900, San Diego, CA 92101-4495, USA
http://www.academicpress.com

Academic Press
Harcourt Place, 32 Jamestown Road, London, NW1 7BY, UK

Library of Congress Catalog Number: 00-107106
International Standard Book Number: 0-12-394690-5 (Set)

International Standard Book Number, Volume 1: 0-12-394961-0

Printed in the United States of America
00 01 02 03 04 IP 9 8 7 6 5 4 3 2 1

Contents

Preface   xi
Contributors   xiii

Chapter 2 Theory of Gelation and Preparation of Gels 27

Preface

The development, production, and application of superabsorbent gels is increasing at a remarkable pace. Research involving functional materials in such areas as medical care, medicine, foods, civil engineering, bioengineering, and sports is already widely documented. In the twenty-first century innovative research and development is growing ever more active. Gels are widely expected to be one of the essential solutions to various problems such as limited food resources, environmental preservation, and safeguarding human welfare.

In spite of the clear need for continued gel research and development, there have been no comprehensive references involving gels until now. In 1996, an editorial board led by the main members of the Association of Polymer Gel Research was organized with the primary goal of collecting a broad range of available information and organizing this information in such a way that would be helpful for not only gels scientists, but also for researchers and engineers in other fields. The content covers all topics ranging from preparation methods, structure, and characteristics to applications, functions, and evaluation methods of gels. It consists of Volume 1, The Fundamentals; Volume 2, Functions; Volume 3, Applications; and Volume 4, Environment: Earth Environment and Gels, which consists of several appendices and an index on gel compounds.

Because we were fortunate enough to receive contributions from the leading researchers on gels in Japan and abroad, we offer this book with great confidence. We would like to thank the editors as well as the authors who willingly contributed despite their very busy schedules.

This handbook was initially proposed by Mr. Shi Matsunaga. It is, of course, due to the neverending effort by him and the editorial staff that this handbook was successfully completed. We would also like to express great appreciation to the enthusiasm and help of Mr. Takashi Yoshida and Ms. Masami Matsukaze of NTS Inc.

Yoshihito Osada
Kanji Kajiwara
November, 1997

Contributors

Editors-in-Chief

Yoshihito Osada, *Professor, Department of Scientific Research, Division of Biology at Hokkaido University Graduate School*

Kanji Kajiwara, *Professor, Department of Technical Art in Material Engineering at Kyoto University of Industrial Art and Textile*

Principal Editorial Members

Tadao Shimomura, *President, Japan Catalytic Polymer Molecule Research Center*

Okihiko Hirasa, *Professor, Department of Education and Domestic Science at Iwate University*

Yoshitsugu Hirokawa, *Technical Councilor, Science and Technology Promotional Office, Hashimoto Phase Separation Structure Project*

xiii

Takao Fushimi, *Examiner, Patent Office Third Examination Office at Ministry of International Trade and Industry*

Tsutomu Matsunaga, *Director, Chemistry Bio-Tsukuba*

Lin Wang, *Senior Scientist, P&G Product Development Headquarters*

Ito Takeshi, *Assistant Manager, Tokyo Office Sales and Development Division of Mitsubishi Chemical Co.*

Seigo Ouchi, *Head Researcher, Kanishi Test Farm at Agricultural Chemical Research Center of Sumitomo Chemical Co.*

Mitsuo Okano, *Professor, Tokyo Women's Medical College*

Masayoshi Watanabe, *Assistant Professor, Yokohama National University Department of Engineering, Division of Material Engineering*

Contributors

Aizo Yamauchi, *President, International Research Exchange Center of Japan Society of Promotion for Industrial Technology*

Yoshihito Osada, *Professor, Department of Scientific Research in Biology at Hokkaido University Graduate School*

Hidetaka Tobita, *Assistant Professor, Department of Engineering, Material Chemistry Division at Fukui University*

Yutaka Tanaka, *Research Associate, Department of Engineering, Material Chemistry Division at Fukui University*

Shunsuke Hirotsu, *Professor, Department of Life Sciences and Engineering, Division of Organism Structures at Tokyo Institute of Technology*

Mitsuhiro Shibayama, *Professor, Department of Textiles, Polymer Molecule Division at Kyoto University of Industrial Art and Textile*

Hidenori Okuzaki, *Assistant, Department of Chemistry and Biology, Division of Biological Engineering at Yamanashi University*

Kanji Kajiwara, *Professor, Department of Technical Art in Material Engineering at Kyoto University of Industrial Art and Textile*

Yukio Naito, *Head of Research, Biological Research Center for Kao*

(the late) Kobayashi Masamichi, *Honorary Professor, Department of Science, Division of Polymer Molecular Research at Osaka University Graduate School*

Hidetoshi Oikawa, *Assistant Professor, Emphasis of Research on Higher Order Structural Controls in Department of Reactive Controls at Reactive Chemistry Research Center at Tohoku University*

Yositsugu Hirokawa, *Technical Councilor, Science and Technology Promotional Office, Hashimoto Phase Separation Structure Project*

Makoto Suzuki, *Professor, Department of Engineering, Division of Metal Engineering at Tohoku University Graduate School*

Ken Nakajima, *Special Research, Division of Basic Science in International Frontier Research System Nano-organic Photonics Material Research Team at Physics and Chemistry Research Center*

Toshio Nishi, *Professor, Department of Engineering Research, Division of Physical Engineering at Tokyo University Graduate School*

Hidemitsu Kuroko, *Assistant Professor, Department of Life Environment, Division of Life Environment at Nara Women's University*

Shukei Yasunaga, *Assistant, Department of Technical Art in Material Engineering at Kyoto University of Industrial Art and Textile*

Mitsue Kobayashi, *Special Researcher, Tokyo Institute of Technology*

Hajime Saito, *Professor, Department of Science, Division of Life Sciences at Himeji Institute of Technology*

Hazime Ichijyo, *Manager of Planning Office, Industrial Engineering Research Center in Department of Industrial Engineering, Agency of Industrial Science and Technology at Ministry of International Trade and Industry*

Masayoshi Watanabe, *Assistant Professor, Yokohama National University Department of Engineering, Division of Material Engineering*

Kunio Nakamura, *Professor, Department of Agriculture, Division of Food Sciences at College of Dairy Agriculture*

Hideo Yamazaki, *Shial, Inc. (Temporarily transferred from Tonen Chemical Co.)*

Koshibe Shigeru, *Shial, Inc. (Temporarily transferred from Tonen Chemical Co.)*

Hirohisa Yoshida, *Assistant, Department of Engineering, Division of Industrial Chemistry at Tokyo Metropolitan University*

Yoshiro Tajitsu, *Professor, Department of Engineering at Yamagata University*

Hotaka Ito, *Instructor, Division of Material Engineering at National Hakodate Technical High School*

Toyoaki Matsuura, *Assistant, Department of Opthamology at Nara Prefectural Medical College*

Yoshihiko Masuda, *Lead Researcher, Third Research Division of Japan Catalytic Polymer Molecule Research Center*

Toshio Yanaki, *Researcher, Shiseido Printed Circuit Board Technology Research Center*

Yuzo Kaneko, *Department of Science, Division of Applied Chemistry at Waseda University*

Kiyotaka Sakai, *Professor, Department of Science, Division of Applied Chemistry at Waseda University*

Teruo Okano, *Professor, Medical Engineering Research Institute at Tokyo Women's Medical College*

Shuji Sakohara, *Professor, Department of Engineering, Chemical Engineering Seminar at Hiroshima University*

Jian-Ping Gong, *Assistant Professor, Department of Scientific Research, Division of Biology at Hokkaido University Graduate School*

Akihiko Kikuchi, *Assistant, Medical Engineering Research Institute at Tokyo Women's Medical College*

Shingo Matukawa, *Assistant, Department of Fisheries, Division of Food Production at Tokyo University of Fisheries*

Kenji Hanabusa, *Assistant Professor, Department of Textiles, Division of Functional Polymer Molecules at Shinshu University*

Ohhoh Shirai, *Professor, Department of Textiles, Division of Functional Polymer Molecules at Shinshu University*

Atushi Suzuki, *Assistant Professor, Department of Engineering Research, Division of Artificial Environment Systems at Yokohama National University Graduate School*

Junji Tanaka, *Department of Camera Products Technology, Division Production Engineering, Process Engineering Group at Optical Equipment Headquarters at Minolta, Inc.*

Eiji Nakanishi, *Assistant Professor, Department of Engineering, Division of Material Engineering at Nagoya Institute of Technology*

Ryoichi Kishi, *Department of Polymer Molecules, Functional Soft Material Group in Material Engineering Technology Research Center in Agency of Industrial Science and Technology at Ministry of International Trade and Industry*

Toshio Kurauchi, *Director, Toyota Central Research Center*

Tohru Shiga, *Head Researcher, LB Department of Toyota Central Research Center*

Keiichi Kaneto, *Professor, Department of Information Technology, Division of Electronic Information Technology at Kyushu Institute of Technology*

Kiyohito Koyama, *Professor, Department of Engineering, Material Engineering Division at Yamagata University*

Yoshinobu Asako, *Lead Researcher, Nippon Shokubai Co. Ltd., Tsukuba Research Center*

Tasuku Saito, *General Manager, Research and Development Headquarters, Development Division No. 2 of Bridgestone, Inc.*

Toshihiro Hirai, *Professor, Department of Textiles, Division of Raw Material Development at Shinshu University*

Keizo Ishii, *Manager, Synthetic Technology Research Center at Japan Paints, Inc.*

Yoshito Ikada, *Professor, Organism Medical Engineering Research Center at Kyoto University*

Lin Wang, *Senior Scientist, P&G Product Development Headquarters*

Rezai E., *P&G Product Development Headquarters*

Fumiaki Matsuzaki, *Group Leader, Department of Polymer Molecule Science Research, Shiseido Printed Circuit Board Technology Research Center*

Jian-Zhang (Kenchu) Yang, *Researcher, Beauty Care Product Division of P&G Product Development Headquarters*

Chun Lou Xiao, *Section Leader, Beauty Care Product Division of P&G Product Development Headquarters*

Yasunari Nakama, *Councilor, Shiseido Printed Circuit Board Technology Research Center*

Keisuke Sakuda, *Assistant Director, Fragrance Development Research Center at Ogawa Perfumes, Co.*

Akio Usui, *Thermofilm, Co.*

Mitsuharu Tominaga, *Executive Director, Fuji Light Technology, Inc.*

Takashi Naoi, *Head Researcher, Ashikaga Research Center of Fuji Film, Inc.*

Makoto Ichikawa, *Lion, Corp. Better Living Research Center*

Takamitsu Tamura, *Lion, Corp. Material Engineering Center*

Takao Fushimi, *Examiner, Patent Office Third Examination Office at Ministry of International Trade and Industry*

Kohichi Nakazato, *Integrated Culture Research Institute, Division of Life Environment (Chemistry) at Tokyo University Graduate School*

Masayuki Yamato, *Researcher, Doctor at Japan Society for the Promotion of Science, and Japan Medical Engineering Research Institute of Tokyo Women's Medical College*

Toshihiko Hayasi, *Professor, Integrated Culture Research Institute, Division of Life Environment (Chemistry) at Tokyo University Graduate School*

Naoki Negishi, *Assistant Professor, Department of Cosmetic Surgery at Tokyo Women's Medical College*

Mikihiro Nozaki, *Professor, Department of Cosmetic Surgery at Tokyo Women's Medical College*

Yoshiharu Machida, *Professor, Department of Medical Pharmacology Research at Hoshi College of Pharmacy*

Naoki Nagai, *Professor, Department of Pharmacology at Hoshi College of Pharmacy*

Kenji Sugibayashi, *Assistant Professor, Department of Pharmacology at Josai University*

Yohken Morimoto, *Department Chair Professor, Department of Pharmacology at Josai University*

Toshio Inaki, *Manager, Division of Formulation Research in Fuji Research Center of Kyowa, Inc.*

Seiichi Aiba, *Manager, Department of Organic Functional Materials, Division of Functional Polymer Molecule Research, Osaka Industrial Engineering Research Center of Agency of Industrial Science and Technology at Ministry of International Trade and Industry*

Masakatsu Yonese, *Professor, Department of Pharmacology, Division of Pharmacology Materials at Nagoya City University*

Etsuo Kokufuta, *Professor, Department of Applied Biology at Tsukuba University*

Hiroo Iwata, *Assistant Professor, Organism Medical Engineering Research Center at Kyoto University*

Seigo Ouchi, *Head Researcher, Agricultural Chemical Research Center at Sumitomo Chemical Engineering, Co.*

Ryoichi Oshiumi, *Former Engineering Manager, Nippon Shokubai Co. Ltd. Water-absorbent Resin Engineering Research Association*

Tatsuro Toyoda, *Nishikawa Rubber Engineering, Inc. Industrial Material Division*

Nobuyuki Harada, *Researcher, Third Research Division of Japan Catalytic Polymer Molecule Research Center*

Osamu Tanaka, *Engineering Manager, Ask Techno Construction, Inc.*

Mitsuharu Ohsawa, *Group Leader, Fire Resistance Systems Group of Kenzai Techno Research Center*

Takeshi Kawachi, *Office Manager, Chemical Research Division of Ohbayashi Engineering Research Center, Inc.*

Hiroaki Takayanagi, *Head Researcher, Functional Chemistry Research Center in Yokohama Research Center of Mitsubishi Chemical, Inc.*

Yuichi Mori, *Guest Professor, Department of Science and Engineering Research Center at Waseda University*

Tomoki Gomi, *Assistant Lead Researcher, Third Research Division of Japan Catalytic Polymer Molecule Research Center*

Katsumi Kuboshima, *President, Kuboshima Engineering Company*

Hiroyuki Kakiuchi, *Mitsubishi Chemical, Inc., Tsukuba Research Center*

Baba Yoshinobu, *Professor, Department of Pharmacology, Division of Pharmacological Sciences and Chemistry at Tokushima University*

Toshiyuki Osawa, *Acting Manager, Engineer, Thermal Division NA-PT at Shotsu Office of Ricoh, Inc.*

Kazuo Okuyama, *Assistant Councilor, Membrane Research Laboratory, Asahi Chemical Industry Co., Ltd.*

Takahiro Saito, *Yokohama National University Graduate School, Department of Engineering, Division of Engineering Research*

Yoshiro Sakai, *Professor, Department of Engineering, Division of Applied Chemistry at Ehime University*

Seisuke Tomita, *Managing Director, Development and Production Headquarters at Bridgestone Sports, Inc.*

Hiroshi Kasahara, *Taikisha, Inc. Environment System Office*

Shigeru Sato, *Head Researcher, Engineering Development Center at Kurita Engineering, Inc.*

Okihiko Hirasa, *Professor, Iwate University*

Seiro Nishio, *Former Member of Disposable Diaper Technology and Environment Group of Japan Sanitary Material Engineering Association*

VOLUME 1

The Fundamentals

CHAPTER 1

Definition and Classification of Gels

Chapter contents

Section 1
Gels: Introduction

AIZO YAMAUCHI

1.1 WHAT ARE GELS?

According to the newly edited version of the *Polymer Dictionary* [3], gels are defined as "polymers and their swollen matters with three-dimensional network structures that are insoluble in any solvents." Also see References [1] and [2] for what constitutes a gel. The dictionary also states that the relationship between the crosslinked structure of gels and characteristics of swelling matter is one in which "linear and branched polymers absorb, swell, and eventually disperse as individual molecules into good solvents. On the other hand, the degree of swelling of crosslinked polymers is limited due to the three-dimensional network structure, although they can swell by the interaction with the solvent." It also notes that gels exist under special conditions not found in solids, liquids, and gases: "swollen gels that have absorbed a large amount of solvent are in the states between solids and liquids and their properties change from viscous liquids to hard solids according to the chemical composition and other factors." Gelation occurs via crosslinking. However, crosslinking does not necessarily require covalent bond formation. It can also be achieved by secondary forces as seen in hydrogen bonding. Thus, gels are crosslinked three-

4

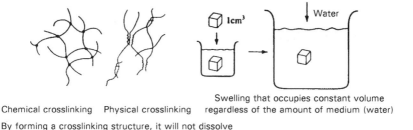

Chemical crosslinking   Physical crosslinking   Swelling that occupies constant volume regardless of the amount of medium (water)

By forming a crosslinking structure, it will not dissolve

(a) Crosslinked structure   (b) Swelling

Sugar cube   Gel   Sugar cube

Dissolves (liquid)   Swelling (gel)   Insoluble (solid)

(c) Intermediate between liquid and solid

Fig. 1 Schematic diagrams of solution, gel, and solid.

dimensional (3D) networks, absorb solvents, and swell to a limited degree without dissolution. In addition, they exist in states that are somewhere between a solid and a liquid. See Fig. 1 for elucidation.

1.2 CLASSIFICATION

A gel consists of both a 3D network structure and a medium. Although it is possible for the medium to be a gas, it is generally a liquid. It will be considered here that gels are 3D network polymers swollen by liquids (solvents). Gels are classified based on the type of crosslinking that creates their 3D networks, as well as whether they are natural or artificial, the shape and size of the gel configuration, and the types of solvents. Table 1 presents many of the classifications of gels to which this work will refer.

1.2.1 Method of Crosslink Formation

Gels can be formed via covalent bonding, the coulombic force, hydrogen bonding, and coordination bonding, and although physical interaction

Table 1   Classification of gels.

Crosslinking system (semi-crosslinking)	Covalent bonding	Crosslinking agent · chemical cross linking
	Coulombic force	Light · radiaton
	Hydrogen bonding	Polyelectrolytes
	Coordination bonding	Natural gels, frozen gels
	Entanglement	Small molecules · ions
		High degrees of polymerization, branched, weak in strength
Structural polymers	Natural gels	Food, protein, polysaccharides, tissues living in the natural world
	Hybrid gels	Medical materials, artificial skin, artificial corneas, artificial pancreas model
	Synthetic gels	Organic polymer, contact lenses, high water absorbent resins, silica gels
Configuration size	Micro-gel	Intramolecular crosslinking: does not enlarge ,
	Macro-gel	Intermolecular crosslinking: normal gel
Solvent	Air	Aerogel, Xerogel
	Water	Hydrogel
	Oil	Lyopic (organo) gel

(such as entanglement) is not a form of chemical crosslinking, gels can be formed this way. Most synthetic matter is of the covalent bonding type, generally polymerized by adding crosslinking agents when polymers are synthesized. The crosslinked structure is strong. Those gels using coulombic force arc poly-ion complexes between polyelectrolytes with opposite charges or polyelectrolytes with multivalent ions, such as calcium.

The gels liquefy and a sol-gel transition occurs when the coulombic force is overcome by changing either pH or ionic strength. The hydrogen bonding system, created when crosslinked structures are formed by hydrogen bonding between the polymers, is seen mostly in natural matter. In these cases, gels will change to sols when hydrogen bonding is broken by the environment, for example, by heating. Coordination bonding is created when crosslinked structures are formed between polar groups of a polymer and the material that is coordinated. Situations that do not involve crosslinking were mentioned previously. Polymer chains become entangled and a gel is created when the molecular weight of the polymer is large or there is branching. In this case, the crosslink point is not specified and the gel configuration is unstable. As bonding strength is weak, in time it will disperse in the solvent and dissolve.

1.2.2 Natural Gels and Synthetic Gels

Typical gelled foods from ancient times are gels that were created using natural materials. Because they are made of natural polymers, there are many forms of polysaccharides and proteins such as in kon-nyaku (a gelatinous cake made of the root of devil's tongue) and tofu. On the other hand, the superabsorbent disposable diapers, perfumes, and contact lenses we use daily are gels made of water soluble, synthetic polymers such as poly(vinyl alcohol) and poly(hydroxyethyl methacrylate). It is possible for them to be made of either a single component or multicomponents. There has been much attention given to the hybrid gels used in medicine and biotechnology due to their improved compatibility with bioorganisms, which allow organic components to assimilate synthetic materials, as well as for special bioorganic functions to be added to gels. As organisms, excluding hard structures like bones and teeth, are composed mostly of water and yet can maintain their shapes in water, it is not an exaggeration to state that human bodies are made mostly of gels. This can be understood if one examines the cornea, lens, or vitreous humor of the eye, or neuronal tissue, all of which are gels. Modern medicine anticipates using hybrid gels capable of fusing, with an organism. These include, but are not limited to, artificial skin [4] and corneas [5].

1.2.3 The Crosslink Structure and Size

Although 3D crosslinked structures are usually shown using a lattice model, in reality, polymers can be rather spread out or coiled up into a ball, depending on the relationship between them and their solvent (one such relationship is the solubility parameter). There are heterogeneous gels in which crosslinked units are locally concentrated and homogeneous gels with evenly distributed crosslinked units. The conceptual diagram for crosslinked structures is shown in Fig. 2. When the size of a gel network is to be given, the average molecular weight between the crosslinked units can be used. The spatial size of a substance can also be determined by measuring the permeation of water and solutes in the medium. In the former, there are calculation methods for degree of swelling that use the Flory formula or the dynamic modulus. The latter is more practical as it is

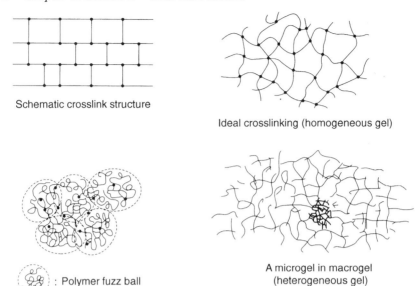

Schematic crosslink structure

Ideal crosslinking (homogeneous gel)

: Polymer fuzz ball

A microgel in macrogel
(heterogeneous gel)

Fig. 2   Schematic diagram of crosslink points.

often important for actual applications of gels, such as the permeability of membranes and their control release properties. There are equations to calculate these quantities by measuring diffusion through membrane of penetrants of known size, such as cells, enzymes, certain proteins and dyes, or the permeability of water under pressure.

Table 2 provides comparisons of network size using the gel membrane of a relatively homogeneous network structure. This network structure was crosslinked by the free radical reaction initiated by γ-ray

Table 2   Size of networks (nm) [6].

Swelling ratio[a]	Pore radius from water permeability	Average distance between crosslink points	
		From swelling ratio	From modulus
8.77	1.7	18	11
11.9	1.9	27	16
15.3	2.5	36	21
17.8	2.7	43	25

[a]Weight of swelling gel/weight of dried gel

irradiation on a poly(vinyl alcohol) aqueous solution. The mean distance between crosslinked units almost matches, but the size calculated from the water permeability is one order of magnitude smaller. The point of this calculation was to obtain the mean radius of the pores when water passes through the hydrated polymer fuzz balls.

1.2.4 Microgels and Macrogels

With respect to the size of gels, there are microgels and macrogels. In a dilute solution, the distance to other molecules is larger, making intermolecular crosslinking more difficult. This creates intramolecular crosslinking of the same polymer chain and crosslinking among several molecules. The spreading of molecules due to crosslinking leads to local concentration fluctuations. Subsequently, the probability of contact with polymer chains becomes less, thereby creating microgels. The compatibility of microgels with solvents diminishes as crosslinking progresses. These microgels sometimes precipitate as ultrasmall particles. However, regardless of whether they precipitate or not, microgels consist of one to several polymer molecules in which each of the polymers holds crosslinked units. This makes them fundamentally different from polymer solutions (sols) that are homogeneously dissolved in the solvent as molecules, without having crosslinked units. These microgels are dispersed in a medium and have been used as the basic material in paints. Painting a surface is easier because the particles adhere and create a strong film after drying. Gels are typically bulks made of, in extreme cases, single, giant molecules; such gels are called macrogels. These giant molecules consist of polymer chains connected by crosslinking.

The distribution of crosslinked units inside macrogels is not uniform. For example, when crosslinking is induced by light, the higher energy flux on the surface than on the inside tends to create higher crosslink density on the outside. This makes it possible to create a skin-core morphology in the gel. When a gel is formed by simultaneous polymerization and crosslinking of a monomer and crosslinking agent, it is necessary to consider carefully the reaction conditions, including concentration and temperature. Especially important is the reactivity ratio of the monomer and crosslinking agent. When their reactivity is very different, a preferential crosslinking reaction takes place at the beginning and end of polymerization and localized crosslinks appear in the macrogel. This is schematically shown in Fig. 3. This is called microgelation within a

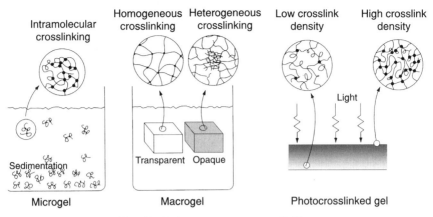

Fig. 3   Heterogeneous crosslinking.

macrogel, and is one reason why gels become heterogeneous. In this case, when the microgels become heterogeneous on the order of the wavelength of light, the light scatters. This creates an opaque, fragile gel even if the homogeneous gel is clear. As the network size is uneven, it is then important to be cautious when the goal is to use the separation by network effect (i.e., preparation of an electrophoresis gel).

1.2.5   Medium

When water is the medium for a gel, it is called a hydrogel. However, most gels, whether natural or synthetic, are composed of water. Therefore, when speaking of gels, we generally refer to water-medium gels. In addition, there are lyopic gels (also called organogels), which are created from an oil-medium such as silicone oil. These have gained attention in recent years as a shock absorption material in tennis shoes. In a broader sense, there are aerogels or xerogels, which use air as the medium; these include silica gels, dried agar and Kouya tofu (freeze-dried tofu). However, these are not normally considered to be gels.

1.3   THREE-DIMENSIONAL CROSSLINK STRUCTURE AND NETWORKS

It is necessary for gels to have intermolecular crosslink structures of polymers, that is, polymer networks. These networks can range in size from a large scale of 10^3–10^6 m (Internet-sized) to human networks used for direct interaction at 1–10 m, all the way down to a network of 10^{-3}–1 m for daily necessities such as nylon stockings, textiles, bamboo baskets, and wire netting. A polymer network is very fine at 10^{-9}–10^{-6} m. The universal function of any network is to incorporate correct matter. On a large scale the Internet has enveloped the earth and, on a small scale, bird cages protect birds by keeping them in. Similarly, gel networks create very small spaces and special environments. In these spaces, various functions are observed in any organism in which gels originate; they include filtering, diffusion, and atomic or molecular order interactions between the polymer chains and the enclosed solute or solvent. Inside the 3D crosslinked structure, there is a microspace where closely packed polymer chains, solvent, and solute coexist.

Although it is not possible to prepare a polymer solution with a high concentration (50–70%) using high molecular weight polymers, it is possible in a gel system to study the interaction of polymers in a high density system with solvent or solute by taking advantage of gels that exist in a state between liquid and solid. Especially in organisms, purposeful and precise structures are formed and high performance can be obtained

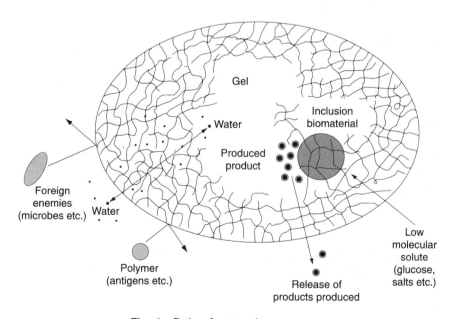

Fig. 4 Role of network structures.

because biopolymer synthesis, crosslinking, and gelation take place simultaneously in organisms. In a hybrid gel with bioorganisms, for example, it is possible to study an artificial pancreas where rejection is avoided by the insertion of a cell colony (i.e., the islets of Langerhans) while insulin is released into the body using gel networks. Accordingly, the gel's precise network structure is one that presents a wall to the outside world and a channel that allows the necessary materials to flow in. This network also serves as a control in an internal microscopic environment (see Fig. 4).

1.4   OTHER USES

The most famous gel is the soft contact lens developed in Czechoslovakia in 1960 using poly(hydroxyethyl methacrylate). In the late 1970s, super-absorbent polymers were developed in the United States and their use in feminine products was worldwide. In 1980, Professor Toyoichi Tanaka discovered phase transition in gels. Today many researchers study those transitions from the basics to applications. Gel applications or functional uses due to phase transitions are found in the chapters that follow and will not be discussed here.

REFERENCES

1   Yamauchi, A., and Hirokawa, Y. (1990). *Functional Gels*, *New Polymeric Materials*. Society for High Polymers, ed., Kyoritsu Publ. p. 3.
2   Ogino, I., Osada, Y., Fushimi, T., and Yamauchi, A. (1991). *Gels*, Sangyo Tosho, p. 3.
3   *New Edition of Polymer Dictionary* (1988). Society for High Polymers, ed., Asakura Shoten, p. 129.
4   Kuroyanagi, Y. (1995). Frontier of artificial skin, *Kobunshi*, **44**: 570.
5   Nakao, H., Matsuda, T., Saishin, M., *et al.* (1993; 1994). Development of hybrid artificial cornea 1,2,3, *Ganki*, **44**: 247; **44**: 1107; **45**: 614.
6   Yamauchi, A. (1977). Composition and structure of gels, *Protein*, *Nucleic Acid and Enzymes*, **22**: (13), 46.

Section 2

Polymer Gels: Crosslink Formations

YOSHIHITO OSADA

2.1 CHARACTERISTICS OF POLYMER GELS

Polymer gels are created from two components, polymer networks and solvents. The polymer network envelops the liquid and prevents it from escaping. In other words, the gel polymer network plays the role of a container that holds a large amount of solvent. Generally, it is thought that gels have the characteristics of both liquids and solids. A very swollen gel acts as a liquid because the diffusion coefficient of small molecules is very high. Gels also can have the properties of a soft solid. This is seen in gels that can be picked up and change shape when they are cut or force is applied to them.

Polymer gels are different from normal solids and liquids and show various characteristics and behaviors. It is known, for example, that the water in gels exists in several different forms: non-freezable water even at very low temperature that exists close to the network and has a strong interaction with the network; bound water that freezes at -10 to $-20°C$; and free water that has the same properties as normal water. There is also the phase transition phenomenon in which a gel is nonlinear. Phase transitions caused by solvent composition, temperature changes, pH

13

changes, ion composition changes in the gel, and electrical fields have been reported [1, 2].

The properties of a polymer gel depend largely on the structure of the polymer network that makes up the gel and the interaction of the network and the solvent. The polymer network's mobility is restricted by its crosslink structure. However, gels possess great mobility because the polymer networks are solvated by a large amount of trapped solvent. The macromolecules that constitute gels spread into a 3D space and exhibit great mobility. Another important characteristic of gels is that they themselves are open materials that are in a nonequilibrium state. Gels are considered open when energy, materials, and information can be exchanged with the outside world. They also function as a place where chemical reactions can take place. Gels respond to outside environments, and exhibit their own characteristics and functions by changing shapes and states. In addition, gels also exhibit electrical conductivity, stereoregularity, and responsivity to external stimuli such as pH, heat, light, and electrical fields depending on the polymer that constitutes the gel. Recent research has focused on creating stimuli-responsive polymers by using the characteristics of polymer gels.

There are various methods of classifying polymer gels. Generally, they are grouped into three categories:

1. classification based on liquids that fill 3D networks;
2. classification based on polymers that form gels; and
3. classification based on the formation method of polymer networks.

Categories 1 and 2 were discussed in subsection 1 and category 3 is discussed in what follows.

2.2   CLASSIFICATION BY CROSSLINK FORMATIONS

The classification of crosslink formation can be largely divided into those formed by a chemical reaction (chemical gels) and those formed by aggregation caused by hydrogen bonding or ionic bonding, and by the physical entanglement of polymer chains (physical gels) [1, 2]. Figure 1 shows an example of crosslinking formation. Generally, gels formed by chemical bonding cannot be dissolved again, and are thus called irreversible gels. Physical gels, on the other hand, create gels in a reversible way, by changes in temperature, composition of solvent, and pH, and are thus

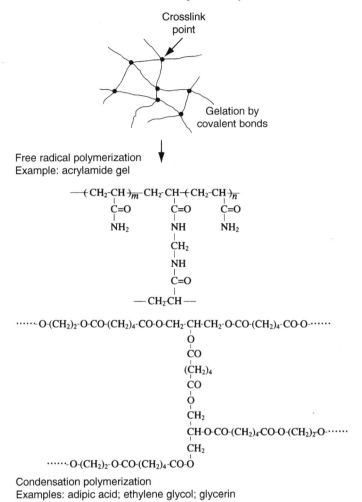

Fig. 1 Examples of chemical gels.

called reversible gels. Many of the natural polymer gels fall into this category. When forming a network using chemical bonding, the branching of the polymer chain becomes crosslinked and that becomes the crosslink point. On the other hand, when crosslinked, these polymer chains seen in natural polymer gels resemble the microcrystals that are seen in physical gels.

When this happens, which is due to a tendency for the crystal to align, the concept of a junction zone (which is a 1D structure) is involved.

When the gel forms, it is possible to go from one edge of the gel to the other by going through a crosslinked unit (or branched area), following another polymer chain going through a different crosslinked unit. This is called percolation. The probability Ω of finding a crosslink point of the polymer chain to which another polymer chain is connected through crosslinking can be given as $\Omega = \varepsilon(x - 1) = \varepsilon x$ where x is the degree of polymerization and ε is the concentration of the crosslinked chemical repeat unit. Generally, the critical branching index α_c is given by $\alpha_c = 1/(f - 1)$ where f is the number of polymer chains that pass through a crosslink point or region. In this situation, problems of polymer concentration and loop formation have not been considered. When these issues are considered, the theory is known as percolation theory and will be discussed in Chapter 2.

2.2.1   Chemical Gels

In the creation of network structures by chemical bonding (covalent bonding), there is a method of: (1) crosslinking at the same time as polymerization; or (2) crosslinking by chemical reaction after linear polymer chains have been synthesized. The latter method can be further divided into the addition polymerization in the presence of divinyl compounds (radical polymerization, anionic polymerization, ionic polymerization, etc.) or the formation of crosslinked structures by polycondensation of multifunctional compounds. In the addition reaction, free radical polymerization is generally utilized. In this free radical polymerization method, initiators are usually used, but light, radiation, and plasmas can also be used.

2.2.1.1   Crosslinking methods during polymerization reactions

2.2.1.1.1   Creating gels using condensation polymerization

Gels made by polycondensation polymerization utilize multifunctional monomers. In the polycondensation reaction, polymers gradually build molecular networks and when their molecular weights become infinite, gels are formed. For example, the polycondensation reaction of bifunctional monomers such as ethylene glycol and adipic acid will form linear polymers. It is possible, however, to create gels that have 3D network structures by adding some monomers that are more than trifunctional (i.e., glycerin). Flory [3], Stockmayer [4] and Case [5] have reported on this type of gelation. According to Flory, the critical condition to obtain 3D network structures during a polycondensation reaction is suggested to be $\alpha_c = 1/(f - 1)$, where α_c is the critical branching index for gelation and f is the number of the monomer's functional group. At the same time, the relationship between α, reaction rate, and reactant ratio has been reported. It is shown as $\alpha = P^2 \rho/\{R - P^2(1 - \rho)\}$, where P is the reaction rate of an acid or alcohol group, R is the reactant ratio, and ρ is the ratio of monomer in branch units against the total number of monomers. These two equations can predict the extent of reaction at the gel point in the polycondensation reaction of bifunctional monomers with multifunctional monomers if the feed ratio of monomers is known. When a trifunctional molecule such as glycerin and a bifunctional molecule such as succinic acid are mixed together, a gel that has 3D networks is created. However, as these types of gels possess high crosslink density, it is difficult for them to dissolve into solvents. Alkyd resin is known as a thermosetting resin. If, for example, glycerin, lauric acid and phthalic acid are condensed together, a gel is made. This gel would not work as a molding compound, but would be used mainly as a coating material. As mentioned here, these are condensation reactions caused by ester bonds, but there are also gels made by amide bonding. For example, when making a reverse osmosis film, a gel film made by the condensation of diamine compounds and tricarboxylic acid is used.

2.2.1.1.2 Creating gels using free radical polymerization

(a) Thermal Polymerization (Polymerization Using Free Radical Initiators) It is possible to obtain gels by polymerizing an appropriate monomer, crosslinking agent, and initiator in a good solvent of the polymer, as in the example of vinyl monomers in the presence of divinyl compounds. Ethylene glycol dimethacrylate and methylene bisacrylamide are often used as crosslinking agents. Free radical initiators start polymerization at a low temperature, so there are times when redox initiators are occasionally used. In these initiators, there is an optimum temperature (high-temperature initiator, normal initiator, low-temperature (redox) initiator and very low-temperature initiator) that creates radicals. Therefore, it is important to select the initiators by matching them with gel synthesis conditions.

When creating a gel with a uniform network structure, it is important to select a combination where the reactivities of the monomer and the crosslinking agent are similar. It is ideal to have monomers with copolymer reactivity ratios $r_1 = r_2 = 1$, as such a combination leads to random polymerization. However, it is not easy to find such a combination. Generally, it would mean looking for the combination of a monomer with a similar reaction ratio and a crosslinking agent, but the above-mentioned relationship is possible only at a low degree of polymerization. When the degree of polymerization is increased, a slight difference between r_1 and r_2 increases the difference in the consumption rates of the monomer and the crosslinking agent. It is necessary to use caution, as this offsets crosslink density and distribution.

(b) Radiation Polymerization   In the radiation normally used in radiation polymerization, there is a γ ray with low output but high permeability, and an electron beam with low permeability but high output possibilities. It is possible to create hydrogels by radiation polymerizing water-based monomers such as acrylic amide, vinyl pyrrolidone, and 2-hydroxyethyl methacrylic acid. The radiation polymerization method is superior in the following ways [6]:

1. chemical initiators are unnecessary, creating a uniform gel;
2. low-temperature polymerization is possible, making the fixing of bioactive substances possible;
3. it has good repeatability; and
4. it is economical.

However, there are monomers that do not polymerize or self-crosslink.

(c) Photopolymerization   There are two methods of polymerization using light. One is to irradiate the monomer in the presence of a crosslinking agent using the light that matches the vinyl monomer's absorption wavelength. A photosensitizer is a material that produces free radicals by using UV or visible light. Those compounds that produce active free radicals are peroxides or azobis-based compounds. The other method is to shine light with an added photosensitizer. The creation of gels using light does not work for mass production, but the reaction proceeds at low temperature. It is, therefore, suitable for immobilizing materials (such as enzymes) that are changed by heating.

(d) Plasma Polymerization   Polymer gels that possess high water absorption and significant metal ion adsorption ability can be made by using water-soluble ultrahigh molecular weight polymers that are created by the plasma-initiated polymerization of aqueous solutions. For example, if a

gel is produced by 2-acrylamide-2-methylpropane sulfonic acid in the presence of a crosslinking agent, the gel obtained absorbs 3000 times the gel's dry weight. This gel also exhibits excellent absorption characteristics for heavy metal ions, such as cobalt and chrome, or a suspension of bentonite. However, the monomer's reactivity under plasma polymerization is unique. Acrylic acid, methacrylic acid, and acrylamide will polymerize but acrylonitrile and styrene will not. See Reference [7] for more information.

2.2.1.2 Methods of crosslinking among polymer chains

2.2.1.2.1 The method for crosslinking using the functional group of polymer

A polymer containing isocyanate ($R-N=C=O$) can undergo polyaddition reactions with such compounds as diols or diamines, crosslink and become a gel. This polyaddition reaction results in urethane and urea bonds. There are other functional groups that can create gels if the crosslinking agent is selected. Several of these possibilities will be noted later. This crosslinking method can achieve the creation of gels with uniform networks, compared to methods in which linking is done simultaneously with polymerization. (see Table 1) [8].

2.2.1.2.2 Radiation crosslinking

By irradiating using γ-rays, it is possible to crosslink poly(vinyl alcohol) [9], poly(methyl vinyl ether), polyethylene, polystyrene, polyacrylate and natural rubber in water. The crosslinking is considered to take place by

Table 1 Examples of a crosslinking agent in relation to a polymer's functional group.

Crosslinking agent	Type of polymer
Dialdehyde compounds	Polymers with amino or hydroxyl group
Amine compounds	Halogen-type polymers, carboxylic polymer ester, isocyanate, epoxy bases, polymers with methylol base
Aziridine compounds	Polymers with carboxyl base
Di or polymethylol phenolic resin	Polymers with nitril base, mercapto base, carboxyl base
Halogen compounds	Polymers with amine, diene
Di or polyisocyanate compounds	Polymers with active hydrogen like $-OH$, $-SH$, $-NH_2$, $-COOH$
Alcohols like diol, polyol, bisphenol	Polymers and cellulose chlorosulfonate bases and isocyanate bases
Diepoxy compound	Polymers with carboxyl bases, hydroxide bases, mercapto base, chlorosulfonate base

splitting water molecules by radiation, extracting the hydrogen from the main chain by the resultant free radicals, and coupling the polymer free radicals formed. Radiation crosslinking is affected by the following: (1) condition of the irradiation (total dose, rate of irradiation, test specimen size, temperature, pressure, etc.); and (2) properties of the polymer (multiple bonding, etc.).

2.2.1.2.3   Photo crosslinking

This method is suited to large amounts of crosslinking and low swelling polymer gels. Examples are known of water soluble polymers, such as poly(vinyl alcohol) and poly(N-vinyl pyrrolidine), being photocrosslinked via diazo resins, bisazides, chromic acid, and photodimerization of polymers having photosensitivity, such as styrenebazolium salt, on the water soluble polymer main chain.

2.2.1.2.4   Plasma crosslinking

It is possible to crosslink the surface of polymers or their thin films, such as polyethylene, poly(tetrafluoroethylene) and nylon, by contact with an inert gas that has been excited by ultrasound. This method, called CASING processing, is suited for creating gels with low solvent affinity and high crosslink density. It is also a crosslinking method suited for adding stickiness to polymer surfaces.

2.2.2   Physical Gels

It is possible to form gels through physical crosslinking among polymer networks using hydrogen bonding, ionic or chelate formation.

A network structure that uses secondary forces is easy to create, but it generally lacks stability due to sol-gel transitions caused by changing temperature, types, ionic strength, or pH. The sol-gel transitions refer to the status of the sol and gel being reversibly transferred to each other or to one of the sides (sols are colloidal dispersions in solvents having fluidity). For example, agar and gelatin aqueous solutions will become gels when the temperature is lowered, but will become sols again when the temperature is increased. Such gels are called reversible gels (see Fig. 2).

2.2.2.1   Crosslinking by hydrogen bonding

When hydrogen bonding is formed among polymer molecules using the freeze-drying method, lyophilization, it forms a gel that has excellent mechanical and water uptake properties [10, 11]. One example of this is the gel formation by laophilization of poly(vinyl alcohol) that has

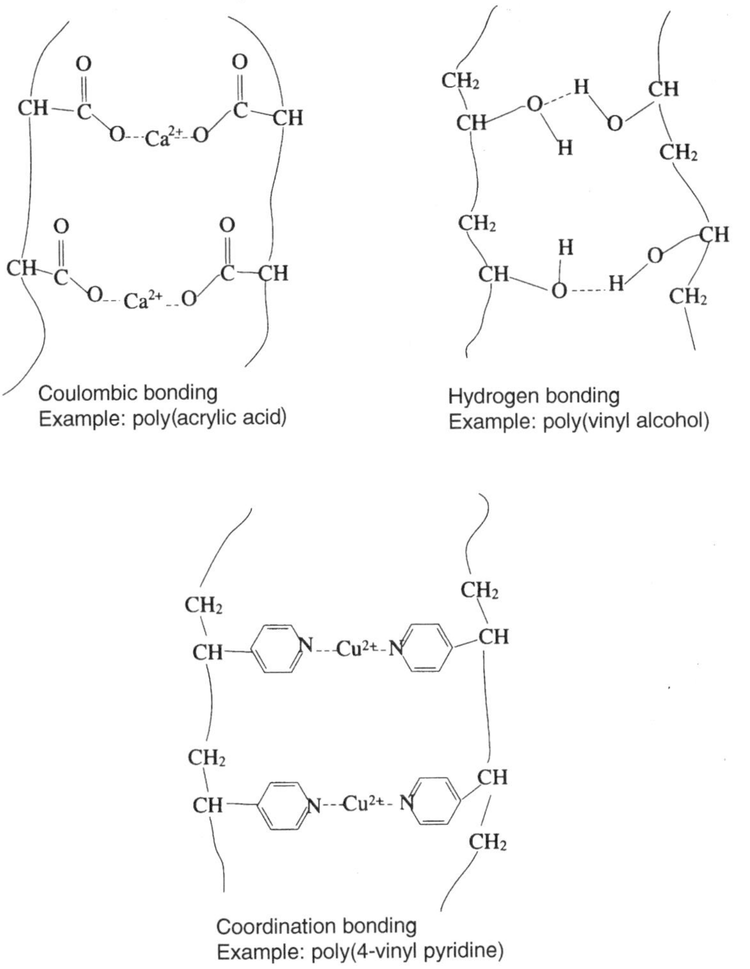

Coulombic bonding
Example: poly(acrylic acid)

Hydrogen bonding
Example: poly(vinyl alcohol)

Coordination bonding
Example: poly(4-vinyl pyridine)

Fig. 2 Examples of a physical gel (From: Gels: Basics and Applications of Soft Materials, Sangyo Tosho).

completely been saponified. In addition, gels can be prepared by forming complexes via hydrogen bonding between two different polymers such as poly-(methacrylic acid)-poly(ethylene glycol) or poly(acrylic acid)-poly-(vinyl alcohol) [12].

2.2.2.2 Crosslinking via ionic bonds

Mixing two types of polyelectrolyte solutions of opposite charges leads to a polyelectrolyte complex gel. The gel formation depends on the solvent types, ionic strength, pH, and polymer concentration. It is possible to tailor the degree of swelling and modulus of the gel by adjusting these parameters via changes in static interaction. When the ratio of polycation and polyanion is 1 : 1, a neutral gel is formed.

Polyelectrolyte complex gels are soluble in an appropriate ratio of the ternary mixture of water, organic solvent and salt so it is possible to form films by coating.

2.2.2.3   Crosslinking by coordination bonding

This is the method for forming gels by bonding poly(carboxylic acids), such as poly(acrylic acid), or strong acid polymers, such as poly(styrene sulfonic acid), by alkaline earth metal ions [13]. This bonding is formed via hydrating water molecules rather than direct bonding between the metal ions and chelates. There is a correlation between the radius of the hydrating ion and bonding strength. The smaller the radius of hydration, the easier the gel forms due to the increased static interaction. Gelation also depends on the polymer's molecular weight, concentration, types of solution that cause crosslinking reactions, and salt concentrations. When the polymer concentration is small, or if the degree of polymerization is not high enough, the polymer may not gel.

2.2.2.4   Crosslinking caused by helix formation

Gelation of agar (from agarose and agropectin), gelatin, agarose, arginic acid, and carrageenan [14] is thought to be caused by helix formation. This is the gelation caused by a heat induced sol-gel transition. For example, when an aqueous solution of gelatin at an appropriate concentration is cooled, the viscosity increases and it gels upon the sol-gel transition at $-25°C$. Crosslinking of this gel is thought to be due to the formation of a thermally stable helix by hydrogen bonding between the NH group of the prolin in gelatin and its neighboring CO group of the peptide in hydroxyprolin (see Fig. 3). Arginic acid, a polysaccharide polyelectrolyte, is made of a block copolymer of two uronic acids and the polymer has three structural units. This gel is made of a so-called egg junction that is formed by bonding two polysaccharide chains via calcium ions. The gelation mechanism of carrageenan, a polysaccharide from red seaweed, involves first forming bonds between polymer chains, and then forming double helix domains by cooling (coil-domain transition).

After this, double helices aggregate through counter ions, such as the K^+ ion, and they are thought to gel by forming crosslinked domains (double-helix association). Another mechanism proposed involves the

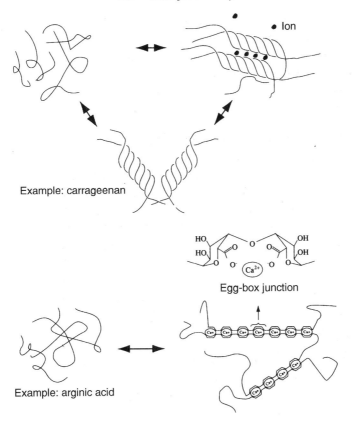

Example: carrageenan

Egg-box junction

Example: arginic acid

Fig. 3 Formation of helices.

formation of aggregates made of a single-chain helix via counter ions. Either way, it is thought to form crosslinked domains by the association of helical structures. Carrageenan also has a sulfonic acid group and is divided into three groups according to its content, ι-, κ- or λ-carrageenan. The properties also differ. Generally, κ-carrageenan is thought to be the easiest to gel. The gelation of agar and agarose is also thought to be caused by the formation of double helices by a parallel arrangement of the helix. The crosslinking of polysaccharides and protein gels is thought to be formed by helices among the polymer chains and the association of these, but detailed analysis on the structure of crosslinked domains has not been done.

Table 2 Formation and classification of crosslink region.

Types of gels	Mechanisms and methods of crosslinking		Examples
	Crosslinking during polymerization	Condensation polymerization	Polyester, polyamide etc.
		Free radical polymerization / Thermal polymerization / Photopolymerization / Radiation polymerization / Plasma polymerization	Copolymerization between various divinyl monomers and divinyl compounds
	Crosslinking after polymerization	Chemical crosslinking	Poly(vinyl alcohol) – aldehyde, etc.
		Photocrosslinking	Poly(vinyl alcohol) – styrilbazolium salt, etc.
		Radiation crosslinking	Acrylic polymer
		Plasma crosslinking	Polyethylene, poly(tetrafluoroethylene), etc.
	Hydrogen bonding	Freeze drying method / Freeze-thaw method / Freeze low-temperature crystallization method / Polymer complex	Poly(acrylic acid), etc. Polyacrylic acid) – poly(vinyl alcohol) Poly(methacrylic acid) – poly(ethylene glycol), etc.
	Ionic bonding	Mixing	Poly(vinyl benzyl trimethyl ammonium chloride) – poly(sodium methacrylic acid), etc.
	Coordination bonding	Chelate reaction	Poly(vinyl alcohol) – Cu^{2+} , poly(acrylic acid) – Fe^{3+} , etc.
	Helix formation	Helix formation among polymers	Agar, gelatin, carrageenan, arginic acid, etc.
	Hydrophobic bonding	Hydrophobic interaction	Egg white alubumin, platelet alubumin

Crosslinking after polymerization

Chemical gels   Physical gels

Synthetic polymer gels, natural polymer gels, hybrid gels

Polymer gels

2.2.2.5 *Crosslinking by hydrophobic bonding*

Gelation of proteins such as egg albumen and serum albumin is thought to be caused by parts of the hydrophobic side chains. These side chains are exposed on the outside of the globular protein molecules by the addition of heat and the formation of a supramolecular assembly. This is done by placing globular particles side by side due to the balance of the hydrophobic interaction and ionic repulsion. (See Table 2 for more information.)

REFERENCES

1 de Rossi, D., Kajiwara, K., Osada, Y., and Yamauchi, A. (1991). *Polymer Gels*, New York: Plenum.
2 de Gennes, P.G. (1979). *Scaling Concepts in Polymer Physics*, Ithaca, New York: Cornell University Press.
3 Flory, P.J. (1973). *Principles of Polymer Chemistry*, Ithaca, New York, Cornell University Press, pp. 347–398.
4 Stockmayer, W.H. (1943). *J. Chem. Soc.* **63**: 3083.
5 Case, L.C. (1957). *J. Polym. Sci.*, **26**: 333.
6 Yamauchi, A. (1972). *Protein Nucleic-acid Enzyme*, **22**: (13), 1418.
7 Nagata, Y. (ed.) (1900). *Plasma Polymerization* Tokyo Kagaku Dojin.
8 Yamashita, S., and Kaneko, T. (eds.) (1982). *Handbook of Crosslinking Agents*, Taisei Publ.
9 Hatakeyama, T., Yamauchi, A., and Hatakeyama, H. (1984). *Eur. Polym. J.*, **20**: 61.
10 Nambu, M. (1982). Japan Tokkyo Kokai, 82–130543.
11 Watase, M., Nishinari, K., and Nambu, M. (1983). *Polym. Commun.* **243**: 52.
12 Tsuchida, H., and Ookawara, S. (eds.) (1974). "Lecture Series on Polymer Experiment, 7. Functional Polymers, Chapter 9," Kyoritsu Shuppam.
13 Polymer Complex Study Group Ed. (1989). "Fundamentals on Polymer Complexes," Gakkai Shuppan Center.
14 Rees, D.A. (1969). *Adv. Carbohydr. Chem. Biochem.*, **24**: 267.

CHAPTER 2

Theory of Gelation and Preparation of Gels

Chapter contents

Section 1
Theory of Gelation

HIDETAKA TOBITA

1.1 INTRODUCTION

Carothers [1], known as the inventor of nylon, proposed the first theory of gelation from the viewpoint of infinite molecular weight. However, his theory is now regarded as obsolete because he defined a gel as the point at which the number average molecular weight becomes infinite. On the other hand, Flory [2–4] proposed the concept of "infinite molecular weight with respect to the weight average molecular weight," which is the concept still in use today. Stockmayer (see Reference [5] by Stockmayer and Zimm), who was impressed with Flory's simplified approach, refined the theory mathematically and established the theory of gelation [6–8]. This theory is now widely regarded as the Flory–Stockmayer model. There have been many gelation theories proposed. They can be classified as the "classical theory" based on the statistical theory of dendritic structures [2–4, 6–13], the "kinetic theory" [6, 14–23] based on the infinite number of simultaneous differential equations describing the growth of polymers with various degrees of polymerization and the "percolation theory," which is based on 3D structure formation [24–

29

29]. It should be noted that the meaning of "classical" should not be regarded as "old" but rather "standard".

1.2 RANDOM CROSSLINKING

We will consider here cases in which linear chain molecules are randomly crosslinked. Therefore, we will deal with the crosslinked state with maximum probability given the various chain lengths and number of crosslinks.

To clarify further, a crosslinked polymer is schematically shown in Fig. 1. In crosslinked polymers, the primary polymer is the polymer in which all crosslinks are attached. Therefore, in the case of random crosslinking, the polymer chains used as raw material are the primary polymers. The primary polymer is connected with another primary polymer via a crosslink. At this point, two crosslink points are formed with one crosslink because the crosslink point is defined as a unit where three chains merge. The crosslink density (ρ) is defined as the number of crosslink points divided by the number of chemical repeat units. Therefore, the crosslink density of the crosslinked polymer in Fig. 1 is $\rho = 4/12 = 0.333$. Also, the crosslink density of the primary polymer A is $\rho = 1/4 = 0.25$. The definition adopted here was proposed by Flory [8] and is the one most widely used. However, it is important to understand this definition because there are occasions where the crosslink density is defined by the number of crosslinks (crosslink points/2) as in the case of crosslinking by radiation. The probability of each chemical repeat unit being the crosslink point in the random crosslinking system is constant and is ρ in Flory's definition of crosslink density. Thus far the

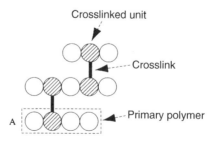

Fig. 1 Schematic diagram of crosslinked polymer.

discussion revealed that the difference between the definition of gelation by Carothers and Flory lies in the difference between the concept of "number" average degree of polymerization and "weight" average degree of polymerization.

1.2.1 Average Degree of Polymerization

The readers may have wondered why there is such a subtle difference. Let us start by explaining the physical difference in the distribution of degree of polymerization and the average degree of polymerization based on the "number" and "weight."

Taking the linear polymer as an example, the number average distribution of the degree of polymerization ($N(n)$) (n indicates the degree of polymerization) can be obtained by random sampling of the chain ends among an infinite number of molecules as shown in Fig. 2a. This probability of the degree of polymerization is the number average degree of polymerization ($\bar{P}_n$). On the other hand, placing each monomer unit in each lattice unit in a 2D lattice as shown in Fig. 2b and sampling the lattice randomly, the weight average distribution of the degree of polymerization ($W(n)$) can be obtained. Its probability is the weight average degree of polymerization ($\bar{P}_w$). As the probability of choosing a larger molecule is higher using weight average as compared with number average, the weight average degree of polymerization is always greater than the number average degree of polymerization for polymers other than monodisperse ones.

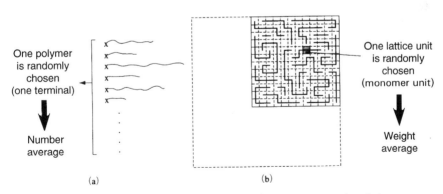

One polymer is randomly chosen (one terminal)

Number average

(a)

One lattice unit is randomly chosen (monomer unit)

Weight average

(b)

Fig. 2 Concept of the distribution of the number average and weight average degrees of polymerization.

Next, let us select a polymer based on the weight average concept by choosing a chemical repeat unit randomly in a random crosslink system. Assuming that the chemical repeat unit in Fig. 3 is selected, the probability of the degree of polymerization of the primary polymer chain including this chemical repeat unit is equivalent to the weight average degree of polymerization $\bar{P}_{wp}$ of the primary polymer. This primary polymer is called the zeroth generation polymer. The probability of the crosslink point of this primary polymer can be expressed as $\bar{P}_{wp}\rho$, where ρ is the crosslink density. Therefore, on average, there are $\bar{P}_{wp}\rho$ primary polymers in the first generation polymer. Since the probability of having the crosslink point in any unit is constant in the first generation primary polymers (the unit used to connect with the zeroth generation polymer is the unit that is randomly chosen in the primary polymers), these primary polymers are chosen on a weight basis. Thus, the probability for the degree of polymerization of the primary polymer is $\bar{P}_{wp}$.

Thus, the total probability of the degree of polymerization of the polymer that belongs to the first generation is $(\bar{P}_{wp}\rho)\bar{P}_{wp}$. Since each first generation primary polymer uses one chemical repeat unit to connect with the zeroth generation polymers, the probability of finding the crosslink point between the first and second generations is $(\bar{P}_{wp}\rho)(\bar{P}_{wp} - 1)\rho$ and the total degree of polymerization of the polymers that belong to the second generation is $(\bar{P}_{wp}\rho)^2(\bar{P} - 1)$. Accordingly, the degree of polymerization of the polymer that includes the first selected chemical repeat

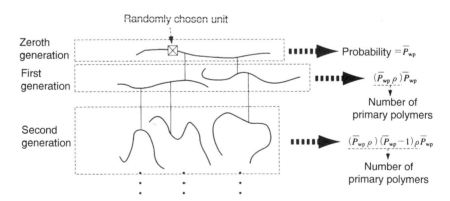

Fig. 3 Calculation of the weight average degree of polymerization in a randomly crosslinked system.

unit, that is the weight average degree of polymerization of the polymers in the system, can be expressed as follows:

$$\bar{P}_{\text{w}} = \bar{P}_{\text{wp}} + (\bar{P}_{\text{wp}})^2 \rho \sum_{i=0}^{\infty} \{(\bar{P}_{\text{wp}} - 1)\rho\}^i = \frac{\bar{P}_{\text{wp}}(1 + \rho)}{1 - (\bar{P}_{\text{wp}} - 1)\rho} \quad (1\text{a})$$

$$\frac{\bar{P}_{\text{wp}}}{1 - \bar{P}_{\text{wp}}\rho} \quad (\text{for } \bar{P}_{\text{wp}} \gg 1 \text{ and } \rho \ll 1) \quad (1\text{b})$$

Therefore, the weight average degree of polymerization approaches infinity at $\rho = 1/\bar{P}_{\text{wp}}$ and in the reacting system gels.

On the other hand, the number average degree of polymerization is the value obtained by dividing the total number of chemical repeat units (in this system, it is constant) by the number of polymer molecules. As the number of molecules reduces by one as one crosslink is formed, the number average degree of polymerization can be expressed as follows:

$$\frac{1}{\bar{P}_{\text{n}}} = \frac{1}{\bar{P}_{\text{np}}} - \frac{\rho}{2} \quad (2\text{a})$$

Hence,

$$\bar{P}_{\text{n}} = \frac{\bar{P}_{\text{np}}}{1 - (\bar{P}_{\text{np}}\rho)/2} \quad (2\text{b})$$

where $\bar{P}_{\text{np}}$ is the number average degree of polymerization of the primary polymer. From Eq. (2), it is apparent that the number average degree of polymerization is finite even at gelation. In other words, if a polymer is selected by weight basis, the probability of choosing the molecules with infinite molecular weight agrees with the weight fraction (nonzero value). However, no matter how one chooses the polymer by number basis, as long as $\bar{P}_{\text{n}}$ is finite, one cannot choose the infinite molecular weight polymers, meaning the probability of choosing such molecules is zero. Although this merely indicates that there is an infinite number of molecules in the system, it offers an interesting paradox in terms of the meaning of words. The existence of an infinite number of molecules does not reflect upon the number average degree of polymerization. This is why the definition of a gel offered by Carothers is not acceptable.

Let us interpret the meaning of Eq. (1) once more. Equation (1b) indicates that gelation takes place when the weight average number of the crosslink point becomes one. Some readers may not expect this. One might wonder why one crosslink point can continue to connect primary polymers and why two crosslink points per molecule are not needed. It should be noted that only the gelled molecules need to have *2 crosslink points per primary polymer at gelation*. A homogeneous crosslink structure formed by random crosslinking means that the probability of having a crosslink point for any chemical repeat unit is equal. The number of crosslink points exhibits bimodal distribution (fluctuation). Therefore, the crosslink density and degree of polymerization of the primary polymers are not constant for all molecules. They are a function of the size of the crosslinked polymers. As will be seen in subsection 2.3 here, the crosslink density at gelation equals twice ($\rho_{\text{gel}} = 2\rho$) that of the (average) crosslink density of the total system. Furthermore, the number average degree of polymerization of the primary polymer that is incorporated into the gel agrees with the weight average degree of polymerization of the total system ($\bar{P}_{\text{np}\cdot\text{gel}} = \bar{P}_{\text{wp}}$) [30]. Therefore, from Eq. (1b), the following equation is derived:

$$\bar{P}_{\text{np}\cdot\text{gel}} \times \rho_{\text{gel}} = 2 \,(\text{at gelation}) \tag{3}$$

Consequently, it becomes apparent that the concept of *the necessity of two crosslink points per primary polymer at gelation* is based only on the largest molecules in the system. Equations (1)–(3) can be used as long as the gel is a random crosslinked system regardless of the distribution of the degree of polymerization of the primary polymer chains. Thus, when the number average degree of polymerization of the primary polymer is the same, the larger the polydispersity index ($\bar{P}_{\text{wp}}/\bar{P}_{\text{np}}$) the larger the weight average degree of polymerization, which accelerates gelation (low crosslink density at gelation).

Figure 4 shows the most probable distribution of the number average degree of polymerization, $\bar{P}_{\text{wp}} = 200$, and weight fraction of the gel (W_{gel}) for the random crosslinked primary polymer chains. This most probable distribution, called the Flory distribution, is a typical distribution function for polycondensed systems. This distribution is obtained when the probability of connecting with the neighboring unit is constant and is $\bar{P}_{\text{wp}}/\bar{P}_{\text{np}} = 2$. In addition, in free radical polymerization the instantaneous value of the distribution of the degree of polymerization follows this most probable distribution when the effect of coupling can be neglected. It can be seen from the figure that, while the weight average degree of polymerization suddenly increases and approaches infinity near the gel

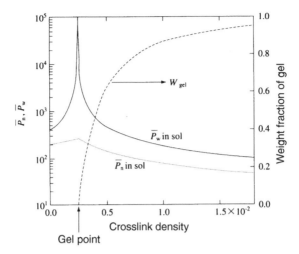

Fig. 4   Change of the average degree of polymerization and weight fraction of gel in random crosslinking.

point, the number average degree of polymerization remains relatively small even at the gel point. Also, the gel that is formed at the gel point incorporates sol molecules as though it were a sponge and the weight fraction of the gel increases rapidly. At this point, since the larger the sol molecule, the greater the probability of sol molecules to be incorporated into the gel, the greater is the reduction in the weight average degree of polymerization of the sol molecules beyond the gel point.

1.2.2   Branching and Crosslinking

Often branching and crosslinking are not separately classified and cross-linked polymers are frequently included in the genre of branched polymers. However, it is necessary to separate these two from the standpoint of gel formation.

Figure 5 shows an example of the structure of a randomly branched primary polymer by showing how one terminal of the primary polymer is randomly connected. The branch structure shown in the figure is generally called long branching. However, it is necessary to recognize the fact that many relatively short branches do coexist. In such long branching, only a T-shape connection is formed. However, if the other terminal can form branches, an H-shape connection can also be formed. It can be shown that

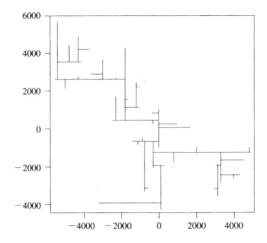

Fig. 5   An example of a homogeneous branched structure.

the weight average degree of polymerization can be expressed by the following equation by applying the sampling method that was stated in subsection 2.1 [31]:

$$\bar{P}_{\mathrm{w}} = \frac{\bar{P}_{\mathrm{wp}}}{(1 - \bar{P}_{\mathrm{np}}\rho_{\mathrm{b}})^2} \tag{4}$$

where ρ_{b} is the branch density. Equation (4) is valid regardless of the distribution of the degree of polymerization. For the system that branches only at the end of the chain, the branch density ρ_{b} must always be less than $1/\bar{P}_{\mathrm{np}}$ because the branch is formed only at one end. Hence, Eq. (4) indicates that it is not possible to form gels with infinite molecular weight.

In practice, in order to form the homogeneous branch structure as shown in Fig. 5, it is necessary to combine special reactants and reaction methods [32]. For example, the long chain branching formation by chain transfer mechanism in free radical polymerization will not form homogeneous branching. Also, the average degree of polymerization is different from that predicted by Eq. (4). Nevertheless, the fact that gels will not form cannot be overcome even by a combination of special reactants and reaction methods. In other words, the T-shape connection alone will not form gels. Only when an H-shape connection is formed, can gels be made. An example of this is the free radical polymerization where termination includes inhibition of coupling. In this case, the H-shape connection can be formed, thus resulting in gel formation.

Termination of inhomogeneity alone will not lead to gel formation [23, 33, 34]. Therefore, from the viewpoint of gelation, it is appropriate to classify the T-shape connection for branching and H-shape connection for crosslinking. The oft-stated expression *gels are formed by repeated formation of branching* lacks technical accuracy. It is easy to judge whether or not the given reaction system gels by examining the possibility of H-shape connection formation; in other words, the possibility of reaction among the molecules that are selected via weight average basis.

1.2.3 Distribution of the Degree of Polymerization and Crosslink Structure

When the distribution of the degree of polymerization of the primary polymer chain is given by a simple distribution function, it is possible to derive the distribution function of the degree of polymerization analytically [14, 15, 30, 31, 35–39]. For the weight average degree of polymerization of the primary polymer chains with the most probable distribution that is crosslinked randomly, the distribution of the weight average degree of polymerization for the fraction that has k crosslinks is given by the following equation:

$$W_k(r) = \frac{r}{(\bar{P}_{np})^2} \exp\left\{-\left(\frac{\bar{P}_{np}\rho + 1}{\bar{P}_{np}}\right)r\right\} \left(\frac{(P_{np}\rho)^k}{(k+1)!(2k+1)!}\right) \left(\frac{r}{P_{np}}\right)^{3k} \quad (5)$$

The total distribution of the degree of polymerization can be obtained by summation as follows:

$$W(r) = \sum_{k=0}^{\infty} W_k(r) \quad (6)$$

As shown in the W_k function, the fraction with k crosslinks can be given by the Schulz–Zimm distribution. The total distribution is given by its sum and takes the form of a supergeometric function.

Figure 6 shows the change of the degree of polymerization of random crosslinked primary polymer chains that obey the most probable distribution. The probability density function that expresses the distribution of the degree of polymerization is plotted in the graph using semi-log scale as the probability variable in a manner similar to that employed in

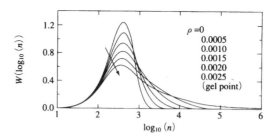

Fig. 6   The variation of the distribution of the degree of polymerization for random crosslinking.

size exclusion chromatography (SEC). It can be seen that the distribution becomes broader (in particular the tail in the high molecular side lengthens) as gelation nears. Although only the results up to the gel point are shown here, Eqs. (5) and (6) are applicable beyond gelation as long as sol molecules are considered.

A detailed evaluation can be given of the crosslink structure that is formed by random crosslinking using Eq. (5) [30]. For example, the number and weight average degree of polymerization of the fraction that possesses k crosslink points are expressed as follows:

$$\bar{P}_{n,k} = \frac{(3k + 1)\bar{P}_{np}}{1 + \bar{P}_{np}\rho} \tag{7}$$

$$\bar{P}_{w,k} = \frac{(3k + 2)\bar{P}_{np}}{1 + \bar{P}_{np}\rho} \tag{8}$$

Hence, each fraction's degree of polymerization merely increases linearly against the number of crosslink points k. The polydispersity index is given by

$$\frac{\bar{P}_{w,k}}{\bar{P}_{n,k}} = \frac{3k + 2}{3k + 1} \tag{9}$$

It indicates that the larger the number of crosslink points, the smaller the polydispersity index. For example, the fraction with the number of crosslink points $k = 3$ shows the polydispersity index

$$\frac{\bar{P}_{w,k}}{\bar{P}_{n,k}} = 1.1$$

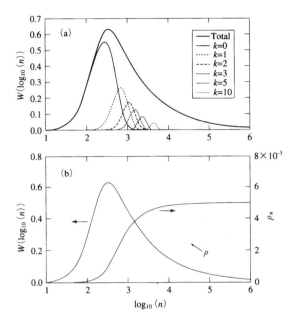

Fig. 7   The relationship between the distribution of the degree of polymer-ization or crosslink density and the degree of polymerization.

Figure 7 displays the calculated results of the degree of polymeriza-tion at exactly the gel point ($\bar{P}_{n,k} = 200$). It can be seen that many linear polymers ($k = 0$) still exist even at the gel point. Also, each fraction overlaps with others heavily, and the distribution will not show a skewed shape. With improvements in modern analytical techniques, skewed distributions and sometimes bimodal distributions are occasionally reported [40, 41]. Such skewed distribution cannot be formed in the genre of Flory's ideal dendritic model; however, it becomes important in a real system with nonideal parameters such as structural dependence of the crosslinking reaction (including cyclization) and degree of polymeriza-tion.

For example, skewed distribution can be predicted theoretically by restricting the space of reaction such as in suspension polymerization and limiting the reaction among large molecules [42]. This case corresponds to the special situation of the degree of polymerization dependence where reaction among large molecules becomes more difficult.

Illustrated in Fig. 7(b) is the distribution of the degree of polymer-ization of all polymers and the probability of crosslink density as a function of the degree of polymerization, which is the average crosslink density of the polymer with a specific degree of polymerization [30]. The crosslink density increases as the molecules grow larger, although it soon approaches very close to an asymptotic value ($\rho_n|_{n\to\infty}$). The value indicated by the dotted line is the (average) crosslink density ρ of the total system. The value of ρ will be smaller than $\rho_n|_{n\to\infty}$ because the crosslink density of polymers of low degree of polymerization is extremely small. The average crosslink density of the total system is often used to represent the system as long as the gel content is not 100%. However, the crosslink density of large and small molecules does not coincide with ρ. In fact, it is important to recognize that the molecule with crosslink density ρ seldom exists. (The average value is merely a statistical number. For example, as one seldom finds a Japanese person who is exactly the average Japanese, the average does not express the properties of a specific polymer.) As stated before, Fig. 7 expresses the states exactly at the gel point and $\rho = 2.5 \times 10^{-3}$. As discussed in subsection 2.1, it can be seen from Fig. 7 that there is a relationship $\rho_n|_{n\to\infty} = 2\rho$ at the gel point [30, 43] and it can be proven accurately [43].

Finally, homogeneous crosslink structure formed by random crosslinking will be briefly discussed. Figure 8 shows an example of the structure of crosslinked polymers formed slightly before gelation, although cyclization is not considered here. While the structure shown here is exactly that of a homogeneous crosslink structure, it is not certain whether we can see it be homogeneous with the naked eye. It important to distinguish the difference between homogeneity and uniformity. A

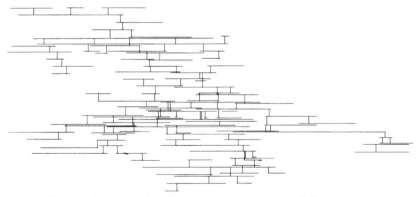

Fig. 8   An example of a homogeneous crosslink structure.

uniform crosslink structure is, for example, the structure with a crosslink point after every five chemical repeat units. In this case, the probability of having a crosslink point at each chemical repeat unit is the repeated sequence of $\{0, 0, 0, 0, 1\}$.

On the other hand, homogeneous structure means that the probability of having a crosslink point for every unit is constant. Therefore, these two structures constitute two extremes in terms of the regularity of the distribution of crosslink points. Human nature is not easily compatible with statistical treatment. One might wish to consider the problems of structural control after attempting to accept that the structure shown in Fig. 8 is homogeneous. A reader is recommended to read an excellent popular science book, *Bully for Brontosaurus* by Gould (1991) [44], where graphic examples of 2D random and regular distributions are illustrated.

1.3   FREE RADICAL COPOLYMERIZATION

The copolymerization of a vinyl monomer and a divinyl monomer is frequently used to prepare a gel. In this section, the basics required to treat a free radical copolymerization from the theoretical point of view will be discussed.

1.3.1   The Process of Crosslink Structure Formation

In free radical polymerization, unlike for the random crosslink polymers discussed in the previous section, primary polymer chains are formed during polymerization. The crosslink points are also formed as the polymers propagate. Furthermore, free radical polymerization is a typical example of classic kinetically controlled systems. It is impossible to disconnect the formed bonds even by accident. Therefore, it can be easily seen that the resultant structure will not necessarily conform to the maximum probability given the primary polymer chains and crosslink density.

In a free radical copolymerization conducted in a homogeneous batch reactor, its crosslink structure formation process will be quite complex as compared with random crosslinking, because both the composition and degree of polymerization of the primary polymer change simultaneously. Here, we will consider kinetically the inhomogeneity of crosslink density that accompanies the compositional changes of the primary chain copolymers.

Figure 9 shows a schematic diagram of the crosslinking process. In the diagram, primary polymer A is the primary polymer generated by reactivity θ. Figure 9 illustrates the process where two primary polymers are crosslinked by reacting the primary polymer B that is growing at reactivity $\psi(\psi > \theta)$ with the unreacted pendant double bonds of A. In this case, from the viewpoint of B, this crosslink point is formed simultaneously with the formation of B. Hence, the density of such crosslink points will be called instantaneous crosslink density (ρ_i). The instantaneous-type crosslink point is formed by reacting with the pendant double bonds on the other primary polymer chains. It is a function of only the primary polymer chain formation process. On the other hand, from the viewpoint of A, the crosslink point on the A chain is formed after molecule A is formed. Thus, this type of crosslink density is called latent crosslink density. Latent crosslink density is a function of both the primary polymer formation process (how many divinyl monomer units are incorporated in the chain and possess the pendant double bonds that can be used for crosslinking), and what occurs later (what fraction of pendant double bonds are consumed and crosslink points are formed). As described earlier, the crosslink point is defined by a triple-branching point. Therefore, the probability of this primary polymer possessing the

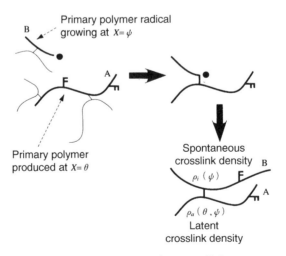

Fig. 9   Formation process of a crosslink structure.

crosslink density $\{\rho(\theta, \psi)\}$ in terms of primary polymer A with reactivity θ is the sum of the instantaneous crosslink points $\{\rho_i(\theta)\}$ and the latent crosslink points $\{\rho_a(\theta, \psi)\}$. Thus, it can be expressed by the following equation. The instantaneous crosslink points are formed as the A molecule is formed while the additional crosslink points are then formed on the A molecule by consuming the pendant double bonds on A:

$$\rho(\theta, \psi) = \rho_i(\theta) + \rho_a(\theta, \psi) \tag{10}$$

These two types of crosslink density, $\rho_i(\theta)$ and $\rho_a(\theta, \psi)$, can be derived by a simple mass balance equation with respect to the crosslink point on the primary polymer molecule formed at each instance [22, 23, 45, 46].

As a natural consequence of the crosslinking reaction process, the density of the primary polymer differs depending on the time of this primary polymer formation. That is, in the case of the copolymerization of vinyl and divinyl monomers, the generally formed inhomogeneous cross-link formation can be regarded as a natural consequence of the mechanism of crosslink formation. This is true except for the "special reaction conditions" by favorable timing of the incorporation of divinyl monomer in the polymer chain (formation of pendant double bonds) and consumption of pendant double bonds (formation of crosslinks). These *special reaction conditions* are used by Flory as *simplified conditions* when the Flory–Stockmayer theory is applied to the copolymerization of vinyl and divinyl monomers. *Flory's simplified conditions* include the following three assumptions: (1) the reactivities of the monomer and the double bonds in the polymer are all equal; (2) any double bond reacts independently; and (3) there will be no intramolecular reactions (cyclization) within the finite size molecules (sols).

Under Flory's simplification conditions, it can be proven that (the probability of) the crosslink density of all the primary polymers is equal by using the kinetic model shown here [47]. That is, under Flory's simplified conditions, the kinetic model that considers the change of the crosslink structure and the model based on random crosslinking become identical. Even for the system under kinetic control, the model that is based on random crosslinking will be strictly applicable. Of course in real systems an inhomogeneous crosslink structure will be formed because of: (1) the difference in reactivity as expressed in copolymerization reactivity ratio between the double bonds of the monomers; (2) the reactivity of the pendant double bonds with the double bonds in the monomers and the dependence of the reactivity on the degree of polymerization and structure; and (3) the existence of a cyclization reaction that depends strongly on the conformation of the polymer chain rather than the average concentration of the functional group.

At this point, the copolymerization reactivity ratio of an ideal monomer pair that fulfills Flory's simplified condition, which states that the reactivity of every double bond is equal, will be briefly mentioned. In many cases, the copolymer reactivity ratio is determined by the composition of the polymer based on the monomer units or the remaining monomers. In this case, the obtained copolymer reactivity ratio is given by the monomer units. The situation is simple for copolymerization between vinyl monomers. However, it requires caution when divinyl monomers are included. Now, we will use subscript 1 for the vinyl monomer and subscript 2 for the divinyl monomer. Divinyl monomers possess two vinyl groups in the molecule $k_{12} = 2k_{11}$ and

$$r_1 = \frac{k_{11}}{k_{12}} = \frac{k_{11}}{2k_{11}} = 0.5 \tag{11}$$

similarly, as $k_{22} = 2k_{21}$,

$$r_2 = \frac{k_{22}}{k_{21}} = \frac{2k_{21}}{k_{21}} = 2 \tag{12}$$

Naturally, if this reactivity ratio is expressed by the vinyl unit, it will be $1:1$. However, the chemical repeat units that are incorporated in the polymer are usually not analyzed with respect to the number of vinyl groups.

As it can be easily imagined by looking at the copolymerization reactivity ratio of $r_1 = 0.5$ and $r_2 = 2$, it is important to pay attention to the fact that the variation of the copolymer composition is very large under Flory's simplified conditions, which form homogeneous crosslink structures in a homogeneous batch reaction system [23]. We often encounter the expression, *the compositional drift of copolymers during the polymerization reaction is the cause of heterogeneous crosslink structure formation.* This is a totally erroneous expression if the copolymer composition is considered in monomer units (of course if the author uses a vinyl unit basis, it is correct). Therefore, it is a confusing expression. Actually, it is not the change of the copolymer composition that harms the formation of a homogeneous crosslink structure. Rather, the right changes in copolymer composition are needed. It is also dangerous to attempt to control crosslink structure only by the reactivity ratio of the monomers. It is essential to understand that, when divinyl monomers are incorporated into polymer chains, only the potential candidates for cross-link points have been formed. Only after the pendant double bonds have reacted are crosslinks generated for the first time. In other words, controlling the crosslink structure of the copolymerization of vinyl and divinyl monomers requires balancing the following two processes: (1) the timing of incorporating the divinyl monomer into the polymer; and (2) the timing to react the incorporated pendant double bonds.

As is apparent from the discussion so far, the crosslink formation process depends strongly on the reaction operation.

As an example, let us consider suspension polymerization. In suspension polymerization, the monomers from the monomer droplets transfer into the polymer particles as long as monomer droplets exist. Thus, regarding the change of monomer concentration in the space of polymerization, that is, polymer particles, this reaction can be considered a semibatch reactor from the viewpoint of the change of monomer concentration. Figure 10 depicts the calculated results of crosslink density distribution of suspension polymerization and homogeneous phase polymerization under Flory's simplified conditions [48, 49]. In Fig. 10, the abscissa indicates the reactivity (θ) at the time of each primary polymer formation and the ordinate shows the probability of crosslink density of each primary polymer. In the figure, for example, the line with $\psi = 0.6$

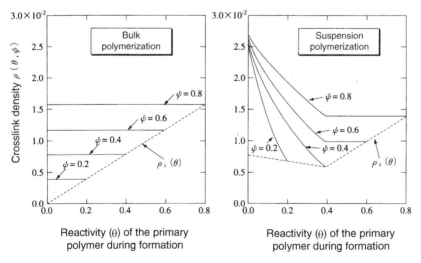

Fig. 10 Comparison of the crosslink density distribution of bulk polymerization and suspension polymerization under Flory's simplified conditions.

indicates the crosslink density of the primary polymers that are formed at each reactivity ($\theta < \psi$) when the current reactivity (ψ) is 0.6. As shown in the figure, the same crosslink density is obtained regardless of the time of reaction in the bulk copolymerization and its relationship will not change even if the reaction proceeds. That is, the batch reaction system produces a homogeneous crosslink structure at any stage of polymerization. On the other hand, in suspension copolymerization systems, the crosslink density varies significantly as a function of the time of primary polymer formation. As a characteristic of this, the crosslink density of the primary polymer formed at an early stage of polymerization is extremely large. Furthermore, the homogeneous crosslink structure will not be formed in the suspension polymerization even if Flory's simplified conditions are used. Here, a breakpoint is seen at a reactivity of 0.4 in the suspension polymerization because the reactivity at which monomer droplets disappear was assumed to be 0.4.

In free radical polymerization, as crosslinking is basically a reaction between a polymer and a polymer free radical, the larger the polymer concentration, the greater the probability of forming crosslinks, that is, the reaction with pendant double bonds. Therefore, a semibatch operation, seedless suspension polymerization, and continuous reactor all enhance crosslinking due to a high polymer concentration as compared with the bulk batch reactor. Figure 11 depicts the variation of average crosslink

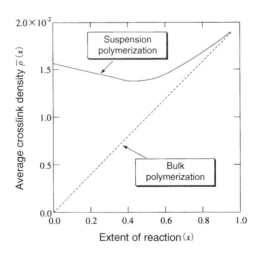

Fig. 11 The variation of average crosslink density that accompanies both bulk and suspension polymerization under Flory's simplified conditions.

density of both bulk and suspension polymerization under Flory's simplified conditions [48, 49]. Also, as seen in suspension polymerization, it should not be surprising that the average crosslink density was reduced during the reaction. This simply indicates that the ratio of the crosslinking rate and the polymerization (propagation) rate is reduced,

1.3.2 Molecular Weight Distribution

When the rate parameters of free radical polymerization are known, it is possible to know: (1) the distribution of the degree of polymerization of the primary polymer chains that are formed at each time; (2) the crosslink density (the distribution of the crosslink density expressed by Eq. (10)) that formed at each time; and (3) information on the probability of a primary polymer chain being connected to other primary polymer chains [50]. In the genre of ring-free models where cyclization is ignored, the bonding state of every primary polymer chain is determined unambiguously by these three conditions. The analytical solution of the molecular weight distribution function under these limiting conditions has not been obtained other than in the average degree of polymerization [51]. Nonetheless, it is possible to analyze numerically using a Monte Carlo computer simulation the statistical nature of the reacting systems [50–54]. This computer simulation also has the advantage of direct observation of individual crosslinked polymers.

Here, the classical theory not limited by spatial limitation is discussed. However, in the area of 2- and 3D percolation processes, computer simulation has been active [25–29]. On the other hand, with the formation of microgels via crosslinking reactions within a microspace, such as suspension polymerization, it is not necessary to consider the critical change of the weight average degree of polymerization, which is characteristic of infinite molecular weight systems [42, 49, 55]. It is easily understood by considering the crosslinking process of the primary polymer to the already formed primary polymers consecutively. In such a case, the weight average molecular weight increases only linearly, and the gel point cannot be determined experimentally. That is, if the gel point is determined based on the insoluble portion of the gel in a good solvent, it is important to note that we only determined the molecular weight at which the polymer becomes insoluble in the used solvent. Although such a situation is extreme, it is important to know that a quite different molecular weight distribution than that of the bulk system is obtained due to the size of the maximum molecules formed in a limited polymer particle size. Thus, it is necessary to introduce a different definition for microgelation in a microspace (such as in suspension polymerization) from that for bulk systems.

1.3.3   Cyclization

It is expected that cyclization is very important in real copolymerization. The difficulty of dealing with cyclization is that it is controlled by the polymer conformation [56]. A long-range correlation problem also needs to be solved. Although there have been a number of models on cyclization reported, unification of the concept has not been successful to date [57–60]. While the reason why the Flory–Stockmayer model ignores cyclization is to simplify analysis, it can also be interpreted based on classical kinetic theory. In a bulk reaction system, there will be an infinite number of molecules and the weight fraction of a molecule other than the gelled molecule is zero. Therefore, by applying the classical kinetic theory where the probability of reaction is expressed by the multiplication of the concentration of functional groups, the probability of a polymer free radical reacting with a pendant double bond will be zero because its weight fraction is zero. This is the fundamental reason why Flory ignored cyclization in his model. As far as the classical kinetic theory is concerned, there will be no cyclization other than in gelled molecules. Although the Stockmayer interpretation calls for inhibition of cyclization even in gelled molecules [6, 63], many modern theories including the ring-free model accept the existence of cyclization in the gelled molecules. However, even in the classical kinetic theory, the sol molecules accept the possibility of cyclization in a reaction within a limited space, such as in suspension polymerization.

The readers are referred to several reviews for information or various problems of real free radical polymerization systems [56, 64–66].

REFERENCES

1   Carothers, W.H. (1936). *Trans. Faraday Soc.*, **32**: 39.
2   Flory, P.J. (1941). *J. Am. Chem. Soc.*, **63**: 3083.
3   Flory, P.J. (1941). *J. Am. Chem. Soc.*, **63**: 3091.
4   Flory, P.J. (1941). *J. Am. Chem. Soc.*, **63**: 3096.
5   Stockmayer, W.H., and Zimm, B.H. (1984). *Ann. Rev. Phys. Chem.*, **35**: 1.
6   Stockmayer, W.H. (1943). *J. Chem. Phys.*, **11**: 45.
7   Stockmayer, W.H. (1944). *J. Chem. Phys.*, **12**: 125.

8   Flory, P.J. (1953). *Principles of Polymer Chemistry*, Chapter IX, Ithaca, New York: Cornell University Press.
9   Gordon, M. (1962). *Proc. Roy. Soc.* (London), **A268**: 240.
10 Gordon, M., and Ross-Murphy, S.B. (1975). *Pure Appl. Chem.*, **43**: 1.
11 Kuchanov, S.I., Koralev, S.V., and Panyukov, S.V. (1988). *Adv. Chem. Phys.*, **72**: 115.
12 Macosko, C.W., and Miller, D.R. (1976). *Macromolecules*, **9**: 199.
13 Durand, D., and Mruneau, C.M. (1982). *Makromol. Chem.*, **183**: 1007.
14 Saito, O. (1958). *J. Phys. Soc. Japan*, **13**: 198.
15 Saito, O. (1972). In *The Radiation Chemistry of Macromolecules*, volume 1, M. Dole, ed., New York: Academic Press, p. 223.
16 Mikes, J., and Dusek, K. (1982). *Macromolecules*, **15**: 93.
17 Donoghue, E., and Gibbs, J.H. (1979). *J. Chem. Phys.*, **70**: 2346.
18 Ziff, R.M., and Stell, G. (1980). *J. Chem. Phys.*, **73**: 3492.
19 Fukutomi, T. (1991). *Macromolecules*, **24**: 616.
20 Thorn, M., Breuer, H.-P., and Honerkamp, J. (1994). *Macromol. Theory Simul.*, **3**: 585.
21 Bamford, C.H., and Tompa, H. (1954). *Trans. Faraday Soc.*, **50**: 1097.
22 Tobita, H., and Hamielec, A.E. (1989). *Macromolecules*, **22**: 3098.
23 Tobita, H., and Hamielec, A.E. (1989). In *Polymer Reaction Engineering*, K.-H. Reichert and W. Geiseler, eds., Weinheim, Germany: VCH, p. 43.
24 Broadbent, S.R., and Hammersley, J.M. (1957). *Proc. Camb. Philos. Soc.*, **53**: 629.
25 Stauffer, D., Conoglio, A., and Adam, M. (1982). *Adv. Polym. Sci.*, **44**: 103.
26 Bansil, R., Herrmann, H.J., and Stauffer, D. (1984). *Macromolecules*, **17**: 998.
27 Boots, H.M.J. (1986). In *Integration of Fundamental Polymer Science and Technology*, L.A. Kleintjens, and P.J. Lemstra, eds., London: Elsevier Applied Science, p. 204.
28 Grest, G.S., Kremer, K., and Duering, E.R. (1992). *Europhys. Lett.*, **19**: 195.
29 Bowman, C.N., and Peppas, N.A. (1992). *Chem. Eng. Sci.*, **47**: 1411.
30 Tobita, H. (1995). *J. Polym. Sci. Polym. Phys. Ed.*, **33**: 1191.
31 Tobita, H. (1996). *Macromol. Theory Simul.*, **5**: 129.
32 Tobita, H. (1997). *Polymer*, **38**: 1705.
33 Zhu, S. and Hamielec, A.E. (1994). *J. Polym. Sci., Polym. Phys. Ed.*, **32**: 929.
34 Tobita, H., and Hatanaka, K. (1995). *J. Polym. Sci., Polym. Phys. Ed.*, **33**: 841.
35 Kimura, T. (1962). *J. Phys. Soc. Japan*, **17**: 1884.
36 Saito, O. (1992). *Statistical Properties of Polymers*, Chapter 5, Tokyo: Chuo University Press.
37 Tobita, H., Yamamoto, Y., and Ito, K. (1994). *Macromol. Theory Simul.*, **3**: 1033.
38 Tobita, H. (1996). *Macromolecules*, **29**: 3010.
39 Tobita, H. (1996). *Macromol. Theory Simul.*, **5**: 1167.
40 Hasegawa, Y., Aoto, H., and Matsumoto, A. (1993). *Konbunshi Ronbunshu*, **42**: 2964.
41 Degoulet, C., Nicolai, T., Durand, D., and Busnel, J.P. (1995). *Macromolecules*, **28**: 6819.
42 Tobita, H., and Yamamoto, K. (1994). *Macromolecules*, **27**: 3389.
43 Flory, P.J. (1947). *J. Am. Chem. Soc.*, **69**: 30.
44 Gould, S.J. (1991). *Bully for Brontosaurus*, Chapter 17, New York: W.W. Norton.
45 Flory, P.J. (1947). *J. Am. Chem. Soc.*, **69**: 2893.
46 Tobita, H., and Hamielec, A.E. (1992). *Polymer*, **33**: 3647.
47 Tobita, H., and Hamielec, A.E. (1990). In *Integration of Fundamental Polymer Science and Technology*, volume IV, P.J. Lemstra and L.A. Kleintjens eds., London: Elsevier Applied Science, p. 33.

48 Tobita, H. (1992). *Macromolecules*, **25**: 1671.
49 Tobita, H., and Nomura, M. (1996). *Kobuhshi Kako*, **45**: 543.
50 Tobita, H. (1993). *Macromolecules*, **26**: 836.
51 Tobita, H., and Zhu, S. *Polymer* (in press).
52 Tobita, H. (1993). *Macromolecules*, **26**: 5427.
53 Tobita, H. (1994). *Macromolecules*, **27**: 5413.
54 Tobita, H. (1996). *Network Polymer*, **17**: 189.
55 Tobita, H. (1995). *Acta Polym.*, **46**: 185.
56 Dusek, K. (1982). In *Polymerization*, R.N. Haward ed., London: Applied Science Publishers, p. 143.
57 Jacobson, H., and Stockmayer, W.H. (1950). *J. Chem. Phys.*, **18**: 1600.
58 Kilb, R.W. (1958). *J. Chem. Phys.*, **62**: 969.
59 Semlyen, J.A. (1976). *Adv. Polym. Sci.*, **21**: 41.
60 Dotson, N.A., Macosko, C.W., and Tirrell, M. (1992). In *Synthesis, Characterization, and Theory of Polymer Network and Gels*, S.M. Aharoni ed., New York: Plenum Press, p. 319.
61 Friedman, B., and O'Shaughnessy, B. (1993). *Macromolecules*, **26**: 4888.
62 Suematsu, K., Okamoto, T., Kohno, M., and Kawazoe, Y. (1993). *J. Chem. Soc. Faraday Trans.*, **89**: 4181.
63 Tanaka, F. (1994). *Physics of Polymers*, Tokyo: Shokabo Publ., p. 103.
64 Ooiwa, M. (1976). *Crosslinking and Gelation, Copolymerization* (2), Tokyo: Baifu-kan Publ., p. 400.
65 Matsumoto, A. (1992). *Polymerization Accompanying Gelation, Synthesis and Reaction of Polymers* (1), Tokyo: Kyoritsu, p. 94.
66 Matsumoto, A. (1996). *Adv. Polym. Sci.*, **123**: 41.

Section 2
Evaluation of Gel Point

JOH TANAKA

2.1 INTRODUCTION

As often defined in dictionaries, gels are sols such as in collodial solution, which solidify by losing fluidity. In order to obtain gels, one would start with a solution and pass it through a process of solidification by adding special conditions. Therefore long-term discussions have centered on where the borderline between sols and gels lies. This relates to *gel preparation* and, thus, it is always discussed any time gels are studied or developed, regardless of differences in nuance. In other words, this relates to the question, What are gels?

Judging from the literature on the sol/gel boundary (i.e., the gel point), they can be classified into three groups. They are those that deal with the gel point as a function of solution concentration, temperature, or reaction time. When gels are considered as infinite networks that have a 3D connection of solute molecules and contain solvent, the minimum necessary amount of solute molecules to form infinite networks is the gel point with respect to the solution concentration. For those treated as a function of temperature, they are mostly solutions that form gels by increasing or decreasing the temperature. They include those familiar in our daily life, such as agar, gelatin, and egg white. In these cases, one often observes a phenomenon in which the gels formed revert to sols by reversing thermal history (thermoreversible sol/gel transition). For example, in agar gelatin, higher temperature sols are cooled and, when they solidify, this is their gel point. The temperature at which gels start to flow when heating occurs is called the melting point. It is also possible to observe differences between the gel point and the melting point. The formation process of synthetic polymer gels, such as the polymerization of multifunctional monomers or the reaction between a linear polymer and crosslinking agent, can be regarded as an example of a gel point being treated as a function of reaction time. The gel point is the time when the gel starts forming after the time the reaction started. The gel point from the viewpoints of solution concentration, temperature, and reaction time is not independent but mutually related. Various characteristics of the correlations depend on the chemical entity.

Methods often used for determining the gel point include the inverted test tube method, fallen ball method, viscoelasticity measurement, and differential scanning calorimetric (DSC) method. In what follows experimental studies on the gel point and melting point from the viewpoint of the lost fluidity of sols will be individually introduced, with an emphasis on experimental procedures.

2.2   INVERTED TEST TUBE METHOD

This method is the simplest one used to study the gel point of a solution or the melting point of a gel. This method is often used to learn the condition of gelation qualitatively [1–3]. However, it lacks generality in that one merely observes whether or not the sample flows upon inversion of a test tube. Therefore, it is necessary to provide both qualitative and quantitative data, such as the amount of sample and the diameter of the test tube when the results are written and disseminated in research papers, etc. When the gel point and melting point of thermoreversible gels are measured, for example, sufficient care must be paid to thermal history, such as the isothermal temperature and heating rate of the sample.

The Eldridge–Ferry study is famous for the measurement of melting point by the inverted test tube method in which qualitative data are involved [4]. In this study, a quantitative evaluation of the relationship between the transition temperature and polymer concentration for the sol-gel transition of gelatin aqueous solutions is reported. The experimental method involves a gel formed by keeping a certain amount of an aqueous solution of gelatin in an ice bath for 24 h. Then, they determined the melting point T_m as the temperature at which the gel dropped to the bottom of the glass tube when it was heated at $12°C/hr$. This method is reputedly able to determine the melting point precisely with the data obtained having been determined to within $\pm 0.2°C$.

Ferry [22] found linearity when $\log c$ and $1/T_m$ were plotted, where c was the concentration of gelatin and T_m was the melting point of the gel expressed in absolute temperature. Similarly, they also found linearity between $\log M$ and $1/T_m$. The equation derived from the van't Hoff equation is used to express this linear relationship:

$$\log c = \frac{\Delta H_m}{2.303 R \cdot T_m} + \text{const} \tag{1}$$

where R is the gas constant and ΔH_m is the heat of the reaction of the equilibrium formation.

$$2 \text{ moles of crosslinking sites} \rightleftharpoons 1 \text{ mole of crosslink} \tag{2}$$

This is based on the concept that a gelatin solution gels by forming 3D networks with many crosslinking reactions of polymer chains [5] and, at the gel point (or melting point), chemical equilibrium is maintained. Upon deriving Eq. (2), they assumed that gelation takes place when the cross-linked regions reach a certain number, and there is no intramolecular (coupling between some parts of the same molecule) crosslinking. This relates to the melting point measurement using the inverted test tube method. It is necessary to disregard the crosslinks that do not contribute to the elasticity of the gel, that is, the intramolecular crosslinks, considering that gels are formed by the loss of fluidity and appearance of elasticity (on the other hand, the elasticity disappears and fluidity appears in the melting of gels). It is possible to calculate the heat of reaction for the polymer chain coupling, and this method is still frequently used [6].

There are studies to correlate the spacing of crystals that exist in a gel and the gel point obtained by the inverted test tube method. Mandelkern's group noticed gel formation when a linear polyethylene in organic solvents, such as toluene, p-zylene, and decalin, was carefully cooled. They performed a series of experiments to investigate the effect of molecular weight and concentration of the polyethylene on the gelation [3]. Although the gelation of the polyethylene solution is caused by crystallization of part of the polymer chain, a dilute solution will not gel even if crystals are formed. They used samples with molecular weights ranging from 3×10^3 to 6×10^6 and estimated the minimum gelation concentration, c^*. As the gel point increases from 25°C to 86°C, the value of c^* varies by a factor of four.

It has been found that the influence of such a gelation temperature also appears on the spacing, which is determined by small angle x-ray diffraction. The spacing increases as the gelation temperature increases. This variation of spacing is also observed for crystals obtained from dilute solutions below c^*. Its dependence on crystallization shows similar behavior at gelation above c^*. Furthermore, the effect of thermal history on gelation was studied and interesting results have been obtained. Using the gels obtained by quenching, the size of the crystallite of polyethylene with various molecular weights is approximately constant. Even in the crystals obtained by quenching dilute solutions of less than c^*, the size of the crystallites is independent of the molecular weight. Considering these experimental results, it seems that the growth process of crystals is continuous with respect to the concentration of polyethylene. Also, similar crystal growth occurs regardless of the concentration being higher or lower than c^*.

Other than this example, they also studied a wide variety of polymers, such as ethylene-butadiene copolymer, ethylene-vinylacrylate copolymer, and their branched polymers. They indicated that, for cases where the junction zone includes crystals, they could not verify functionality, which is equivalent to the number of polymers that form the junction zone.

Later, Tanaka and Nishinari [7] estimated the structure of the junction zone using the experimental results reported by Domszy *et al.* [3]. They proposed the model shown in Fig. 1 for the junction zone formed by the aggregation of polymer chains. They also derived parameters that correspond to the number of aggregating chains and the parameter ζ that corresponds to the number of chemical repeat units forming the crosslink [8, 9]. The parameter s is called junction multiplicity. In this model, gels are formed with crystallites in the junction zone.

In its formation process, the junction zone is thought to be formed by gathering the functional group consisting of ζ chemical units. The change of free energy during the formation, ΔG_0, is expressed as:

$$\Delta G_0 = \zeta(\Delta H_0 - T \cdot \Delta S_0) \tag{3}$$

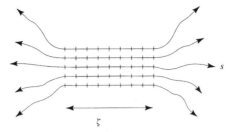

Fig. 1 A junction-zone model that consists of eight polymer chains and ζ segments [7].

where the changes in enthalpy and entropy of an individual chemical repeat unit are ΔH_o and ΔS_o, respectively. Their approach is based on the statistical method [10] and they proposed an equation to correlate the concentration, temperature, and molecular weight at the gel point:

$$\ln c^* = \zeta \frac{\Delta H_o}{k_B T} - \frac{1}{s-1} \ln M + \text{const} \tag{4}$$

In reality, there is a possibility of having distribution on junction multiplicity and variation in the junction zone size. However, in their analysis of crosslinking, they assumed uniform multiplicity and the same junction zone size. Using Eq. (4), it is possible to estimate the values of the parameters s and ζ by plotting c^* against $(10^3/T + \ln M)$ from gel point measurement. Namely, as shown in Fig. 2, s can be calculated from the slope B of the plot, c^* vs T, while keeping M constant. Also, ζ can be estimated from the slope A of the plot c^* vs M, while keeping T constant.

Figure 3 shows the plot, c^* vs $(10^3/T + \ln M)$, of the melting point data of the polyethylene-toluene and polystyrene-carbon disulfide gels. Figure 3(a) illustrates the gel point reported by Domszy et $al.$ [3] using polyethylene-toluene systems. The solid line indicates T dependence of c^* at a constant molecular weight and shows good linearity. Polyethylene is a polymer with a high degree of crystallinity and the heat of fusion of the segment per mole for bulk crystals $(\Delta H_o)_{mol}$ has been reported [11]. Inserting this value and A, which may be obtained from the slope, into the following equation, one obtains $\zeta = 6.51$:

$$\zeta = \frac{10^3 k_B}{|\Delta H_o|} A = \frac{10^3 R}{|(\Delta H_o)_{mol}|} A \tag{5}$$

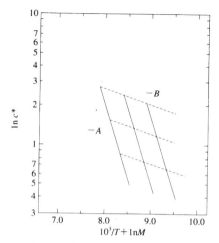

The solid line expresses the relationship between c^* and temperature under constant molecular weight. The dotted line is for the relationship between c^* and molecular weight under constant temperature.

Fig. 2   Schematic diagram of the plot to calculate s and ζ parameters from Eq. (4).

On the other hand, the relationship between c^* and M at a constant gelation temperature is somewhat lacking in linearity. Nonetheless, inserting the slope in the following equation gives $s = 4.0$ which roughly estimates 26 chemical repeat units in the junction zone:

$$s = \left(\frac{1}{B}\right) + 1 \qquad (6)$$

The foregoing calculation ignored the change in free energy due to chain folding. The chemical repeat unit used is from a statistical point of view, which most likely corresponds to several monomer units. It is important to consider these points when detailed analysis is desired.

Figure 3(b) depicts the gel point of polystyrene-carbon disulfide systems and the plot at constant molecular weight shows good linearity.

There are several reports on the gelation of polystyrene solutions. However, no unified interpretation has yet been obtained [12, 13]. Although there is a theory that the junction zone contains solvent

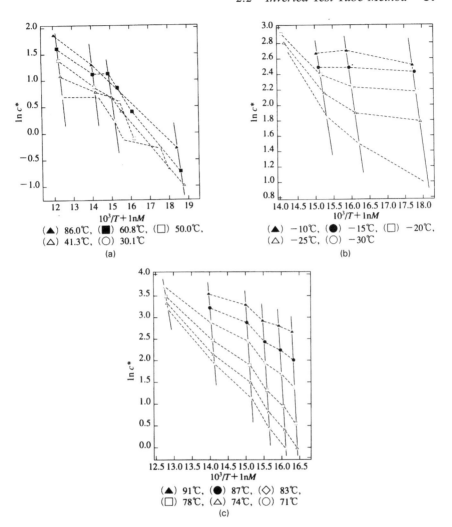

Fig. 3   (a) c^* vs $(10^3/T + \ln M)$ plot [7] (polyethylene-toluene gel), (b) c^* vs $(10^3/T + \ln M)$ plot [7] (polyethylene-carbon disulfide gel), (c) c^* vs $(10^3/T + \ln M)$ plot [7] (poly(vinyl alcohol)-water gel).

molecules, in this section, the segments that contribute to crosslinking are assumed to follow the structure illustrated in Fig. 1. Here again, $\zeta = 3.43$ is obtained by using the heat of fusion of the bulk crystal [11]. The slope at constant T increased as the temperature decreased. The calculated values indicate $s = 16.7$ for higher temperature at $T = -10°C$, and $s = 3.4$ for lower temperature at $T = -30°C$. Based on these results, junction multiplicity is regard as temperature dependent. The melting point plot of the gels from a poly(vinyl alcohol) aqueous system (Fig. 3 (c)) also shows a similar trend and the plot at constant T indicates increased slope with decreased temperature. The calculated results indicate $s = 3.6$ at $T = 91°C$ and $s = 2.1$ at $T = 71°C$, suggesting somewhat fewer polymers compose the junction zone. Such a temperature dependence indicates that the higher the melting temperature of the gel, the thicker the crosslinking, which is consistent with the thermal properties of poly(vinyl alcohol) gels [14, 15].

The c^* vs $(10^3/T + \ln M)$ plot proposed by Tanaka and Nishinari [7] and Tanaka and Stockmayer [9] is the method used to independently calculate the parameter s and ζ. It is an effective approach to characterize the crosslink structure of the thermoreversible gels.

2.3   FALLEN BALL METHOD AND U-SHAPED TUBE METHOD

The fallen ball method is the method used to judge gelation by placing a small ball in a sample and is often used as a qualitative method, as is the inverted test tube method shown in Fig. 4. It is necessary to clarify the diameter, material and density of the ball, and the diameter of the test tube.

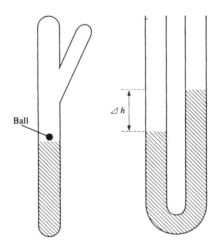

Fig. 4 Schematic diagram of the glass tube used for the fallen ball and U-shaped tube method (gel is prepared in the hatched area).

The U-shaped tube method is used for thermoreversible gels with relatively low modulus ($\sim$100 dyne/cm^2). As shown in Fig. 4, by preparing a gel in such a way that there are differences in meniscus height, the melting point is determined from the change in height. In the experiment reported by Harrison *et al.* [16], the amount of the sample used was as little as 0.3 ml when a capillary 0.25–1 mm in diameter is used at the bottom of the tube and the distance between the two vertical tubes is 20 mm.

They measured the difference in the meniscus Δh by increasing the temperature of the gels, such as methyl acrylate-vinylidene chloride copolymer and determined the melting point from the breakpoint of the log Δh–temperature plot. It was reported that the melting point was not affected by the inner diameter of the capillary or the Δh prior to melting.

Takahashi *et al.* also determined the melting points of various synthetic polymer gels using a U-shaped tube with a capillary having a 1 mm inner diameter [17]. They reported the data with a good reproducibility of $\pm 1°$C when the melting point was measured on the same sample by repeated sol-gel transitions. They also reported that the results were the same despite different initial Δh. From the experimental results, they concluded that crystals are incorporated in the crosslink junction and, thus, they modified Eq. (1) in order to evaluate the number of monomer units constituting the crystals. It was also shown that gelation by increased temperature of a system, such as in a methyl cellulose-water system, and the effect of solvent on the melting point can both be explained [18–20].

2.4 VISCOELASTICITY METHOD

The three methods described thus far are simple methods for measuring the loss of the fluidity of sol and the appearance of elasticity [21]. However, they merely observe apparent phenomena. In other words, as viscosity and elasticity of polymeric materials depend on measurement time, it is necessary to observe the loss of fluidity and appearance of elasticity of materials with long relaxation time, such as gels, over a long period of time. In this sense, dynamic mechanical spectroscopy is useful. Readers are referred to monographs on dynamic mechanical spectroscopy [22–24]. In this section, a few examples of viscoelastic behaviors observed at the gel point will be discussed.

Tung and Dyres [25] measured viscoelastic properties of various thermosetting resins during the gelation process and compared their results with the gel time determined by the American Materials Testing Standards (ASTM). The gel time measurement method of ASTM is used for thermosetting resins. It involves placing a probe stick into the reacting sample, pulling it out at certain time intervals, and determining the gel time when the sample no longer sticks to the wall of the probe [26]. Figures 5 and 6 illustrate the storage moduli (G') and loss moduli (G'') of an epoxy resin and sulfone resin, respectively. For both polymers, G'' is greater than G' at the beginning of the reaction and afterwards at a certain point G' becomes greater than G''. Interestingly, the $G' = G''$ crossover time agrees well with the gel time determined by the ASTM method. Although they observed this phenomenon for as many as 11 samples as shown in Fig. 7, the mechanism was not clear.

What kind of viscoelastic behavior can be observed in samples that are intermittently removed during the gel formation process?

Researchers have explored the possibility of conducting such an experiment using polydimethylsiloxane (PDMS) [27] and examined the G' and G'' behavior at the gel point [28]. Figure 8 shows the results of G' and G'' as a function of time at a constant frequency. A similar trend as reported by Tung and Dynes has been observed. Figure 9 is the frequency dependence of G' and G'' for the samples whose reaction is frozen at the

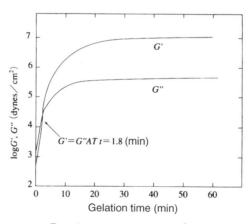

Reaction temperature is 171°C and
measurement frequency is 10 rad/s.

Fig. 5 The time dependence of G' and G'' of an epoxy resin during curing [25].

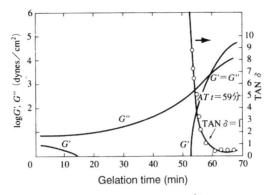

Reaction temperature is 171°C and
measurement frequency is 10 rad/s.

Fig. 6 Time-dependent changes of *G′* and *G″* during curing of sulfone
resin [25].

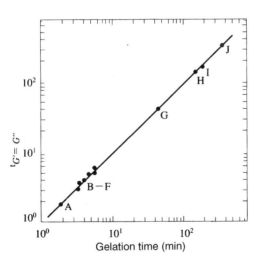

(A)–(F) are epoxy resin, reaction temperature is 171°C
(G) is sulfone resin, reaction temperature is 150°C
(H)–(J) are silicone resin, reaction temperature is 25°C.

Fig. 7   The correlation between the gelation time determined by ASTM and
the time for $G′ = G″$ crossover [25].

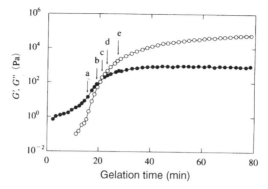

Fig. 8 Time-dependent changes of G' and G'' during crosslinking of poly-dimethylsiloxane [28].

times indicated by the arrows in Fig. 8. In order to avoid large overlap, the lines are shifted along the abscissa. Lines a and b correspond to short reaction times and their G' and G'' show steeper slopes at lower frequencies. The lines d and e are for the long reaction times and G' shows a low-frequency plateau, which seems to indicate a lack of viscoelastic relaxation. In the middle reaction time c, it is $G' = G''$ regardless of the frequency and line c shows good linearity in the log–log plot.

This kind of G' and G'' behavior is regarded as the reflection of the structural variation due to the crosslink network formation via the cross-

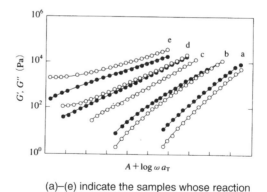

(a)–(e) indicate the samples whose reaction is halted at the times specified in Fig. 8.

Fig. 9 Dynamic mechanical behavior of polydimethylsiloxane for which reaction is halted during curing [28].

linking reaction. Although it may be difficult to explain this causal relationship, the fact that G' no longer shows relaxation at times d and e seems to be consistent with the concept of gelation where elastic response can be observed even for a long time deformation due to the formation of infinite networks.

For line c in Fig. 9, which corresponds to the viscoelastic behavior of the gel point, Winter and Chambon [28] have proposed the following empirical equation:

$$G', G'' \sim \omega^{1/2} \tag{7}$$

where ω is the frequency.

They considered that gels show the same mechanical properties whether or not the crosslink formation is by covalent bonding or physical aggregation. They expected Eq. (7) generally to characterize gelation viscoelastically. They also stated that it may be feasible to correlate with the change of other properties accompanying gelation, such as the strong restriction of the mobility of solute molecules in comparison with sols and the formation of insoluble components.

2.4.1 Conclusions

In this section, several examples of studies on the evaluation of the gel point from the viewpoint of the loss of sol fluidity have been introduced. The majority of these studies focused on the classification of the characteristics of the crosslink structure. In the future, the research focus will likely shift towards an interpretation and correlation of gel properties with respect to parts other than crosslinks and spatial distribution of crosslinks.

Although it was not mentioned in this section, the gel point is regarded as an application example of a critical phenomenon or an example of a phase transition [29–31]. In these studies, it was expected that a universal exponential law exists between the viscous and elastic properties of gels and the equivalent quantity that describes the deviation from the gel point. However, this exponential law is not well understood from the molecular structural point of view. The gel point may appear to be an especially simple property among various properties of gels. However, there are many unanswered problems and future progress in this area is desired.

REFERENCES

1   Iwano, S., Kikuta, N., Sakai, K., and Izumi, Y. (1993). *Polym. Preprints Jpn.*, **42**: E1014.
2   Edwards, C.O., and Mandelkern, L. (1982). *J. Polym. Sci. Polym. Lett. Ed.*, **22**: 355.
3   Domszy, R.C., Alamo, R., Edwards, C.O., and Mandelkern, L. (1986). *Macromolecules*, **19**: 310.
4   Eldridge, J.E., and Ferry, J.D. (1954). *J. Phys. Chem.*, **58**: 992.
5   Veis, A. (1964). *Macromelecular Chemistry of Gelatin*, New York: Academic Press.
6   Watase, M., and Nishinari, K. (1993). *Food Hydrocolloids*, **7**: 449.
7   Tanaka, F., and Nishinari, K. (1996). *Macromolecules*, **29**: 3625.
8   Stockmayer, W.H. (1991). *Macromolecules*, **24**: 6367.
9   Tanaka, F., and Stockmayer, W.H. (1994). *Macromolecules*, **27**: 3943.
10 Fukui, K., and Yamabe, T. (1967). *Bull. Chem. Soc. Jpn.*, **40**: 2052.
11 Mandelkern, L. (1964). *Crystallization of Polymers*, New York: McGraw-Hill Book Co., p. 119.
12 Tan, H.M., Moet, A., Hiltner, A., and Baer, E. (1983). *Macromolecules*, **16**: 28.
13 Gan, J.Y.S., Francois, J., and Guenet, J. (1986). *Macromolecules*, **19**: 173.
14 Shibatani, K. (1970). *Polym. J.*, **1**: 348.
15 Ogasawara, A., Nakajima, T., Yamamura, K., and Matsuzawa, S. (1975). *Prog. Colloid Polym. Sci.*, **58**: 145.
16 Harrison, M.A., Morgan, P.H., and Park, G.S. (1971). *Br. Polym. J.*, **3**: 154.
17 Takahashi, A., Sakai, M., and Kato, T. (1980). *Polym. J.*, **12**: 335.
18 Takahashi, A., Nakamura, T., and Kagawa, I. (1972). *Polym. J.*, **3**: 207.
19 Takahashi, A. (1973). *Polym. J.*, **4**: 379.
20 Kato, T., Yokoyama, M., and Takahashi, A. (1978). *Colloid Polym. Sci.*, **256**: 15.
21 Osaga, Y., and Ross-Murphy, S.B. (1993). *Sci. Am.*, **268**: 82.
22 Ferry, J.D. (1964). *Viscoelasticity of Polymers*, H. Sofue, K. Murakami, and M. Takahashi, Trans., Tokyo: Tokyo Kagaku Dojin (originally published in English).
23 Nakagawa, T. (1960). *Rheology*, Tokyo: Iwanami Publ.
24 Ozaki, K., and Masuda, T. (eds.) (1992). *Lecture Series: Rheology*, Kobunshi Kankokai.
25 Tung, C.Y.M., and Dynes, P.J. (1982). *J. Appl. Polym. Sci.*, **27**: 569.
26 (1970) *Annual Book of ASTM Standards*, pp. D2471-71.
27 Valles, E.M., and Macosko, C.W. (1979). *Macromolecules*, **12**: 521.
28 Winter, H.H., and Chambon, F. (1986). *J. Rheology*, **30**: 367.
29 Stauffer, D., Coniglio, A., and Adam, M. (1982). *Adv. Polym. Sci.*, **44**: 103.
30 de Gennes, P.G. (1984). *Physics of Polymers*, S. Kubo, ed., Yoshioka Publ.
31 Horie, K. (1982). *Netsukokasei Jushi (Thermosetting Polymers)*, **3**: 78.

Section 3
Theory of Swelling

TOSHIYA HIROTSU

3.1 STATIC PROPERTIES

The most noted characteristic of gels is that of swelling. Responding to external conditions, gel networks swell by absorbing solvent or shrink by expelling the solvent until thermodynamic equilibrium is reached. This equilibrium is called swelling equilibrium. Equilibrium is determined by interaction between the gel and solvent and responds sensitively to various conditions such as temperature, solvent composition, pH, hydrostatic pressure, and external electric field, even if it is the same network. The degree of swelling also increases significantly by fixing ions on the networks. Furthermore, depending on the network and solvent, a volumetric phase transition may occur. The volume will change as much as 10–100 times at the phase transition point, responding to even a slight change of external conditions. There are no solids other than gels that show such phenomena. In this section an explanation of the fundamentals of thermostatical theory elucidates the phenomenon of swelling equilibrium. The volumetric phase transition will also be reviewed as a typical example of extreme sensitivity of the swelling equilibrium to temperature.

65

3.1.1 Fundamental Theory of Swelling

3.1.1.1 Excluded volume and θ temperature

From the viewpoint of statistical mechanics, a similar solution to the one used for a nonideal gas applies. In other words, if the intermolecular potential of a gas is replaced with the effective potential among solute molecules, which is the interaction transmitted through solvent molecules, the partition function will be in the same form. In the case of a polymer solution, each segment can be regarded as a gaseous molecule. Using this similarity, the theoretical approach developed for a nonideal gas can be applied to polymer solutions. It is easier to understand the excluded volume and θ temperature from this point of view [1, 2].

Any molecule possesses a region where other surrounding molecules cannot invade. This region is called the excluded volume. The fundamental reason for the swelling phenomenon of gels is this excluded volume. If the two-body interaction potential is expressed as $U(r)$, the excluded volume V_{ex} is given by:

$$V_{ex} = 4\pi \int_0^\infty \left(1 - \exp\left[\frac{-U(r)}{k_B T}\right]\right) r^2 \, dr \tag{1}$$

Using the Lennard-Jones type potential where the expulsion term is replaced by a rigid sphere:

$$U(r) = \begin{cases} \infty & (r < r_0) \\ -u_0(r_0/r)^6 & (r \geq r_0) \end{cases} \tag{2}$$

where u_0 and r_0 are constant, the excluded volume at temperature T is expressed as:

$$V_{ex} = \frac{4\pi}{3} r_0^3 \left(1 - \frac{u_0}{k_B T}\right) \tag{3}$$

where k_B is the Boltzmann constant. In this case, the excluded volume is small at low temperature and large at high temperature.

In Eq. (3), if temperature T equals u_0/k_B, V_{ex} is zero. This appears as though the forces cancel each other out in the gas as a whole despite the fact that each molecule impinges on the others with the force dependent on the distances between molecules. It is a strange situation. Equation (1) for the excluded volume is the same equation that gives the second virial coefficient b_2 in the virial series of nonideal gases. Therefore, the zero excluded volume is equal to the disappearance of the second virial coefficient. The temperature at which b_2 of a gas disappears is called the Boyle temperature. It is also known as the temperature at which the ideal gas law applies to actual gases. As described, this means that the gas as a whole appears as though there are no intermolecular interactions at this temperature.

In a polymer solution, the temperature at which $V_{ex} = 0$ is called the θ temperature, or more generally the θ point. At this point, polymer chains behave as though they were ideal Gauss chains. At $V_{ex} > 0$ the polymer chains spread wider than the ideal chains and at $V_{ex} < 0$ they shrink. In a gel where chains are crosslinked, the change of the spreading of an individual chain is observed as macroscopic volumetric changes of the networks.

3.1.1.2 *Free energy of gels*

Because a gel can be regarded mechanically as a solid and thermodynamically as a solution, the thermodynamics of gels can be formulated as an expansion of that of polymer solutions; however, there is a macroscopic deformation of networks that does not exist in solutions and calculation of its deformation entropy is somewhat challenging. Here this calculation will be briefly discussed and readers are referred to the original literature for more details.

If the Gibbs free energy that accompanies the deformation of a gel is written as ΔG, it is $\Delta G = \Delta G_m + \Delta G_e$ where ΔG_m is the contribution from the mixing of networks and solvent and ΔG_e is from the elasticity. If the networks contain ions, their contribution must be taken into account. This will be discussed later so for the time being neutral networks will be considered.

3.1.1.2.1 Free energy of mixing, ΔG_m

With respect to ΔG_m the results of polymer solutions may be used as they are. If the entropy of mixing is written as ΔS_m and the enthalpy of mixing as ΔH_m, then $\Delta G_m = \Delta H_m - T \, \Delta S_m$ where ΔS_m is the change of entropy when n_1 solvent molecules and n_2 polymer chains mix and form a homogeneous solution [3]. At this point, unlike in the case of mixing of small molecules, calculation issues are involved because the sizes of solvent and solute molecules are extremely different. As well, a number of polymer conformations must be taken into account. However, the results based on a lattice model can be expressed in the following simple form:

$$\Delta S_m = -k_B[n_1 \ln(1 - \phi) + n_2 \ln \phi] \tag{4}$$

where k_B is the Boltzmann constant and ϕ is the volume fraction of the solute. Equation (4) takes the form of replacing the mole fraction of the ideal entropy of mixing into a volume fraction. Next, we will consider change of enthalpy upon mixing. Also, using the lattice model, the change of internal energy of mixing is generated by replacement of the contact of the same species, such as solvent–solvent and segment–segment, with dissimilar species such as solvent–segment. Therefore, the parameter ε, which expresses the change of contact energy, will be introduced as follows:

$$\varepsilon = u_{12} - \frac{(u_{11} + u_{12})}{2} \tag{5}$$

where u_{ij} is the energy of i, j pair formation, and i (or j) $= 1$ and 2 are solvent and solute molecules, respectively. If $\bigcirc$ is solvent and $\bullet$ is solute, a schematic representation of mixing is $\bigcirc\!-\!\bigcirc + \bullet\!-\!\bullet \rightarrow 2\bigcirc\!-\!\bullet$, and the change of internal energy accompanying this reaction is 2ε.

By multiplying the number of contacts p between solvent and solute by ε, the change of contact energy for the entire system can be obtained. The quantity p can be given by the multiplication of valence number z, the number of solvent molecules n_1, and the volume fraction of polymer ϕ. Thus, the enthalpy of mixing is

$$\Delta H_m = p\varepsilon = zn_1\phi\varepsilon = k_B T n_1 \phi \chi \tag{6}$$

where χ is defined as:

$$\chi = \frac{z\varepsilon}{k_B T} \tag{7}$$

and is the quantity expressing the contact energy per solvent molecule divided by $k_B T$. It is called the polymer-solvent interaction parameter or the χ parameter. Using ΔS_m and ΔH_m obtained here, the free energy of mixing is expressed by the following equation:

$$\Delta G_m = k_B T[n_1 \ln(1 - \phi) + n_2 \ln \phi + \chi n_1 \phi] \tag{8}$$

3.1.1.2.2   Free energy of elasticity ΔG_e
Because the elasticity of polymer networks is of an entropy form, the elasticity term can be written as $\Delta G_e = -T \, \Delta S_e$ defining ΔS_e as the change of entropy due to deformation including swelling of networks [3]. We will introduce the deformation ratio that is expressed as the ratio of the length before and after elongation as $\alpha = L/L_0$ where L_0 and L are the

length before and after deformation. In a general deformation, α is a tensor. However, α is regarded as scalar because we will consider only isotropic swelling, although the following discussion can be readily extended to anisotropic deformation. For swelling, it is customary to choose the time at which the network is formed as the reference state (L_0) of the definition of α. This is because the network is regarded as completely relaxed due to the lack of force among chains at this time.

To calculate the change in entropy of network swelling, it is necessary to know the change of entropy to form the network with the deformation which is expressed by the deformation parameter α from the N_c polymer chains at the reference state. This can be expressed by the sum of the change of entropy ΔS_1, which is generated by placing the N_c uncrosslinked and undeformed chains into the position of deformed networks, and the change of entropy ΔS_2, which is generated by the crosslink formation within a pair of such chains.

First, we will obtain ΔS_1. For this, it is necessary to have a statistical parameter expressing the individual chains comprising the networks (see Footnote 1 on p. 71) in addition to the parameter α that defines the macroscopic deformation of networks.

For such a quantity, the end-to-end distance $\boldsymbol{R}$ of the chain will be used. Also, the microscopic deformation of an individual chain is assumed to be proportional to the macroscopic deformation expressed by α. This is called the affine deformation assumption. It expresses the main characteristics of the rubber elasticity theory even with its simplicity. However, it is also known that there are experimental observations that cannot be explained by this assumption. Also, the number of closed loops rather than the number of chains is essential because real networks contain chains that do not contribute to elasticity. Although it is necessary to discuss these subjects for rigorous rubber elasticity theory, they will not be referred to as we will explain only the fundamental elasticity theory.

The calculation of the distribution changes of $\boldsymbol{R}$ uses a random-walk model on a 3D lattice. Expressing the Cartesian coordinates as (x, y, z) and assuming each chain as a Gauss chain, the distribution probability of the components of $\boldsymbol{R}$ is expressed as:

$$P(x, y, z)\mathrm{d}x\,\mathrm{d}y\,\mathrm{d}z = C \exp\{-\beta^2(x^2 + y^2 + z^2)\}\mathrm{d}x\,\mathrm{d}y\,\mathrm{d}z \qquad (9)$$

where C is the normalization constant, $\beta = \{3/(2nb^2)\}^{3/2}$, n is the number of random walks, and b is the distance of each walk. Equation (9) gives the probability of each component of R being between x and $x + dx$, y and $y + dy$, z and $z + dz$, respectively.

Assuming that the collection of N_c chains is deformed with the deformation parameter α, if the number of chains whose end-to-end distance vector is between (x_i, y_i, z_i) and $(x_i + dx, y_i + dy, z_i + dz)$ is N_i after deformation, then the end-to-end distance vector of these N_i chains should have been between $(x_i/\alpha, y_i/\alpha, z_i/\alpha)/\alpha$ and $[(x_i + dx)/\alpha, (y_i + dy)/\alpha, (z_i + dz)/\alpha]$ where the subscript i denotes the end-to-end distance vector rather than the chains. What we wish to obtain is the probability of the deformation expressed by (x_i, y_i, z_i) for the N_i chains prior to deformation (see Fig. 1). This can be obtained similarly to the entropy change for the volumetric change of an ideal gas. With

$$N_i = C \exp\{-\beta^2(x^2 + y^2 + z^2)/\alpha^2\}dx\ dy\ dz/\alpha^3$$
$$P_i = C \exp\{-\beta^2(x^2 + y^2 + z^2)\}$$

The probability Ω_1 is given by:

$$\Omega_1 = \left(\frac{N_c!}{\prod_i N_i!}\right) \prod_i P_i^{N_i}$$

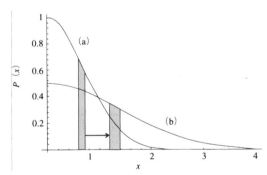

By stretching twice towards the X direction, the distribution function $P(X)$ of the X component of the terminal vector of the polymer chain changes from curve (a) to curve (b). The chains included in the special small region change as indicated by the arrow.

Fig. 1   Conceptual diagram for the calculation of deformation entropy of networks.

where the elements within the parenthesis are the consequences of the equivalence among chains with the same R. Taking logarithms, using Steeling's approximation, and, further, integrating the sum over i,

$$\Delta S_i = \ln \Omega_1 = -\left(\frac{3N_c}{2}\right)(\alpha^2 - 1 - 2\ln\alpha) \tag{10}$$

is obtained.

Next, the change of entropy ΔS_2, which accompanies crosslink formation, will be considered. Crosslinking the chain end of each pair is nothing but entering into a microvolume δ by pairing the terminal segments. Assuming the total volume of a solution to be V, the probability of connecting N_c chains[1] in such a manner is:

$$\Omega_2 = (N_c - 1)(N_c - 3)(N_c - 5)\cdots(1)\left(\frac{\delta}{V}\right)^{N_c/2}$$

$$\doteqdot \left(\frac{N_c}{2}\right)! 2^{N_c/2}\left(\frac{\delta}{V}\right)^{N_c/2}$$

Substituting V with $\alpha^3 V_0$, and eliminating the term that has nothing to do with deformation, we obtain:

$$\Delta S = \ln \Omega_2 = -\left(\frac{3}{2}\right)N_c \ln \alpha \tag{11}$$

From Eqs. (10) and (11), the entropy change for the deformation from the reference state ($\alpha = 1$) is

$$\Delta S = -\left(\frac{3k_B}{2}\right)N_c(\alpha^2 - 1 - \ln\alpha) \tag{12}$$

Accordingly, the change of free energy for the swelling of gels is

$$\Delta G = n_1 k_B T[\ln(1-\phi) + \chi\phi] + \left(\frac{3}{2}\right)N_c k_B T(\alpha^2 - 1\ln\alpha) \tag{13}$$

Because $n_2 = 1$ for networks, the term that contains n_2 is neglected.

3.1.1.3 *Polymer-solvent interaction parameter*

Let us re-evaluate the meaning of χ because it is the central quantity that determines the swelling equilibrium with the free energy derived in the

[1] In the problem of networks, the word chain means the portion from one crosslink point to the other. In this sense, it is sometimes called a partial chain. However, in this chapter, we will simply call it a chain.

previous section. The quantity ε introduced in Eq. (5) has been regarded as the change of energy upon contact. However, in cases in which the solute has an environment-dependent internal structure, or the solvent molecules surrounding the solute have special intermolecular structures, such as water, it is necessary to consider entropy change brought about by local structural changes [1, 3]. Thus, letting the changes of enthalpy and entropy be δ_h and δ_s, respectively, $\chi = (\delta_h - T\delta_s)/k_B T$.

In a system that possesses an upper critical solution temperature, δ_h and δ_s are both positive, χ decreases as the temperature increases, and, thus, the concentration ϕ decreases. That is, the excluded volume increases as the temperature increases and the volume of the gel also increases. On the other hand, in the system that possesses a lower critical solution temperature, the opposite phenomenon will be observed. Also, if δ_h and δ_s are constant, χ should not depend on the polymer concentration. However, the χ's determined from the activity measurement of various polymer solutions have been found to exhibit concentration dependence [4, 5]. This fact suggests that the change of free energy due to the contact between solvent and segment contains contributions not only from 2-body interactions but also from more than 3-body interactions. This also has a strong relationship with the mechanism of volumetric phase transition that will be discussed later.

Next, the relationship between χ and excluded volume will be described. The similarity between a polymer solution and a gas can be considered if the osmotic pressure π of the solution corresponds to the pressure P of the gas and the volume of the solution to the volume V of the gas. Then the van't Hoff equation, $\pi M/c = RT$, corresponds to the constitutive equation of an ideal gas, $PV = RT$, where c is the mass concentration in a unit volume and M is the molecular weight.

From the free energy of mixing of gels derived in the previous section, we obtain the osmotic pressure π as follows:

$$\pi_m = -\left(\frac{N_a}{v_1}\right)\left(\frac{\partial\,\Delta G_m}{\partial n_1}\right)$$
$$= -\left(\frac{N_A k_B T}{v_1}\right)\left[\left(1 - \frac{1}{x}\right)\phi + \ln(1 - \phi) + \chi\phi^2\right] \tag{14}$$

where N_A is Avogadro's number, χ is the number of segments included in a polymer chain and v_1 is the molar volume of the solvent. Expanding the log term of this equation and considering that $\phi/xv_1 = c/M$, we obtain,

$$\frac{\pi}{c} = RT\left[\frac{\phi}{x} + \left(\frac{1}{2} - \chi\right)\phi^2 + \frac{1}{3}\phi^3 + \cdots\right]$$

Comparing this equation with the virial expansion,

$$P_v = k_BT\left[1 + \frac{b_2}{v_1} + \frac{b_3}{v_2} + \cdots\right]$$

we find that the second virial coefficient b_2 and $(\frac{1}{2} - \chi)$ are proportional to each other. Given the specific volume of a polymer w, the proportionality constant is w^2/v_1. That is $\chi = \frac{1}{2}$ corresponds to $b_2 = 0$ and this is the θ point. When χ varies by temperature or solvent composition, the solvent is a good solvent for $\chi > 0.5$ and a poor solvent for $\chi < 0.5$.

3.1.2 Swelling Equilibrium

3.1.2.1 Neutral networks in pure solvents

We will now consider swelling equilibrium based on the free energy as derived previously. Similar to diffusion equilibrium, swelling equilibrium is achieved when the chemical potentials of the solvent inside and outside of the gel are equal [6]. The networks of gels are multifunctional. They themselves are solutes, function as semipermeable membranes, and also act as pressure actuators to increase the chemical potential of the solvent by the elastic force. When the equilibrium of a gel is considered, it is easier to understand it if the equality of chemical potential is replaced by the mechanical balance of the surface. In other words, equilibrium will be achieved when the normal osmotic pressure arising from the difference in solute concentrations π_m and the osmotic pressure from the elasticity of the networks on the surface of a gel.

Thus π_m can be expressed by Eq. (14) when $x \to \infty$ and π_e can be obtained as follows using ΔG_e:

$$\pi_e = -\left(\frac{N_A}{v_1}\right)\left(\frac{\partial \Delta G_e}{\partial n_1}\right) \tag{15}$$

Accordingly, with $\pi_m = \pi_e = 0$, the condition for the equilibrium is

$$\phi + \ln(1 - \phi) + \chi\phi^2 - \left(\frac{N_cv_1}{N_AV_0}\right)\left[\frac{\phi}{2\phi_0} - \left(\frac{\phi}{\phi_0}\right)^{1/3}\right] = 0 \tag{16}$$

where subscript 0 indicates quantities in the reference state at the gel point.

The solution of Eq. (16) is easier to understand in graph form. Figure 2 is the plot with $\pi = -\pi_m$ and $\pi = \pi_e$ as a function of ϕ, and the crossing point of the two curves is the equilibrium. Although the change of equilibrium with various values of χ is shown, in this section, χ is regarded as a constant and independent of concentration.

3.1.2.2   Effect of ion

The structure and properties of polyelectrolyte gels are both theoretically and experimentally poorly understood compared with neutral gels. It is not surprising because there are many unsolved problems regarding both gels and solutions. In polyelectrolyte solutions, the long-range coulombic forces among ions make these problems necessarily multi-body ones [6, 7]. Even in a pure solvent, the system consists of three components, polymer ions, counter ions and solvent, and the equilibrium among them needs to be considered.

Cases for which many studies have been done on electrolyte gels involve those gels with small fixed-charge density of the network. Their ionic effects can be treated as perturbations to neutral networks. In this case, static repulsion is small due to a long average ionic distance and counter ion osmotic pressure based on Donnan's effect becomes dominant. In addition to the aforementioned π_m and π_e, ionic osmotic pressure π_i is added and the balance among these three determines the equilibrium.

The Donnan effect generates π_i through the difference between the concentrations of the counter ions inside and outside of the gel. Its magnitude is determined by the density of the ionic functional group on

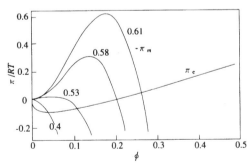

The f coordinate that corresponds to the crosslinking point between $-\pi_m(\phi)$ and $\pi_e(\phi)$. The numbers beside the $-\pi_m(\phi)$ curves indicate χ.

Fig. 2 Graphic display of swelling equilibrium.

the chain and the degree of dissociation of the group. Letting f be the fixed charge of a polymer chain in an electron charge unit, π_i is obtained as

$$\pi_i = f\,RT \tag{17}$$

According to the electroneutrality principle, f can be regarded as the number of counter ions because counter ions and fixed ions compensate each other. As the equilibrium condition is $\pi_m = \pi_e + \pi_i = 0$, instead of Eq. (16) we obtain:

$$\phi + \ln(1-\phi) + \chi\phi^2 - \left(\frac{N_c v_1}{N_A V_0}\right)\left[\left(f+\frac{1}{2}\right)\left(\frac{\phi}{\phi_0}\right) - \left(\frac{\phi}{\phi_0}\right)^{1/3}\right] = 0 \tag{18}$$

In a widely used, method to prepare gels with small fixed electron density a small amount of ionic monomer is added to a neutral monomer when the gel is prepared. It can also be prepared by synthesizing the networks of homopolymer from a weak electrolyte monomer and subsequent control of the solvent pH. Examples of the former are acrylic amide-acrylic acid and NIPAAm-acrylic acid gels. An example of the latter involves adding NaOH to the surrounding liquid of the acrylic acid gel and the degree of dissociation of the carboxylic acid is controlled. Figure 3 depicts the experimental results of swelling by changing the concentration of the fixed electron density of NIPAAm-acrylic acid copolymers. With the addition of a 1% ionic group, the equilibrium degree of swelling at 25°C is increased by as much as 20%, indicating a remarkable ionic effect[2].

3.1.3   Volumetric Phase Transition

Volumetric phase transition is the phenomenon in which the equilibrium degree of swelling or volume shows large discontinuous change in response to the external conditions, such as temperature or solvent composition. Although it was theoretically predicted many years ago, Tanaka *et al.* discovered it experimentally for the first time on acrylamide

---

[2] Swelling curves of gels are usually shown with temperature on the abscissa. However, when the experimental results of material properties are plotted, it is customary to depict the variables on the ordinate and the measured value on the abscissa. As the degree of swelling is determined by measuring the volume as a function of temperature, the graphs are plotted with temperature, which is the variable, on the ordinate in this section.

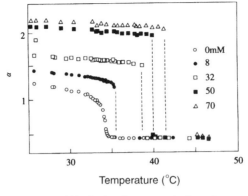

mM indicates the concentration of
acrylic acid at gel preparation.

Fig. 3   Acrylic acid-equilibrium swelling curves of NIPAAm acrylic acid copolymers.

gel [7]. Figure 4 shows the volumetric phase transition of the most widely studied gel, N-isopropylacrylamide (abbreviated as NIPAAm) [8–10].

It should be noted that only when the volume is discontinuous should the term *phase transition* be used. Also, the region where the change is from continuous to discontinuous is the critical phase transition. In principle, one cannot claim phase transition without first experimentally proving discontinuity, such as coexistence of two phases. Although it is not an easy task to show critical phase transitions or even a sufficiently close-to-phase transition, the majority of the literature on gels is vague on these issues and the terms phase transition or critical point are often used without detailed evaluation.

Considering just the free energy of mixing, the borderline between good and poor solvents is at $\chi = 0.5$. As the contribution of elasticity or ions exists in gels, swelling does not necessarily change to shrinking at $\chi = 0.5$. Nonetheless, significant volume change takes place roughly around this value. To clarify when phase transition takes place, the phase stability will be considered.

For this problem it is also easier to replace the stability of gels with the mechanical stability of networks. The requirement for mechanical stability of a homogeneous, isotropic material is given by bulk modulus $K > 0$, and the corresponding stability requirement of networks is given

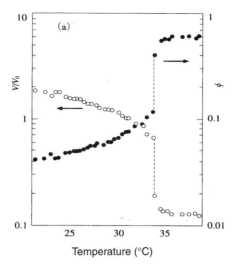

This graph expresses both the volume
ratio (V/V_o) and the volume fraction of the
network (ϕ). The broken line indicates the
primary phase transition. Here binodality
is observed.

Fig. 4   Equilibrium swelling curves for neutral NIPAAm gel.

by osmotic bulk modulus, $K_{os} > 0$. Expressing this condition using the
previously derived free energy, it is given as follows:

$$K_{os} = \left[\phi \left(\frac{\partial \pi}{\partial \phi} \right)_T \right]_{\pi=0}$$

$$= -\left(\frac{N_A}{v_1} \right) \left[\phi \left(\frac{\partial^2 \Delta G}{\partial \phi n_1} \right)_T \right]_{\pi=0}$$

The value of ϕ at $K_{os} = 0$ is the stability limit of the gel, that is, the
spinodal line. Figure 5 shows the calculated spinodal lines of the NIPAAm
gel at given χ_1 and χ_2 values. The hatched area is the unstable region and
point C is the critical point. When the gel enters the hatched region due to
a sudden temperature change, there will be phase separation by spinodal
decomposition.

   If the equilibrium swelling line is (a), phase transfer will not take
place. Volumetric phase transition occurs in the case of (c) where the gel

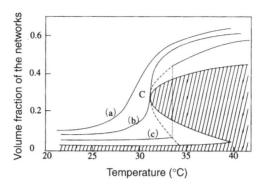

The hatched area indicates an unstable region, the broken line is the spinodal line, and the dotted line is the volumetric phase transition. Curves (a), (b) and (c) are examples of equilibrium swelling curves

Fig. 5 The phase diagram calculated using χ_1 and χ_2 determined experimentally for NIPAAm gel.

passes through the unstable region and enters into a new phase. Next, let us consider the requirement for phase transition.

The liquification of a real gas occurs as the first-order phase transition below the critical temperature. Based on the van der Waals theory, this is related to the fact that there are both converging and quasiconverging solutions near the transition to satisfy the equilibrium requirement rather than only one solution. In Fig. 2, there is only one solution for any c values, and, in this case, volumetric change is continuous and no phase transition will take place. Then, as observed in real polymer solutions, what is the situation if χ is an increasing function of ϕ? Figure 6 shows examples of only the first term with respect to $\chi = \chi_1 + \chi_2 \phi$ and ϕ. In a certain range of c, there are three crossing points. From the calculation of free energy, the middle solution corresponds to an unstable solution, while both ends correspond to stable and quasistable solutions. The experimentally obtained χ_1, χ_2 and N_c values for NIPAAm gels [10] are used in calculation of this graph. Introducing the ϕ dependence into χ corresponds to considering higher than the third virial coefficients. It also indicates that multibody interaction among segments cannot be ignored. Microscopically, this reflects the structural formation of water molecules around a hydrophobic group [11].

According to the theoretical study by Erman and Flory [12], $\chi_1 > \frac{1}{3}$ must be satisfied in the equation of ϕ dependence of χ to observe the first-order phase transition. There are only a few examples other than the NIPAAm gel and its family that are known for neutral gels that can be used to observe the phase transition as a function of temperature in pure solvents. This indicates that such a large ϕ dependence is exceptional.

The effect of ions on the phase transition is extremely strong. It is also apparent from the swelling curves of the copolymer gels made from NIPAAm and acrylic acid. Upon adding only 1% of an ionic monomer, very strong effects, such as enhanced discontinuity and shifting the phase transition to higher temperature, can be seen. This effect can be explained by the osmotic pressure of counter ions as described in subsection 1.2.2. Figure 7 indicates that the equilibrium conditions of polyelectrolyte gels can be determined from the crossing point of $-\pi_m(\phi)$ and $\pi_i(\phi) + \pi_e(\phi)$ curves. In this figure, the ϕ dependence of χ is ignored and, due to the effect of π_i, there are three crossing points, although $-\pi_m(\phi)$ is the same as in Fig. 2. Even though there is no ϕ dependence, a discontinuous transition can take place due to the effect of ions. Introducing fixed charges on the networks makes it easier for both a first-order transition and an increased degree of swelling to occur.

There are several notable phenomena in the phase transition of gels in mixed solvents. One of them is the reentrant phase transition phenomenon. This is the phenomenon in which a successive change, swelling-shrinking-swelling, occurs as a function of solvent composition. Although

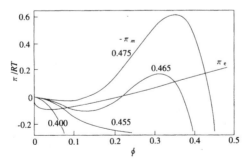

Compared to Fig. 2, the shape of $-\pi_m (\phi)$ changes.
Sometimes, three crossing points might appear.
At this point, it indicates the occurrence of the
primary phase transition.

Fig. 6 Swelling equilibrium when the lowest (primary) ϕ dependence is considered.

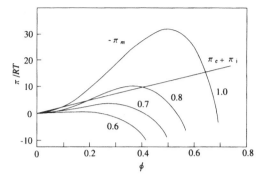

In this figure, the *f* dependence of *c* is ignored. Nonetheless, three crossing points exist. The ordinate is expanded by 50 times compared with 2 and 5 times when compared with Fig. 2 and Fig. 5, respectively. The degree of equilibrium swelling is increased by the ionic effect.

Fig. 7   Cases where the effect of ions is considered ($f = 10$).

it is presumed that structure formation due to the specific interaction between the mixed solvent and networks may contribute to this, a theoretical treatment will be quite complex because the composition of the solvent inside and outside of the gel is not necessarily the same. Figure 8 illustrates an example of re-entrant phase transition.

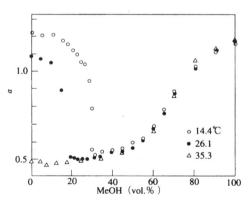

Swelling behavior of neutral NIPAAm gel in water-methanol mixed solvents at various temperatures,
(S. Hirotsu, *J. Chem. Phys.*, **88**: 427 (1988)).

Fig. 8   An example of recoverable phase transition in mixed solvents.

Finally, volumetric phase transition has been observed in all the known examples to date that possess LCST, namely, the lower temperature side is the swelling phase and the higher temperature side is the shrinking phase. Although this does not necessarily mean that the opposite will not occur, at least it is an essential principle known to date regarding the mechanism of volumetric phase transitions. From this there is no doubt that the origin of the transition is structural formation by specific interactions between solvent and segment, because the fact that the shrinking phase has higher entropy than the swelling phase cannot be explained simply by taking networks into account. Considering the structural formation of water around hydrophobic groups, χ includes multibody interaction, and thus the density dependence of χ can be understood and the fact that both δ_h and δ_s will be negative can be explained.

Accordingly, the origin of volumetric transition can be explained at least qualitatively by phenomenological discussion. However, there are many unsolved problems, such as the molecular transition mechanism, dynamic properties, and the behavior of polyelectrolyte gels. Future research may yield many interesting discoveries.

3.1.4 Conclusions

There are many interesting phenomena for the static properties of gels other than those subjects discussed here. For example, among the subjects of recent interest are heterogeneity of networks accompanying crosslink formation and microscopic phase separation of polyelectrolyte gels. Unfortunately, due to space limitations, these subjects were not discussed. The important issues discussed thus far will now be summarized.

1. The swelling equilibrium of the simplest gels made of homopolymer networks and pure solvent can be semiquantitatively explained by the empirical thermostatistical theory of gels advanced mostly by Flory's group.
2. Within this theory, the most important parameter to determine the swelling equilibrium of gels is χ.
3. Volumetric phase transition typically occurs when the concentration dependence of χ is large.
4. When a small amount of fixed charges are introduced into the networks, the degree of swelling increases significantly. This can be

explained by the osmotic pressure of counter ions based on the Donnan effect.
5. Many unexplained polyeletroclyte gel properties remain. There are also many unsolved problems in the swelling behavior or the phase transition of gels in mixed solvents.

3.2   KINETIC THEORY OF SWELLING

MITSUHIRO SHIBAYAMA

Familiar examples of swelling of gels are found in ordinary toys. Koya-tofu also exhibits similar properties—it is able to recover its original size in water. Important points that can be learned from such gels are that the diffusion of water into the gel causes swelling (diffusion limited) and the shape of dried (shrunken) gels is maintained during swelling (isotropic swelling). In this section, we will first consider the principle of the swelling of gels and then: (1) its phenomenology; (2) the kinetic theory of swelling; and (3) application and examples of the kinetic theory.

3.2.1   Phenomenology of Swelling of Gels

When the aforementioned Koya-tofu is immersed in water, water gradually penetrates the tofu, and after the water has completely penetrated (or while penetrating), swelling can be observed. The mechanism of this phenomenon is first, the penetration of water, subsequent plasticization, and then swelling. Such phenomena can be seen in the process of water absorption and swelling of ion exchange resins and polymer absorbents. On the other hand, changing the environment (such as solvent, temperature, salt concentration and pH) of a gel that already contains water can cause it either to swell or shrink. Generally, the rate of swelling or shrinking of water-containing gels is much slower than for dried gels. This is because the dried gels are porous and the surface area that is in contact with the solvent is extremely large compared with water-containing gels. It is necessary to keep this in mind to understand the kinetic theory of gels.

The swelling kinetic theory of gels began by simply solving the diffusion equation for the diffusion of solvent into gels. As early as 1965, Dušek had already treated the swelling of ion exchange resins as the boundary condition of the diffusion equation of spheres [13]. Following this, in the 1970s, theoretical ideas on the structure of gels were developed by Tanaka *et al.*, that is, the continuous body is composed of network molecules [14, 15]. Moreover, an interesting pattern may appear on the surface when gels swell. There are various studies of this phenomenon from the pattern formation point of view; however, we will not deal with this here.

3.2.2 Swelling Kinetic Theory of Gel Networks

Both swelling and shrinking of gels differ fundamentally from the diffusion of gas and liquid. Figure 1 shows extreme examples. The figure on the top schematically shows a straight line drawn with ink onto a filter paper, later wet by water. The bottom figure shows the swelling of a gel in the form of a stick. In the case of ink, diffusion starts from each point of the initial ink pigment (within the dotted line). On the other hand, the gels swell homothetically. This is because the molecular chains in gels are connected to each other and random diffusion cannot take place; however, they try to minimize the deformation energy as a whole. In the case of a spherical gel, it is not necessary to consider strain in any direction other than for the radius because other strains will counteract each other. However, for those gels having anisotropic

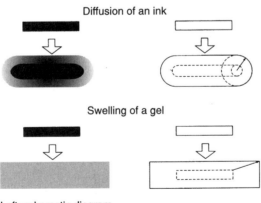

Left: schematic diagram
Right: the position of a specific point after swelling
(dotted line indicates the position prior to swelling)

Fig. 1 Comparison between the diffusion of an ink and swelling of a gel.

shapes, swelling by both cooperative diffusion and the process of strain relaxation must be discussed. Thus, we will first discuss the swelling kinetic theory of a spherical gel taking into consideration the connectivity of networks.

3.2.2.1 *Cooperative diffusion of gel networks*

In order to discuss the movement of the constitutive molecules of gels, it is necessary to describe gels as continuous bodies based on continuum mechanics. Let us consider the process for a point r on the gel network to move to point r' as shown in Fig. 2. The vector defined by the following equation is called the deformation vector:

$$u = r' - r \tag{1}$$

The equation of motion for the movement of a microscopic volume element of the gel is given by

$$F = \nabla \cdot \tilde{\sigma} - f \frac{\partial u}{\partial t} \tag{2}$$

where σ is the stress tensor ($\equiv \sigma_{ik}$, subscript ik indicates each element of a Cartesian coordinate), and f is the coefficient of friction. As force F can be given as a product of the mass of the microscopic volume element and acceleration,

$$F = \rho \frac{\partial^2 u}{\partial t^2} \tag{3}$$

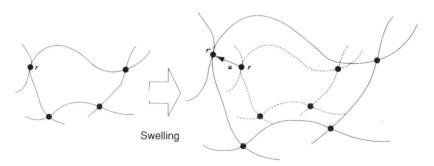

Swelling

Solid line indicates gel network and black dots show crosslink points; r and r' indicate position vector before and after the swelling, respectively, and u is the deformation vector.

Fig. 2   Swelling of gel networks.

where ρ is the density of the microscopic volume element. Substituting Eq. (3) into Eq. (2), and also using the following relationship between the stress tensor and strain tensor, u_{ik}, of an isotropic material:

$$\sigma_{ik} = 2\mu\left(u_{ik} - \frac{1}{3}\nabla \cdot u\delta_{ik}\right) + K\nabla \cdot u\delta_{ik}$$

$$u_{ik} \equiv \frac{1}{2}\left(\frac{\partial u_i}{x_k} + \frac{\partial u_k}{x_i}\right) \tag{4}$$

we obtain the equation of motion for gels,

$$\rho\frac{\partial^2 u}{\partial t^2} = \mu\nabla^2 u + \left(K + \frac{1}{3}\mu\right)\nabla(\nabla \cdot u) - f\frac{\partial u}{\partial t} \tag{5}$$

where K is the bulk modulus and μ is the shear modulus.

The component that constitutes networks, namely the partial polymer chain (for example, the partial chain between crosslinks), may not be able to move alone but may move along with other partial chains influencing each other. Therefore, the diffusion of a partial chain is not self-diffusion but cooperative diffusion. The rate of this cooperative diffusion must be much slower than the self-diffusion of the solvent itself. Considering the movement of networks, the left-hand side of Eq. (5), the momentum term, is negligibly small compared with each term of the right-hand side of the equation. Therefore,

$$f\frac{\partial u}{\partial t} = \mu\nabla^2 u + \left(K + \frac{1}{3}\mu\right)\nabla(\nabla \cdot u) \tag{6}$$

Equation (6) will be the fundamental equation to deal with the kinetics of swelling and shrinking, and also to discuss the dynamics of gels by polarized dynamic light scattering.

3.2.2.2 *Spherical symmetric gels*
In this section, the swelling of spherical symmetric gels will be discussed. In Eq. (6), let

$$D \equiv \frac{(K + \frac{4}{3}\mu)}{f} \tag{7}$$

and taking into account that $\boldsymbol{u}$ depends only on the radius component r of the spherical coordinate, we obtain

$$\frac{\partial \boldsymbol{u}}{\partial t} = D \frac{\partial}{\partial r}\left\{\frac{1}{r^2}\left[\frac{\partial}{\partial r}(r^2 \boldsymbol{u})\right]\right\} \tag{8}$$

Tanaka and Fillmore solved Eq. (8) as a diffusion equation and defined D as the diffusion coefficient (TF theory) [15]. However, it should be cautioned that Eq. (8) is different from the ordinary diffusion equation (7).

For example, the diffusion equation of a component (low molecular weight particle) in a spherical container with respect to concentration c is

$$\frac{\partial c}{\partial t} = \frac{1}{r^2}\frac{\partial}{\partial r}\left(r^2 D \frac{\partial c}{\partial r}\right) = D\left(\frac{\partial^2 c}{\partial r^2} + \frac{2}{r}\frac{\partial c}{\partial r}\right) \tag{9}$$

Using the function $C(r, t) \equiv r^2 u(r, t)$, Eq. (8) becomes

$$\frac{\partial C}{\partial t} = Dr^2 \frac{\partial}{\partial r}\left(\frac{1}{r^2}\frac{\partial C}{\partial r}\right) = D\left(\frac{\partial^2 C}{\partial r^2} - \frac{2}{r}\frac{\partial C}{\partial r}\right) \tag{10}$$

and their difference is apparent. Therefore, in a strict sense, D is not the diffusion coefficient. Therefore, we will call Eq. (8) the swelling equation rather than the diffusion equation and D the effective diffusion coefficient.

When the shear modulus μ can be ignored in comparison to bulk modulus K, that is, when

$$D \cong \frac{K}{f} \tag{11}$$

Eq. (8) can be solved relatively easily. Here, the initial condition is that the stress at time 0 is homogeneous throughout the gel (the strain at the center of the sphere is 0, the maximum at the gel surface, and distributes proportionately throughout the gel), that is,

$$u(r, 0) = \frac{\Delta a_0}{a_\infty} r \tag{12}$$

where a_∞ is the gel's final radius, and Δa_0 is the change of the gel radius ($\Delta a_0 \equiv a_\infty - a_0$ and a_0 is the initial radius of the gel). The boundary condition is that the stress at the gel surface is 0 after the swelling equilibrium. Letting the gel radius be $a(t)$ at time t, the solution under these conditions is

$$\frac{a(t)}{a_0} = \frac{1 - u(a_\infty, t)/a_\infty}{1 - \Delta a_0/a_\infty}$$

$$= \left(1 - \frac{2\Pi_0}{3K}\right)^{-1} \left\{1 - \frac{2\Pi_0}{\pi^2 K} \sum_{n=1}^{\infty} \frac{1}{n^2} \exp\left(-\frac{t}{\tau_n}\right)\right\} \qquad (13)$$

where Π_0 is the initial difference in osmotic pressure along the changes in the environment and it has the relationship

$$\Pi_0 = 3K \frac{\Delta a_0}{a_\infty} \qquad (14)$$

with the bulk modulus and radius change that is normalized by a_∞, $\Delta a_0/a_\infty$. Also, τ_n is the nth mode relaxation time and is given by

$$\tau_n = \frac{a_\infty^2}{n^2 \pi^2 D} \qquad (15)$$

As the real relaxation is dominated by the longest relaxation time τ_1, the effective diffusion coefficient can be given by

$$D_{\text{eff}} \equiv \tau_1^{-1} \left(\frac{2a_\infty}{2\pi}\right)^2 = D \qquad (16)$$

Furthermore, when $t \gtrsim \tau \equiv \tau_1$, the relative radius change is given by

$$\frac{a_\infty - a(t)}{a_\infty - a_0} \cong \frac{6}{\pi^2} \exp[-t/\tau] \qquad (17)$$

Many swelling kinetic studies of gels are done based on Eq. (17). Figure 3 shows the radius change of a spherical gel as a function of time using Eqs. (13) and (17). Although the approximation by Eq. (17) is not appropriate in the region of $0 < t < \tau$, the solution of Eq. (17) can sufficiently evaluate the effective diffusion coefficient D.

The main characteristic of the swelling of gels is the large dependence of swelling rate on gel size. Because from Eq. (16),

$$\tau = \frac{a_\infty^2}{\pi^2 D} \qquad (18)$$

the swelling relaxation time of gels becomes longer in proportion to the square of the final radius a_∞.

The solid line in Fig. 4 indicates the (longest) relaxation time τ as a function of a_∞ when $D = 10^{-7}\,\text{cm}^2/\text{s}$. When $a_\infty = 0.0001\,\text{cm}\ (= 10\,\mu\text{m})$,

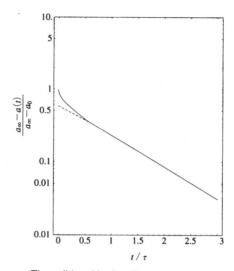

The solid and broken lines are
theoretical functions from Eq. (13)
and approximated function from Eq. (17).

Fig. 3 The normalized time, t/τ, dependency of the normalized radius, $a(t)/a_\infty$, of a gel.

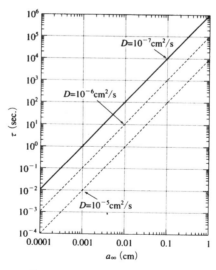

Fig. 4 The ultimate radius a_∞ dependence of swelling relaxation time τ of gels.

τ is approximately 1 s and for $a_\infty = 0.01$ cm ($= 100\,\mu$m), τ is approximately 100 s. Also, when D increases an increment from 10^{-6} cm^2/s to 10^{-5} cm^2/s, τ decreases to $1/10$ (dotted lines in the figure).

This property is important when a gel is used as an actuator. To increase response speed, it is necessary to sufficiently reduce the size or diameter of the gel.

3.2.2.3 *Anisotropically shaped gels*

In anisotropically shaped gels, such as rods or flat plates, the strain caused by diffusion during swelling becomes anisotropic and shear energy will accumulate. In real swelling of gels, shape is determined by minimizing this shear energy. The shear energy F_{sh} is given as follows:

$$F_{sh} = \mu \int_v \left[\left(u_{xx} - \frac{T}{3} \right)^2 + \left(u_{yy} - \frac{T}{3} \right)^2 + \left(u_{zz} - \frac{T}{3} \right)^2 \right] dV \qquad (19)$$

where T is the sum of the diagonal elements of the strain vector, u_{ik}. Also, the integration is performed over the entire gel volume V. The condition of the minimization of strain energy for a spherical gel is

$$\delta F_{sh} = 0 \qquad (20)$$

Although a spherical symmetric gel always satisfies the condition of Eq. (20), the swelling equation, Eq. (6), must be solved while taking into consideration Eq. (20) for a rod-like gel with high aspect ratio or flat plate gel [16]. Li and Tanaka divided such swelling processes into microscopic regions and solved the equation assuming that the network diffusion and strain relaxation mutually take place in each region [17]. At first, only the diffusion of networks takes place and the shear energy accumulates. After completion of diffusion in the microscopic region, strain relaxation occurs. Since strain relaxation is an elastic deformation, it can take place instantaneously. This series of processes takes place in each microscopic region and, as a result, macroscopic description becomes possible. While details are referred to in the original publication [17], here we will discuss only the more important results. Figure 5 shows the effective diffusion coefficient, $D_e(r)$, at relative position r/a normalized by the effective diffusion coefficient of a spherical gel D_0. This graph is plotted when r from the a spherical or rod-like gel with radius a is taken from the center of the gel and for a disk-shaped gel with a thickness of $2a$ it is taken from the thickness. An interesting point is the effective diffusion coefficient at $r = a$, that is at the surface of the gel. While a spherical gel can swell in

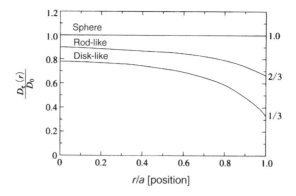

Fig. 5. The dependence of the effective diffusion coefficient $D_e(r)$ of spherical, rod-like and disk-like gels on the relative distance r/a from the center of the gel [25]. (At the surface of the gel $r/a = 1$ and the relative values of the effective diffusion coefficient are $1:2/3:1/3$).

xyz (3) directions, a rod-like gel is limited to swelling two-dimensionally (assuming the *x* axis as the main axis of the rod, it is towards the *xy* (2) directions), and a disk-shaped gel swells one-dimensionally (assuming that the *z* axis is normal to the disk, it swells towards the *z* axis) [18]. Therefore, the swelling of an anisotropic-shaped gel depends on the allowed dimension of the solvent [19]. The effective diffusion coefficients of disk-shaped gels and rod-like gels are $1/3$ and $2/3$ that of spherical gels, respectively (they do not differ to the extent of an order of magnitude).

For example, the swelling of a flat plate is controlled by its thickness and the rate of swelling is greater for thinner plates. As a result, swelling in the *xy* direction (normal to thickness direction swelling) is very rapid.

3.2.3 Critical Relaxation and Phase Separation

Thus far, continuous change of the gel size from state A to B by swelling or shrinking has been described. In many cases, the free energy function of gels $G(\phi)$ is a simple function of the volume fraction ϕ of the polymer component in the gel. However, there are occasions in the swelling and shrinking of gels where the differential of $G(\phi)$ becomes 0 or negative, at which a volumetric phase transition can be observed. As the thermodynamic driving force that causes swelling or shrinking becomes very small in the vicinity of the volumetric phase transition, the so-called critical relaxation phenomenon appears. As a result, the rate of swelling or shrinking becomes extremely small. Tanaka *et al.* [20] evaluated an effective diffusion coefficient of N-isopropylacrylamide, which shows a volumetric transition near 34°C, by performing swelling and shrinking experiments at various temperatures. As shown in Fig. 6, gels shrink by a temperature jump to the vicinity of the critical temperature. It is observed that the closer the temperature to the critical temperature, the slower the rate. This is related to the reduction of $D_{kinetics}$. On the other hand, the effective diffusion coefficient $D_{dynamic}$ that is obtained by dynamic light scattering also becomes extremely small at the critical temperature, which is the critical relaxation.

Passing the critical point and entering a state of shrinking, the gel destabilizes and phase separation takes place. The phase-separated gels tend to have a frozen structure because the molecular mobility in the

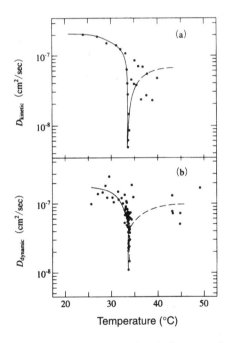

Fig. 6 The temperature dependence of (a) $D_{kinetics}$ obtained from swelling/shrinking and (b) the cooperative diffusion coefficient, $D_{dynamic}$, of N-isopropylamide gel (both quantities show sudden reduction at the critical temperature).

polymer rich phase is markedly inhibited. Accordingly, the kinetics of swelling and shrinking in the vicinity of the critical point will be very complex. In such a case, it is necessary to discuss separately the kinetics of swelling and shrinking in the swollen state and those of the phase separated in the shrunken state [21, 22].

3.2.4 Application and Examples of Kinetics

3.2.4.1 Ion exchange resins

As described in the introduction, the kinetic study of gel swelling started with the analysis of swelling of ion exchange resins by Dušek from the phenomenological and application points of view [13]. An ion exchange resin based on the diffusion equation (Eq. (9)) was recently analyzed, and the possibility of ion exchange by separating the mutual diffusion of counter ions into the diffusion within (partial diffusion) and at the surface of the ion exchange resin (layer diffusion) was proposed [23, 24]. The authors also found that the time variation of the degree of swelling strongly affected solvent and resin compatibility using the dried prepolymer of the ion exchange resin (polystyrene crosslinked by divinylbenzene (DVB): PS-DVB) with various solvents, such as dichloroethylene (TCE), and 1,2,3-trichloropropane (TCP) (see Fig. 7) [25]. This example indicates that, when a dried gel is immersed in a solvent, it swells after passing through the induction period where little swelling takes place, and the rate of swelling depends significantly on the properties of the solvent. The solubility parameter of these solvents in units of $(cal/cm^3)^{1/2}$ is δ (EDC) = 9.1 (trans), δ (CB) = 9.5, δ (DCB) = 10.0, δ (TCE) = 9.3 [26], and there is a trend that the farther away from the solubility parameter of polystyrene (PS), δ (PS) $\cong$ 9.1, the longer the threshold period (TCE is an exception). This indicates that penetration of a solvent in a dried gel depends strongly on the compatibility (or solubility of the polymer) between the solvent and polymer (gel). Therefore, the threshold period without volume change is considered the dissolution process of the polymer that composes the gel. Of course, there is the process of penetration of the solvent in the pores of the gel. However, it is a capillary phenomenon and as such it is regarded as a very rapid process compared with the dissolution process.

Ikkai and Shibayama studied the swelling behavior of a cationic ion exchange resin (sulfonated PS-DVB) and its precursor (PS-DVB resin) in water and toluene as a function of crosslink density, and compared it with the structural analysis results obtained from small angle neutron scattering

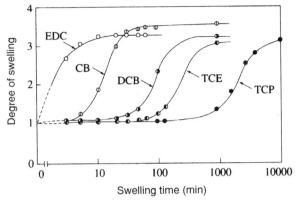

EDC: dichloroethylene   CB: chlorobenzene
DCB: dichlorobenzene   TCE: tetrachloroethylene
TCP: 1, 2, 3-trichloropropane

Fig. 7 The swelling behavior of a dry gel (polystyrene resin that is cross-linked by divinyl benzene) when it is immersed in various halogenated solvents [23].

[27]. Figure 8 shows (a) the dependence squared of swelling relaxation time τ on the final diameter d, and (b) the CD dependence of the apparent diffusion coefficient D_{app} (as the dried ion exchange resin is porous, the apparent diffusion coefficient is used rather than the effective diffusion coefficient in which the effective surface area is taken into consideration). According to this diagram, we can see that τ satisfies Eq. (18) and increases proportionally to d^2, D_{app} is decided only by CD, independent of d, and the larger the CD, the smaller the diffusion coefficient.

3.2.4.2 Acrylamide gels

Tanaka *et al.* demonstrated that the effective diffusion coefficient determined by swelling experiments and the cooperative diffusion coefficient obtained from dynamic light scattering for a spherical gel of acrylamide agree within experimental error [14]. Furthermore, Peters and Candau modified the TF theory using an analysis that involved shear modulus, but it has been ignored in the TF theory [28]. Tokita and Tanaka found that the rate of water passage through a temperature-sensitive N-isopropylacrylamide becomes very large near the volumetric phase transition temperature of 33.6°C [29, 30]. This has been explained as being due to the reduction of the friction coefficient f by the movement of gel networks at the volumetric phase transition.

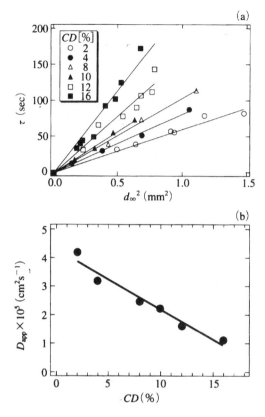

Fig. 8 (a) The dependence d_{∞}^2 of ultimate diameter of the resin with swelling relaxation time τ (upper diagram), and (b) the crosslink density, *CD*, dependence of apparent diffusion coefficinet, D_{app}, of the cationic ion exchange resin having various *CD* [27].

3.2.4.3 Ion complex gels

Shibayama *et al.* [19, 31] studied the cooperative diffusion coefficient of chemically crosslinked poly(vinyl alcohol) (PVA) containing ion complex crosslinking by using the kinetics of swelling and shrinking and dynamic light scattering. The PVA gels are prepared by formation of a complex with boric acid ions. The gels shrink (at low concentration) or expand (at high concentration) depending on boric acid ion concentration. Figure 9 shows the swelling and shrinking behavior (Step I) of the PVA gel when it is immersed in a solution that contains boric acid ions (a mixed solution of

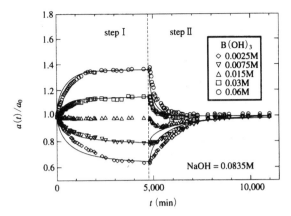

Step I: swelling or shrinking of PVA gel with boric acid
Step II: shrinking or swelling of PVA gel after boric acid
has been removed. The solid line indicates the
theoretical curve.

Fig. 9 Swelling/shrinking behavior of chemically crosslinked poly(vinyl alcohol) (PVA) gel accompanying the formation of a boric acid complex [19].

boric acid and sodium hydroxide) as well as the reverse process (Step II) when the gel that has reached equilibrium is immersed in a sodium hydroxide solution that does not contain boric acid ions. The solid line in the figure is the result of theoretical calculations based on Eq. (13). There is a slight deviation from the theoretical curve compared with the ordinary swelling and shrinking behavior due to the contribution of complex formation, especially at an early stage of Step II. Nonetheless, swelling and shrinking behavior is predicted relatively well using the theoretical curve. Figure 10 illustrates the boric acid concentration dependence of the effective diffusion coefficient at Steps I and II.

The effective diffusion coefficient D was about 10^{-7} cm^2/s, which is similar to that of the acrylamide gels. Also, from the fact that the effective diffusion coefficient of the swelling and shrinking of the reverse process is approximately the same, the swelling and shrinking process caused by complex formation can be considered to be a reversible process with the same effective diffusion coefficients [19].

3.2.5 Conclusions

The kinetic theory of gel swelling has been explained from the viewpoints of theory, experiments and applications, all of which are summarized in

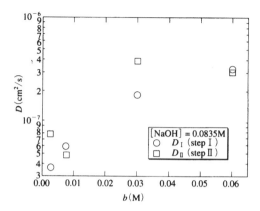

Fig. 10 Boric acid concentration dependence of the effective diffusion coefficient during Step I and Step II processes.

the following: (1) the deformation of gels is a homothetic deformation (isotropic swelling); (2) the swelling of gels is similar to a diffusion process and the swelling relaxation time is proportional to the square of the gel size; (3) the effective diffusion coefficient that characterizes swelling can be estimated by swelling experiments, and this value almost agrees with the cooperative diffusion coefficient that can be obtained from dynamic light scattering; and (4) effective diffusion coefficients depend on the dimension of swelling, and those of a very thin gel (swelling dimension of 1) and a very thin rod-like gel (swelling dimension of 2) are 1/3 and 2/3 those of ordinary 3D gels (swelling dimension of 3).

The kinetic theory of swelling has been developed by ignoring the shear modulus and assuming the constancy of the effective diffusion constant during the swelling process. It has greatly contributed to the analysis of swelling behavior of gels. These two assumptions made theoretical treatment and predictions easy and allowed good agreement with experimental results. However, the shear modulus cannot be ignored in comparing with the bulk modulus, and this assumption cannot be applied especially for a gel in a good solvent as pointed out by Peters and Candau [28] and Onuki [32]. Also, the effective diffusion coefficient is not a constant during the swelling process; instead it is a function of the degree of swelling and the position within the gel. In particular, the assumption of the constancy of the effective diffusion coefficient becomes a serious problem for swelling of dried gels and volumetric transition to the shrinking phase. Therefore, the kinetic theory of the swelling of gels described in this section is semi-quantitative and is an area in which future progress is strongly desired. Those who are interested in an exact analytical solution should refer to the review by Onuki [32].

REFERENCES

1   Doi, M., and Ko, A. (1992). In *Iwanami Lecture Series, Modern Physics 19: Polymer Physics*, Iwanami Publ.
2   Tanaka, F. (1994). *Physics of Polymers*, Shoka Do.
3   Flory, P.J. (1953). *Principles of Polymer Chemistry*, Ithaca, New York: Cornell University Press, Chap. 12.
4   Eichinger, B.E., and Flory, P.J. (1968). *Trans. Faraday Soc.*, **64**: 2053.
5   Flory, P.J. (1970). *Disc. Faraday Soc.*, **49**: 7.
6   Flory, P.J. (1953). *Principles of Polymer Chemistry*, Ithaca, New York.: Cornell University Press, Chap. 13.
7   Tanaka, T., Fillmore, D., Sun, S.T., Nishio, I., Swisslow, G., and Shah, A. (1980). *Phys. Rev. Lett.*, **45**: 1636.
8   Hirotsu, S., Hirokawa, Y., and Tanaka, T. (1987). *J. Chem. Phys.*, **87**: 1392.
9   Hirotsu, S. (1994). *Phase Transitions*, **47**: 183.
10 Hirotsu, S. (1987). *J. Phys. Soc. Jpn.*, **56**: 233.
11 Hirotsu, S., and Kaneki, A. (1988). In *Dynamics of Ordering Processes in Condensed Matter*, S. Komura, and H. Furukawa, New York: Plenum Publishing, pp. 481–486.
12 Erman, B., and Flory, P.J. *Macromolecules*, **19**: 2342.
13 Dusek, K. (1965). *Chem. Commun.*, **30**: 3804.
14 Tanaka, T., Hocker, L.O., and Benedek, G.B. (1973). *J. Chem. Phys.*, **59**: 5151.
15 Tanaka, T., and Fillmore, D.J. (1979). *J. Chem. Phys.*, **70**: 1214.
16 Tanaka, T., Sun, S.-T., Hirokawa, Y., Katayama, S., Jucera, J., Hirose, Y., and Amiya, T. (1987). *Nature*, **325**: 796.
17 Li, Y., and Tanaka, T. (1990). *J. Chem. Phys.*, **92**: 1365.
18 Landau, L.D., and Lifshitz, E.M. (1972). *Elastic Theory*, T. Sato and Y. Ishibashi, Trans., Tokyo: Tokyo Tosho (originally published in English).
19 Shibayama, M., Uesaka, M., Nomura, S., and Shiwa, Y. (1996). *J. Chem. Phys.*, **105**: 4350.
20 Tanaka., T., Sato, E., Hirokawa, Y., Hirotsu, S., and Peetermans, J. (1985). *Phys. Rev. Lett.*, **55**: 2455.
21 Sekimoto, K. (1993). *Phys. Rev. Lett.*, **70**: 4154.
22 Tomari, T., and Doi, M. (1994). *J. Phys. Soc. Jpn.*, **63**: 2093.
23 Yamamizu, T., Mizuno, O., Tada, K., and Takeda, K. (1991). *Kobunshi Ronbunshu*, **48**: 75.
24 Kawakami, F., Tada, K., and Takeda, K. (1992). *Kobunshi Ronbunshu*, **49**: 181.
25 Kawakami, F., Tada, K., and Takeda, K. (1992). *Kobunshi Ronbunshu*, **49**: 189.
26 Brandrup, J., and Immergut, E.H. (eds) (1989). *Polymer Handbook*, 3rd edition, New York: Wiley.
27 Ikkai, F., and Shibayama, M. (1996). *J. Polym. Sci. Polym. Phys. Ed.*, **34**: 1637.
28 Peters, A., and Candau, S.J. (1988). *Macromolecules*, **21**: 2278.
29 Tokita, M., and Tanaka, T. (1991). *J. Chem. Phys.*, **95**: 4613.
30 Tokita, M., and Tanaka, T. (1991). *Science*, **253**: 1121.
31 Shibayama, M., Takeuchi, T., and Nomura, S. (1994). *Macromolecules*, **27**: 5350.
32 Onuki, A. (1993). *Adv. Polym. Sci.*, **109**: 63.

Section 4
General Theory of Gel Preparation

HIDENORI OKUZAKI

4.1 INTRODUCTION

Gels are *in the state where a large amount of solvent is enclosed in polymer networks*, and polymer networks are formed by crosslinking. Therefore, *to make gels* is nothing other than forming a *crosslink structure*. Crosslink structure can be divided into formation by covalent bonds or intermolecular physical bonds [1].

Crosslink structure by a covalent bond is formed by the energy of heat, catalysts, light, radiation, plasma, and electric fields. There are methods to crosslink during polymerization or to crosslink polymer chains after polymerization. The former method is easy and can be used in a wide variety of monomers. On the other hand, by using the latter method, gels can be processed into any shape, such as fibers or films. Also, it is possible to crosslink the polymers while maintaining the higher order structure or orientation formed in solution.

Network structure by intermolecular physical forces is formed by hydrogen bonding between polymer chains, static bonding, complex bonding, hydrophobic bonding, and van der Waals bonding. Many of the natural polymer gels, such as polysaccharides and proteins, belong to

98

this category. These gels are usually prepared by mixing or cooling solutions. A sol-gel transition takes place by varying temperature, pH, or ionic strength. Usually, the mechanical strength and chemical stability of the gels obtained are weaker than those in the gels prepared by covalent bonding. However, by introducing crystallizable side chains into the polymer structure, it is possible to make gels with strong 3D network structures where microcrystals form crosslink points.

4.2 GEL PREPARATION BY COVALENT BONDING

4.2.1 The Method to Crosslink Simultaneously during Polymerization

4.2.1.1 Thermal polymerization

Among the reactions that form crosslinks by adding heat are vinyl polymerization, polyaddition polymerization, polycondensation polymerization, and addition condensation reaction [2]. Condensation polymerization has been widely used for the synthesis of polyesters and polyamides (Fig. 1). In polymerizing monomers with more than trifunctional groups, either alone or with other monomers, branching and crosslinking occur and 3D networks are formed [3, 4]. In particular, alkyd resin made of polyester is used for coating [5]. This reaction is reversible and the equilibrium constant is 400 for polyamides and 10 for polyesters. To increase the degree of polymerization and synthesize gels with better efficiency, it is necessary to: (1) use high purity monomers; (2) precisely match the number of functional groups; and (3) eliminate small molecular weight reaction byproducts such as water, alcohol and hydrogen chloride, from the reacting system by heating or evacuating.

Polyurethane [6], polyurea, and epoxy resin [7] are obtained by polyaddition polymerization. Formation of crosslinks depends on reaction conditions. It is complex because, in many cases, both the combination of various addition reactions and condensation contribute, and the number of functional groups and reactivities change.

Formaldehyde also reacts with aromatic hydrocarbons or amines at various ratios and forms relatively small molecular weight reaction intermediates [6]. These are multifunctional and easily form crosslinks by heating with appropriate hardeners. Phenolic resin, urea resin, and melamine resin are typical examples and 3D networks are formed by addition condensation reactions that repeats the addition and condensation

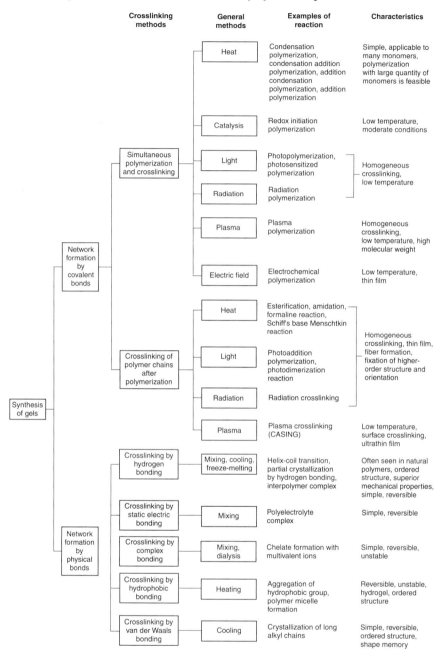

Fig. 1   The synthetic method of gels and their properties.

reactions. However, except for urethane, these systems are seldom used for gel preparation.

The most general method for preparing gels is free radical polymerization of a vinyl monomer with a divinyl compound that is used as a crosslinker. The vinyl monomers include acrylic acid and its esters, acrylic amide, styrene, and vinylacetate. For a crosslinking agent, usually methylene bis-acrylamide, ethylene glycol dimethacrylate, or divinyl benzene is used. The composition and properties of gels vary significantly depending on the combination of monomers and crosslinking agents.

If we express a vinyl monomer as M and a divinyl compound as B-B, the basic reaction of the crosslink formation process follows the free radical copolymerization theory [3, 8]:

$$\sim\!\!\!\sim\!M\cdot + M \xrightarrow{k_{MM}} \sim\!\!\!\sim\!M\cdot$$

$$\sim\!\!\!\sim\!M\cdot + B\text{-}B \xrightarrow{k_{MB}} \sim\!\!\!\sim\!\underset{\underset{B}{|}}{B}\cdot$$

$$\sim\!\!\!\sim\!\underset{\underset{B}{|}}{B}\cdot + B\text{-}B \xrightarrow{k_{MB}} \sim\!\!\!\sim\!\underset{\underset{B}{|}}{B}\cdot$$

$$\sim\!\!\!\sim\!\underset{\underset{B}{|}}{B}\cdot + M \xrightarrow{k_{BM}} \sim\!\!\!\sim\!M\cdot$$

$$r_1 = k_{MM}/k_{MB}, \; r_2 = k_{BB}/k_{BM}$$

where $\sim\!\!\!\sim\!M\cdot$ and $\sim\!\!\!\sim\!\underset{\underset{B}{|}}{B}\cdot$ are the growing terminal of the monomer and crosslinking agent, respectively, and r_1 and r_2 are the reactivity ratio of the monomer and crosslinking agent, respectively. Since the crosslinking takes place completely randomly when $r_1 = r_2 = 1$, an ideal homogenous gel can be obtained (see Fig. 2). However, it is difficult to find the combination to satisfy this relationship, and thus it is necessary to choose a monomer and crosslinking agent with similar chemical structure and reactivity. In addition, the reactivity of the remaining vinyl group is expected to be different upon the reaction of the first vinyl group of the

(a) Homogeneous gel   (b) Heterogeneous gel

Fig. 2 Network structure of gels (schematic diagram).

two in the crosslinking agent. To obtain a homogeneous gel, it is essential to select the solvent, monomer concentration, and polymerization temperature so that the solution will not be turbid and cause precipitation. For free radical polymerization, it is necessary to use initiators, such as azobisisobutylnitrile. Because the oxygen in the solution acts as a radical scavenger, it is necessary to deaerate the solution sufficiently or replace the dissolved gas with an inert gas like nitrogen.

4.2.1.2   Catalytic polymerization (redox initiated polymerization)

Because peroxides cause a redox reaction and readily generate free radicals in the presence of an appropriate reducing agent, gels can be formed below room temperature [9]. Examples of combinations of oxidizing agent/reducing agent include hydrogen peroxide/ferric salt and benzoyl peroxide/dimethylaniline. However, caution is necessary. If an appropriate temperature is not used, the generated free radicals will not be effectively used up in polymerization and they can then cause an explosion. The selection of solvent is also important because peroxides undergo redox decomposition in solvents such as ketones, ethers, and amines.

4.2.1.3   Photopolymerization

In photopolymerization, polymerization and crosslinking are performed by using light that corresponds to the absorption wavelength (180–220 nm) of the vinyl group of monomers and crosslinking agents [10]. On the other hand, the method to generate free radicals by adding photosensitizers, such as peroxides and azobis compounds, and to polymerize monomers and crosslinking agents, is called photosensitized polymerization. Crosslinking reactions are controlled by changing the concentration of monomer and crosslinking agent, intensity of light, and irradiation time. It is desirable to investigate the absorption spectra of the monomer and crosslinking agent prior to their use. As the activation energy of photopolymerization is as small as 3–5 kcal/mol and the reaction proceeds at low temperature, it is possible to achieve inclusion fixation of a biocatalyst in a gel [11].

4.2.1.4   Radiation polymerization

Vinyl monomers polymerize by γ-ray and electron beams that are higher in energy than visible radiation [12, 13]. Although it is possible to observe ionic polymerization, the majority of cases are by a free radical mechanism. It is possible to obtain homogeneous gels because: (1) it is unnecessary to use a free radical initiator or catalyst; (2) the reaction proceeds at low temperature; and (3) the reproducibility of irradiation is good. The degree of crosslinking is controlled by the kind and concentration of the vinyl monomer and crosslinking agent, solvent, irradiation dose, time, and temperature.

4.2.1.5 Plasma polymerization

In low-temperature plasma by glow irradiation, there are various active species, such as electrons, ions, free radicals, excited molecules, and photons, at a wide energy distribution. In particular, because plasma electrons possess high energy, they can be used for polymerization of vinyl monomers. Figure 3 shows a typical instrument for plasma polymerization [14].

An ampoule or flask that is connected to a vacuum line is placed between external parallel plate electrodes, a liquid or solid monomer fills the container, and plasma is generated for 30 s to several minutes after evacuation. The active polymerizing species generated in the plasma comes into contact with the surface of the monomer phase, diffuses into it, and initiates polymerization. Only the initiation reaction takes place in the gaseous phase, and growth and termination reactions take place in the condensed monomer phase. The characteristics of this polymerization are: (1) ultrahigh molecular weight polymer can be obtained [15] as well as strong, homogeneous gels despite having a small amount of crosslinking agent; (2) the obtained gel shows high water absorption [16] and excellent ability to adsorb metallic ions [17]; and (3) biomaterials such as enzymes

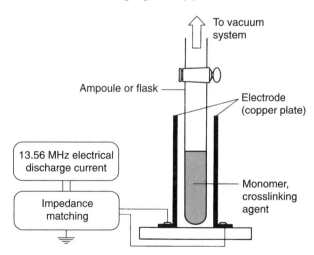

Fig. 3 Apparatus for plasma-initiated polymerization.

can be fixed in the gel without denaturing it because polymerization takes place at low temperature. The effect of monomer reactivity and solvent in the plasma polymerization are unique. Readers are referred to a monograph [14] for details.

4.2.1.6   Electropolymerization

Vinyl monomers can also be polymerized by electrolysis. In electropolymerization, there are cases in which vinyl monomers directly become free radicals, anions or cations, and where added salts are activated [18]. Oxidation-reduction potential of the vinyl monomer and added salt are good indices for judging the occurrence of these different mechanisms. As electrolysis relates only to initiation of polymerization, similar gels can be formed by adding crosslinking agents, such as divinyl compounds. However, there will be active species formed near the electrode surface and thus termination tends to occur. Therefore, it is generally difficult to obtain high molecular weight polymers.

In the method to crosslink simultaneously with polymerization, the majority of combinations are between a vinyl monomer and a divinyl compound. Heat, catalyst, light, radiation, plasma, and electric fields merely act as initiators for initiation reactions of free radical polymerization. However, these external energies interact uniquely with monomers and solvents. As a result, the state of crosslinking and properties of gels strongly reflect this fact. Table 1 summarizes the gel preparation methods that use free radical polymerization.

4.2.2   Method to Crosslink Polymer Chains Already Formed

4.2.2.1   Chemical reactions

Those synthetic, biological and natural polymers that possess hydroxyl, carboxyl and amine reactive functional groups on the side chain or the chain ends can be crosslinked by aldehyde, hydroxyl, carboxylic acid, N-hydroxymethyl and epoxy groups. Examples of various polymers and crosslinking agents are shown in Table 2. In comparison to the method of simultaneously crosslinking during polymerization, this method can: (1) produce homogeneous crosslinking; (2) prepare gels with various properties depending on the type of crosslinking agents; (3) be applied for thin films and fibers; and (4) be applied to bio and natural polymers like polysaccharides and proteins. In addition, stereoregular gels can be

Table 1 Characteristics of gel preparation methods by free radical polymerization.

	Thermal polymerization	Catalytic polymerization	Photopoly-merization	Radiation polymerization	Plasma polymerization	Electrochemical polymerization
Monomer, crosslinking agent	Various vinyl compounds	Various vinyl compounds	When the polymer is unstable in light, it is not possible	When the polymer is unstable under radiation, it is not possible	Specific, only with limited compounds	Various vinyl compounds
Additives	Initiators	Redox initiators 10°C to room temperature	Photosensitizer	Unnecessary	Unnecessary	Electrolyte
Polymerization temperature	Room temperature to 100°C	Selection of initiators, change of temperature	Low temperature Selection of photosensitizer, change of light flux	Low temperature	Room temperature	Room temperature
Adjustment of reaction	Selection of initiators, change of temperature	Selection of initiators, change of temperature	Special light source	Change of the dose	Change of radiation time and temperature	Change of electrolyte and current
Apparatus and operation	Simple	Simple		Special radiation source	Glow source, short time treatment	Electrode source, simple
Properties of gels	Heterogeneous—homogeneous	Homogeneous	Homogeneous	Homogeneous	Homogeneous, high water absorbence, absorption of metals	Heterogeneous, filmlike

Table 2   Chemical crosslinking of polymer chains.

Linear polymers	Reactive functional groups	Examples of crosslinking agent
Poly(acrylic acid)	−COOH	Ethylene glycol
Poly(methacrylic acid)		Glycerine
Poly(glutamic acid)		Ethylene diamine
Poly(vinyl alcohol)		Dionium salt
HEMA*	−OH	Glutaldehyde
Hyaluronic acid		Glutaric acid
Starch		N,N′-di(hydroxymethyl) urea
Polyarylamine	−NH₂	Ethylene glycol diglycidyl ether
Poly(ethylene imine)		Dibromohexane
Polylysine	−NHR	Di-iodoethane
Collagen		
Poly(vinyl pyridine)		
Poly(vinyl imidazole)	−NRR′	

*HEMA: poly(2-hydroxyethyl)methacrylate

prepared by crosslinking the higher order and orientated structures that are formed in a solution.

The γ-benzyl-L-glutamate (PBLG) forms an α-helix in an appropriate solvent and exhibits cholesteric liquid crystal behavior.

By crosslinking this polymer with ethylene diamine or triethylene tetramine, a liquid crystalline gel can be prepared [19]. A magnetic liquid crystal gel that exhibits swelling anisotropy can be obtained by crosslinking oriented PLBG under a magnetic field.

4.2.2.2   Photocrosslinking [20]

The technique that uses light to crosslink and insolubilize polymers has been applied to photosensitive resins or photoresists. This technique is largely divided into photoaddition and photodimerization reactions. Examples for photoaddition reactions include tetraazonium salts and bisazide compounds. When light is shone on a multifunctional aromatic diazonium salt (such as tetraazonium salt), nitrogen gas is immediately generated and free radicals are formed, leading to crosslinking and insolubility of the polymer [21, 22]. Bisazide compounds decompose by light and produce nitrane. Nitrane is an isoelectric reaction intermediate like carbene, and crosslinks polymers by addition, insertion or hydrogen extraction reactions [23].

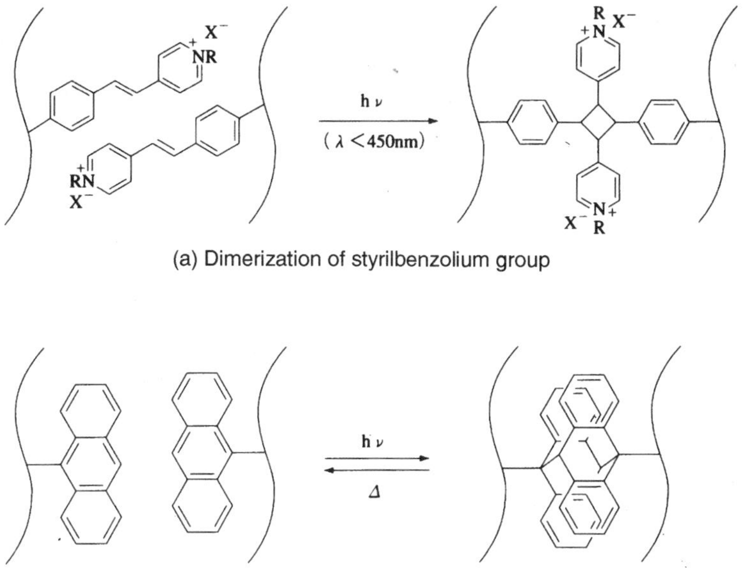

(a) Dimerization of styrilbenzolium group

(b) Dimerization of anthraryl group

Fig. 4 Formation of crosslinking by photodimerization.

The crosslinking of polymers by photopolymerization is achieved by introducing functional groups that photodimerize, such as styrilbazorium [24], styrylbenzolium salt [20], and anthracene thimine [25], onto polymer chains (see Fig. 4). The characteristics of this method are: (1) crosslinking takes place homogeneously; (2) enzymes and cells can be fixed as crosslinking takes place at low temperature [26]; and (3) patterning of gels by photoresist is possible. The degree of crosslinking is controlled by functional groups in the polymer chain, the concentration of the polymer, and the wavelength, intensity, and irradiation time of the light.

4.2.2.3 Radiation crosslinking

When gels are prepared by radiation crosslinking [12, 13], a polymer can be radiation crosslinked in a solid state followed by swelling with an appropriate solvent or a polymer solution can be directly irradiated and gelled. The polymers that are radiation crosslinked in the solid state are used in polymer forms, electric wire, cable and tire, and are swollen in an appropriate solvent to become radiation crosslinked gels. However, cross-linking and chain scission of polymers take place competitively and, thus, the number of polymers that can be used is limited. In general, reactivity to radiation and the chemical structure of a polymer are closely correlated. For example, if the main chain structure is $[-CH_2-CHR-]_n$ it crosslinks and if the structure is $[-CH_2-C(CH_3)R-]_n$, then it decomposes.

Polymers that crosslink in the solid state can also be used in solution. Particularly in water, the hydrogen abstraction created by reactive POH radical due to the decomposed water indirectly yields polymer radicals [12]. The rate of crosslinking in solution is greater than in solids due to a high mobility of the generated polymer free radicals.

Shown in the following chemical formula is the process of cross-linking when a poly(vinyl alcohol) aqueous solution is irradiated by γ-rays [27]:

A gel fiber of 200 μm diameter can be manufactured by γ-ray irradiation of an aqueous solution of poly(vinyl methylether) (PVME) [28]. This fiber is considered a candidate for chemomechanical material application because it is porous and has a phase transition at approximately $38°C$.

4.2.2.4   *Plasma crosslinking*
One of the methods of crosslinking polymers by plasma is CASING (crosslinking with activated species of inert gas) [14, 29]. Although inert gases such as helium and argon do not cause chemical reactions in a plasma, it is possible to sever chemical bonds on polymer surfaces. The resultant polymer free radicals recombine among them and crosslinks form. For example, plasma crosslinking of polyethylene proceeds as follows:

$$He \xrightarrow{\text{plasma}} He^* + He^+ + e^- + h\nu + \text{etc.}$$

$$RCH_2CH_2CH_3 + He^* \longrightarrow RC\cdot HHCH_2CH_3 + He + H^*$$

$$RC\cdot HCH_2CH_3 + H^* \longrightarrow RCH=CHCH_3 + H_2$$

$$\begin{array}{c} RC\cdot HCH_2CH_3 \\ RC\cdot HCH_2CH_3 \end{array} \longrightarrow \begin{array}{c} RCHCH_2CH_3 \\ | \\ RCHCH_2CH_3 \end{array}$$

The surface chlorine or fluorine atoms are eliminated by briefly irradiating poly(vinyl chloride) or poly(tetrafluoro ethylene), thereby leading to dense crosslinking [30]. Generally, CASING treatment will not change wettability but it does improve adhesion. As crosslinking takes place only on the surface, this method is appropriate for preparing ultrathin gels. The degree of crosslinking is controlled by the power and irradiation time of plasma and the type and concentration of the inert gas.

4.3 GEL FORMATION BY INTERMOLECULAR PHYSICAL BONDING

In the previous section, crosslinking by covalent bonds was described. However, it is possible to crosslink by intermolecular physical bonding, such as hydrogen bonding, static bonding, coordination bonding, hydrophobic bonding, and van der Waals bonding.

In general, these interactions form junction zones like microcrystals, helices, ion complexes, and micelles (see Fig. 5).

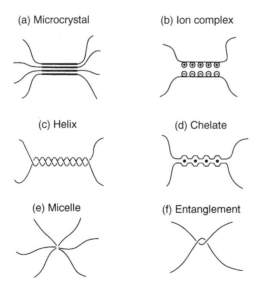

Fig. 5 Formation of crosslinking by intermolecular linkages.

4.3.1 Synthetic Polymers

4.3.1.1 Crosslinking by hydrogen bonding

Poly(vinyl alcohol) (PVA) will gel by microcrystals acting as crosslink points. These microcrystals are formed by hydrogen bonding between crosslink points (see Fig. 5(a)). In general, PVA gels with superior mechanical properties can be prepared by methods such as the freeze treatment method [31] partial drying by freezing [32], the repeating freeze/thaw method [33], and the freezing low temperature crystallization method [34]. The crosslink formation mechanism of the freezing low temperature crystallization method is considered as follows [35]:

1. The free water in a polymer solution freezes by cooling and polymer chains are excluded.
2. Then, the local concentration of PVA increases and hydrogen bonding between polymer chains is formed leading to the formation of crystallites.
3. Repeated thawing and freezing grow microcrystals and strong 3D networks are formed.

Generally, the obtained gels are cloudy due to microscopic heterogeneity. However, it is also possible to prepare transparent gels using a mixed solvent of dimethyl sulfoamide and water.

In addition, crosslinking can be formed by hydrogen bonding between different polymers. Polymer complexes can be formed between proton donor polymers like polyarylamine and poly(vinyl alcohol) and proton acceptor polymers like poly(ethylene glycol) and poly(vinyl pyrrolidone) through hydrogen bonding [36]. The characteristics of polymer complexes include: (1) bonding is selective; (2) directionality exists; (3) composition based on functionality is approximately 1 : 1; and (4) they are reversible. The preparation of gels is relatively simple. However, appropriate polymer concentration, ionic strength, and pH need to be selected.

4.3.1.2 Crosslinking by static bonding

When polyelectrolytes with differing charges are mixed under appropriate conditions, molecular assemblies via static interactions are formed (Fig. 5(a)) [37], and these will become the crosslink points to form polyelectrolyte complex gels [38]. The characteristics of crosslinking by static bonding are: (1) the bond strength is as high as 10–100 kcal/mol; (2) it is

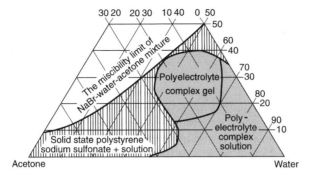

Fig. 6 The ternary phase diagram of polybenzyl-trimethyl-ammonium (PVBMA)-polystyrene sodium sulfonate (PNaSS) complex.

isotropic; and (3) the composition ratio of anion to cation is approximately $1:1$. When polyelectrolyte complex gels are prepared, important parameters include chemical structure and concentration of the polymer electric charge density, pH, ionic strength, composition and dielectric constant of the solvent, and temperature and rate of mixing. By controlling these parameters appropriately, gels for the desired purpose and applications can be prepared. Figure 6 shows an example of a polyelectrolyte complex gel made from (poly(vinyl-benzyl-trimethyl-ammonium chloride) (PVBMA) and poly(styrene sodium sulfonate) (PNaSS).

4.3.1.3   Crosslinking by coordination bonding

Synthetic polymers with side chains that can form coordination complexes, such as polycarboxylic acids, polyols and polyamines, will form crosslinks by adding multivalent metallic ions [39]. For example, when NH_2 is added to a copper sulfate solution of poly(vinyl alcohol), the polymer becomes insoluble [40].

When a metallic ion is added to a high concentration polyoxazolin with bipyridil groups, a gel will form in several seconds [41]. The stability of the obtained gels is on the order of $Ru^{3+} > Fe^{3+} > Co^{3+} > Ni^{2+}$ and no gels are obtained for Cu^{2+} and Ag^+. These gels show a thermoreversible nature. They become sol at high temperature and revert to gel again at low temperature.

4.3.1.4   Crosslinking by van der Waals bonding

The block copolymer of acrylic acid and acrylic acid stearyl is an example of crosslink formation via van der Waals forces. The side chain stearic group exhibits a crystal-amorphous transition at approximately 50°C. The amorphous polymer mixes with water or dimethyl sulfoxide, and the stearyl side chain aggregates and crystallizes by cooling, leading to a crosslinked polymer [42]. These gels are generally cloudy and inhomogeneous, but the properties of the gel change significantly with monomer composition, temperature, and composition of solvent. The mechanical strength increases by copolymerizing with divinyl monomers. The structure formation by the stearyl group is less complex. Shape memory [43], artificial valves, or switches that are temperature-actuated are the applications currently under consideration.

4.3.1.5   Other crosslinking

Other gels include those with molecular entanglement, anomalous viscosity, and interpenetrating polymer networks (IPN). For example, when a several % aqueous solution of poly(methacrylic acid) is stirred, the viscosity gradually increases and forms a gel. After stopping agitation and keeping the agitation steady for some time, the gel returns to a sol. This phenomenon is reversible and is known as antithixotropy or rheopexy [44]. It is an interesting example of dynamic gels that are formed in a nonequilibrium state.

On the other hand, IPN gels are a kind of polymer complex obtained by polymerizing the monomers that are included in 3D networks. As these gels possess a tight network structure, any drugs carried in the gels exhibit minimal leakage. Such gels are being evaluated for drug delivery systems where drug delivery is controlled by external stimuli that include temperature and electric fields [45].

4.3.2   Bio and Natural Polymers

4.3.2.1   Crosslinking by hydrogen bonding

Polysaccharides that crosslink by hydrogen bonding include starch, agar, and carrageenan [46]. They are widely used in the photographic industry, for food and medicine, and culture substrates. The main components of agar are agarose and agaropectin. This aqueous solution increases in viscosity when the solution heated above 90°C is cooled and gels at about 30°C [47]. This is because the hydroxyl groups of the polysaccharides form hydrogen bonds and then a double helix (Fig. 5(c)), which is considered to form crosslinking via complex formation.

On the other hand, unlike agar, carrageenan contains a large concentration of sulfate groups, and can be divided into κ-, τ-, and λ-, depending on their concentration. The κ-type exhibits the strongest ability to form gels. Another proposed mechanism is the double helix one described here, which involves a single chain helix leading to a gel (see Fig. 7). As well, the modulus of the gel increases significantly by potassium or cesium ions. This is considered to be due to the lessening of static repulsion of the sulfate group by the ions entering the carrageenan molecules [48]. Starch, made from amylose and amylopectin, dissolves to form a viscous liquid at 60–70°C, and gels at room temperature.

Collagen and gelatin are proteins that crosslink via hydrogen bonding. Collagen exists in bones, cartilage, tendons, skin, and fish scale and is a fibrous protein that forms right-handed helices made of three polypeptides. It has excellent biocompatibility and is used in cosmetics, medical materials, culture substrates, and food packaging [49]. Gelatin is a collagen that is made of irreversibly separate molecules

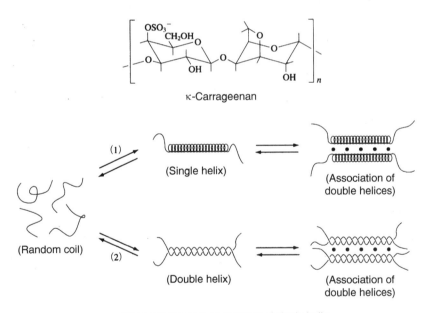

κ-Carrageenan

(1) Formation and association of single helix
(2) Formation and association of double helix
• Indicates cations

Fig. 7 The chemical structure and crosslink formation mechanism of κ-carrageenan.

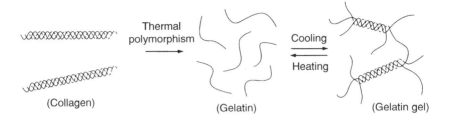

Fig. 8 Thermal polymorphism of collagen and crosslink formation of gelatin.

and is a protein having a molecular weight of 100,000. When a gelatin/ water solution is cooled, it gels at about 25°C. This is because the NH and CO groups in the polypeptides hydrogen bond, partially reproduce tropocollagen, and form a crosslink structure (see Fig. 8) [50]. However, the obtained gel readily exhibits sol-gel transition with temperature and solvent changes and is mechanically weak.

4.3.2.2 Crosslinking by coordination bonding

Hydrocolloids such as arginic acid salt [51], pectin [52], and konyak mannan [53] become gels with the addition of divalent metallic ions. For example, by adding dropwise a sodium arginic acid solution into a calcium salt solution, a gel is readily formed. This is considered to take place because egg box-type crosslinking is formed by the carboxylic group of arginic acid incorporating the calcium ion (see Fig. 9) [54]. By removing calcium ions with a chelating agent like ethylene diamine tetraacetate, the gel returns to a sol. This process occurs reversibly.

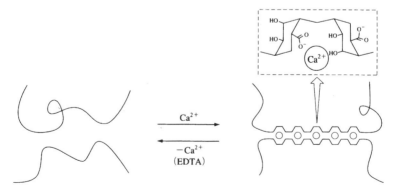

Fig. 9 Crosslink formation model of calcium arginic acid.

4.3.2.3   *Crosslinking by hydrophobic bonding*

An aqueous solution of methyl cellulose and hydroxylpropyl cellulose will gel upon heating [55]. The viscosity of the solution reaches a maximum with the introduction of several alkyl side chains with 6, 12, and 16 carbons onto hydroxyethyl cellulose. The optimum concentration of the alkyl side chain decreases as the length increases [56].

This is considered to be due to crosslink formation by micelles as a consequence of the aggregation of hydrophobic side chains (see Fig. 5 (e)). Here, the driving force of hydrophobic bonding is the positive entropy change accompanying the release of structural regular water molecules that are restricted by the hydrophobic alkyl side chains. Therefore, this is a unique crosslink structure for hydrogels.

Table 3   Synthetic methods of bio and natural polymers.

Examples of gelling polymers	General methods	Formation of crosslinks
Polysaccharides		
Starch	Cooling	Microcrystal formation by hydrogen bonding
Agar		Helix formation by hydrogen bonding
Carrageenan	Cooling	
Gellan		
Arginic acid		
Pectinic acid	Addition of multivalent ions	Crosslinking by coordination bonding (egg-box junction)
Carboxymethyl cellulose		
Konyaku mannan		
Methyl cellulose		
Hydroxypropyl cellulose	Heating	Micelle formation by hydrophobic bonding
Xanthan gum		
Hyaluronic acid	Cooling	Entanglement of molecules (high molecular weight)
Culdran	Heating	Irreversible aggregation
Protein		
Gelatin		
Collagen	Cooling	Helix formation by hydrogen bonding
Albumin		
Soy protein	Heating	Association of partially denatured globular proteins
Casein		
Fibrin		
Elastin	Swells in acid and alkaline solutions	Covalent bonding
Keratin		

4.3.2.4 Other crosslinks

Xanthan gum is used as a thickening and/or suspension agent in foods and pharmaceutical products. This gelling agent has long side chains and gels simply by entanglement of its molecules (see Fig. 5(f)) [54]. As well, hyaluronic acid is extremely hygroscopic and gels in water by molecular entanglement if it has high molecular weight [57]. Curdlan, a polysaccharide that consists of $\beta 1 \to 3$ glycoside bonds, is insoluble in water. However, as the temperature of the suspension is increased, it gels at 54°C after increasing its viscosity, and, at 80°C, it becomes an irreversible gel.

Globular proteins, such as albumen, soy protein and casein, partially denature while maintaining their globular shape. When they are heated or disulfide bonds are broken, these complexes form network structures [58]. Hard proteins, such as fibrin, elastin, and keratin, possess covalent crosslinks and are difficult to dissolve in water. However, they swell in acid or base solutions to some extent. Table 3 summarizes the examples of bio and natural polymers that gel via intermolecular physical bonding.

REFERENCES

1   Hagino, I., Osada, Y., Fushimi, T., and Yamauchi, A. (1991). *Gel-Fundamentals and Applications of Soft Materials*, Sangyo Tosho.
2   Chemical Society of Japan (1992). *Experimental Chemistry Lecture Series 28: Polymer Synthesis*, Tokyo: Maruzen, p. 425.
3   Flory, P.J. (1955). *Polymer Chemistry*, vol I and II, S. Oka, and K. Kanamaru. Trans. Tokyo: Maruzen (originally published in English).
4   Okamura, S. *et al.* (1970). *Introduction to Polymer Chemistry*, Kyoto: Kagaku Dojin, p. 211.
5   Braun, D., Cherdron, H., and Kern, W. (1968). *Experimental Methods in Polymer Chemistry*, Y. Iwakuni. Transl., Asakura Publ. p. 181 (originally published in English).
6   Otsuka, T., and Kinoshita, Y. (1972). *Experimental Methods in Polymer Synthesis*, Kyoto: Kagaku Dojin, p. 299.
7   Kakiuchi, H. (1970). *Epoxy Resins*, Shokodo, p. 64.
8   Polymer Chemistry Society (1972). *The Fundamentals in Polymer Chemistry*, Tokyo: Kagaku Dojin, p. 131.
9   Otsu, T., and Kinoshita, M. (1972). *Experimental Methods in Polymer Synthesis*, Kyoto: Kagaku Dojin, p. 131.
10 Cox, A., and Kemp, T.J. (1975). *Fundamental Photochemistry*, K. Honda. Transl., Kyoritsu Publ. p. 63.
11 Okamura, S., Nakajima, A., Onogi, S., Kawai, H., Nishijima, Y., Higashimura, T., and Ise, N. (1970). *Introduction to Polymer Chemistry*, Kyoto: Kagaku Dojin.
12 (1966). *Polymer Technology Lecture Series 15: Radiation Polymer Chemistry*, Chijin Sokan.
13 Makunouchi, K. (1989). *Proc. 4th Polymer Gel Study Group Lecture Series*, p. 20.
14 Osada, Y. (1986). *Plasma Polymerization*, Tokyo: Kagaku Dojin, p. 213.

15 Osada, Y., Bell, A.T., and Shem, M. (1973). *J. Polym. Sci., Lett. Ed.*, **16**: 309.
16 Osada, Y., and Takase, M. (1983). *J. Chem. Soc. Jpn.*, 439.
17 Osada, Y., and Mizumoto, A. (1985). *Macromolecules*, **18**: 302.
18 Yoshikawa, J. (1986). *Polymer Synthesis*, Kyoto: Kagaku Dojin, p. 147.
19 Kishi, R., Shishido, M., and Tazuke, S. (1990). *Macromolecules*, **23**: 3779.
20 Ishimura, K. (1989). *Proc. 4th Polymer Gel Study Group Lecture Series*, p. 14.
21 Tsunoda, T., and Yamaoka, T. (1964). *J. Appl. Polym. Sci.*, **8**: 1379.
22 Yamaoka, T., and Morita, H. (1988). *Photosensitive Polymers*, Kyoritsu Publ.
23 Smolinski, G., Snyder, L.C., and Wasserman, E. (1963). *Rev. Mod. Phys.*, **35**: 576.
24 Mimsk, L.M., Smith, J.G., van Deusen, W.P., and Wricht, J.F. (1959). *J. Appl. Polym. Sci.*, **2**: 302.
25 Moghaddam, M.J., Inaki, Y., and Takemoto, K. (1986). *Polym. Prepr. Jpn.*, **35**: 495.
26 Uedaira, H., Yamauchi, A., Nagasawa, J., Ichijyo, H., Suehiro, T., and Ichimura, K. (1984). *Sen-i Gakkaishi*, **40**: T-317.
27 Hatakeyama, T., Yamauchi, A., and Hatakeyama, H. (1984). *Eur Polym. J.*, **20**: 61.
28 Taira, S., Morishita, A., Onodera, R., Ichijo, H., and Yamauchi, A. (1989). *Kobunshi Ronbunshu*, **46**: 661.
29 Osada, Y. (1994). *Low Temperature Plasma Material Chemistry*, Sangyo Tosho.
30 Iriyama, Y., and Yasuda, H. (1988). *J. Appl. Polym. Sci., Appl. Polym. Symp.*, **42**: 97.
31 *Tokko Kokai, Jpn.*, **48**-30462 (Kuraray).
32 Watase, M., Nishinari, K., and Nambu, M. (1983). *Polymer Commun.*, **24**: 52.
33 Nambu, M. (1983). *Kobunshi Kako*, **32**: 523.
34 Xuan, C. and Ikada, H. (1983). *Pobal Meeting Record*, **83**: 92.
35 Xuan, C. (1989). *Proc. 4th Polymer Gel Study Group Lecture Series*, p. 8.
36 Abe, Y. (1994). *Polymer Complex*, Kyoritsu Publ.
37 Kambara, S. (1974). *Functional Polymers*, Kyoritsu Publ. p. 489.
38 Bixler, H.J., and Michaels, A.S. (1969). *Encycl. Polym. Sci. Tech.*, **10**: 765.
39 (1989). *Fundamentals of Polymer Complexes*. Polymer Complex Study Group, Gakkai Publ. Center.
40 Deuel, H., and Neukom, H. (1949). *Makromol. Chem.*, **3**: 13.
41 Saegusa, T. *et al.* (1989). *Proc. 38th Polymer Symposium, Jpn.*, p. 47.
42 Matsuda, A., Sato, J., Yasunaga, H., and Osada, Y. (1994). *Macromolecules*, **27**: 7695.
43 Osada, Y., and Matsuda, A. (1995). *Nature*, **376**: 219.
44 Murakami, K. (1991). *Fundamentals of Rheology*, Sangyo Tosho.
45 Katono, H., Sanui, K., Ogata, N., Okano, T., and Sakurai, Y. (1991). *Polym. J.*, **23**: 1179.
46 Yamauchi, A., and Hirokawa, N. (1990). *New Polymeric Materials One point-24: Functional Gels*, Kyoritsu Publ.
47 Watase, M. (1989). *Proc. 4th Polymer Gel Study Group Lecture Series*, p. 26.
48 Smidstrod, O., and Grasdalen, H. (1982). *Carbohydr. Polym.*, **2**: 270.
49 Fujimoto, O. (1994). *Bio Sciences of the Future Series 32: Collagen*, Kyoritsu Publ.
50 Shiraishi, H. (1989). *Proc. 4th Polymer Gel Study Group Lecture Series*, p. 34.
51 Watase, M. (1970). *Nisshokukoshi*, **17**: 148.
52 Watase, M. (1970). *Nisshokukoshi*, **17**: 252.
53 Watase, M. (1970). *Nisshokukoshi*, **17**: 488.
54 Uchida, K. (1988). *Nisshokukoshi*, **31**: 34.
55 (1996). *Colloid Science III-Bio-colloids and Application of Colloids*, Tokyo: Tokyo Kagaku Dojin, p. 43.
56 Sau, A.C. (1987). *Polym. Mater.*, **57**: 497.
57 Yamauchi, A. (1989). *Polymers and Medical Care*, Mita Publ. Kai, p. 356.
58 Clark, A.H., Sauederson, D.H.P., and Suggett, A. (1981). *Int. J. Protein Res.*, **17**: 380.

CHAPTER 3

Structure and Properties of Gels

Chapter contents

Section 1
Structure of Gels

KANJI KAJIWARA

1.1 INTRODUCTION

1.1.1 Preface

What are gels? Hermans [1] described gels to be (1) coherent colloidal suspensions that consist of more than two components; (2) materials that show solid-state mechanical properties; and (3) materials in which the solute and solvent spread continuously throughout the system. This defines tofu, konyak, agar, and jelly, all of which are in a condition in which the sol has frozen or hardened (the origin of the name gel). Such gels are soft, are solid-state materials and contain a large amount of solvent (water). However, 3D molecules, such as crosslinked rubber, epoxies, and silica gel, none of which contain solvents, consist of a single component and thus are not called gels.

According to the Flory–Stockmayer gelation theory [2], the gel point is defined as the point at which the apparent weight average molecular weight becomes infinite by connecting structural units. Thus, it is not necessary for a gel to be a composite of two components. Although there are various definitions of gels, it is safe to state that there is no single definition. In any definition, gels are 3D structures of *infinite structural materials* that occupy the entire given space. Gel formation is the process to form *infinite structure* by intermolecular interaction, such as covalent bonding, i.e., chemical interaction, and hydrogen bonding, hydrophobic bonding, and ionic bonding, i.e., physical interactions. In other words, it indicates that intermolecular interaction extends throughout the space. Therefore, in order to discuss the *structure of gels*, how to define the *infinite structure* within the finite framework becomes the problem.

In the process of gel formation, intermolecular interaction spreads throughout the system. However, interaction spreading is not necessarily homogeneous. According to Flory, gels are classified into four types: (1) an ordered lamella structure that contains gel intermediate structure; (2) a completely random polymer network structure that is formed by covalent bonding; (3) locally ordered polymer networks that are formed by physical association; and (4) amorphous globular structure. In the simplest and ideal system that is made of polymer networks formed by completely random Type 2 covalent bonds, the reaction proceeds three-dimensionally by condensation of multifunctional groups and dendritic polymer networks are constructed. However, as the reaction proceeds, unlike the theoretical assumptions, the reaction will not be homogeneous. As shown in Fig. 1, there is a portion where segment density is dense and rare portions exist separately due to the difference in crosslink density. In addition, in the case of polysaccharides that belong to Type 3, the polymer chains are thought to associate locally and form crosslinks, leading to gel formation (physical gel). As there are multiple intermolecular interactions that become the cause of association, the gel necessarily possesses hierarchy, and various structural levels can be considered.

To analyze an *infinite structure* gel in the finite framework, various levels of structures must be analyzed and, furthermore, these structures need to be reconstructed. Even in a model system where it is considered random, gels are not homogeneous and differ in crosslink density, forming various structural domains. The structure of gels can be understood for the first time by knowing the distribution and bonding states of these domains.

1.1.2 Static Structure

1.1.2.1 The classical model of gel structure

The first gel that was theorized rigorously using a model by Flory and Stockmayer is the one formed by covalent bonding [2]. In this model,

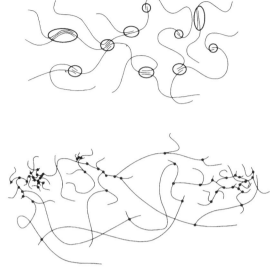

In the physical gel (top), the portion with physical bonding forms clusters (crosslink domain) of finite size.

In the chemical gel (bottom), the crosslink point by covalent bonding can be regarded as having dense and rare parts of crosslink points.

Fig. 1   Models of gel networks.

multifunctional structural units (monomers) react consecutively and grow like tree branches. When the branches fill the entire space, gelation has been achieved [3]. Gordon [4] proposed a formulation as shown in Fig. 2 on the process of f-functional monomers connected randomly by condensation polymerization using the cascade theory of Good [5]. The tree that expresses gelation and the equations that are $1:1$ are given in a dummy variable θ. In Fig. 2, the branching point of the tree ($\bullet$) corresponds to the dummy variable θ and the probability of the tree growing, α, is the probability of further branching. This is the probability that the remaining functional group of the already reacted monomer will react. This formulation indicates that the f-functional random polycondensation polymerization system forms homogeneous gels that lack an intramolecularly crosslinked cyclic structure. The right-hand side equation of Fig. 2 is called a bonding probability series of f-functional random polycondensation polymerization system, and has the following relationship with the

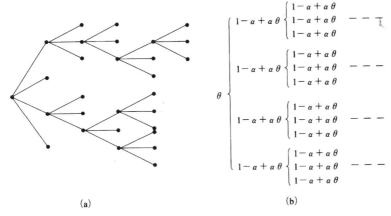

Fig. 2 A tree model ($f = 4$) of f-functional condensation polymerization and its corresponding probability processes.

weight fraction w_x of the branched polymers with a specific degree of polymerization, including the random branching polymers with various modes of branching:

$$W(\theta) = \sum_{x=1}^{\infty} w_x\theta^x = \theta(1 - \alpha + \alpha u(\theta))^f \tag{1}$$

$$u(\theta) = \theta(1 - \alpha + \alpha u(\theta))^{f-1} \tag{2}$$

By graphing the gelation model and, furthermore, having the graph correspond with the equations to describe the gelation process, the calculation of the statistical parameters of the system during the gelation process becomes easy. For example, the weight average degree of polymerization DP_w of the branched polymer immediately prior to the gelation can be readily obtained by differentiating the bonding probability series by θ:

$$DP_w \equiv \left(\frac{\partial w(\theta)}{\partial \theta}\right)_{\theta=1} = \frac{1 + \alpha}{1 - (f-1)\alpha} \tag{3}$$

The cascade theory can also be easily extended to crosslinked systems. The formation of crosslinked gels starts first with crosslinking a linear polymer (primary polymer) by intermolecular covalent bonds and then forming a star polymer with four branches. As the crosslinking via intermolecular covalent bonding proceeds, the amount of branching increases, and gelation takes place when the entire space is occupied by a molecule (see Fig. 3). If each structural unit (monomer) can participate in the crosslinking reaction, the number of functional groups f of f-functional random polycondensation polymerization corresponds to the degree of polymerization of the linear polymer (primary polymer) prior to crosslinking. Accordingly, if f in Eqs. (1)–(3) is replaced with y, which is the degree of polymerization of the primary chain (when there is a distribution in the primary chain, the weight average degree of polymerization y_w is used), then these equations can be used for crosslinked systems as they appear. (However, it is necessary to add the information on the molecular weight of the primary chain to Eqs. (1) and (2) as front factors. Also, this will be seen by multiplying the degree of polymerization of the primary chain on the denominator of Eq. (3). See reference [6] for further details.)

The model that does not contain cyclic groups in a molecule is called the classical model, and while it has been criticized as being too simplistic in describing the gelation process, it can define the gel point precisely. Also, if the statistical values of the system before and after the gel point (gel fraction, average degree of polymerization, average radius of gyration, etc.) are normalized by the distance from the gel point, real systems can be explained satisfactorily. The gel point is determined by the point where the weight average molecular weight becomes infinite, that is, by

$$\alpha_c = \frac{1}{f-1} \tag{4}$$

(crosslinked systems can be expressed by replacing f by y_w), and corresponds to the point where the gel fraction begins increasing from 0. After the gel point is determined, the reaction proceeds (α increases) and the gel fraction increases (sol fraction decreases). The gel fraction is 1

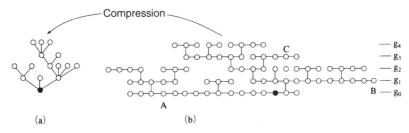

Fig. 3   A branch model of primary chain crosslinking and compression into a tree model.

at $\alpha = 1$ (sol fraction is 0) and all structural units in the system have been incorporated into the gel. Usually, α/α_c is called the degree of crosslinking and

$$\varepsilon \equiv 1 - \frac{\alpha}{\alpha_c} \tag{5}$$

corresponds to the equivalent distance from the critical point α_c (see subsection 1.1.2.2, percolation model).

In the classical Flory–Stockmayer model (FS model), the cyclic structure by intramolecular crosslinking is not considered. However, in real reaction systems, both intramolecular and intermolecular crosslinking occur competitively. In an extreme case, gelation will not occur when intramolecular crosslinking dominates. Because intramolecular crosslinking will not contribute to the gelation, α_c increases in comparison to an ideal situation where no intramolecular crosslinking takes place. Intramolecular crosslinking becomes more pronounced in a dilute solution. Substitution also causes a shift of the gel point. It is possible that if one of the multifunctional groups has reacted, the reactivities of other groups will change. As the reactivity increases (decreases), the gel point increases (decreases). However, as far as the FS model is concerned, only the gel point shifts, and even if intramolecular crosslinking and substitution effects are taken into account, the structure of gels will not fundamentally change as shown in Figs. 2 and 3. Furthermore, the critical behavior near the gel point (Table 1) is the same.

Table 1   Critical indices.

Proportionality equations	Percolation	Classical theory
(1) $DP_w \propto \lvert\varepsilon\rvert^{-\gamma}$	$\gamma = 1.74$	$\gamma = 1$
(2) $\langle S^2 \rangle_z^{1/2} \propto \lvert\varepsilon\rvert^{-\nu}$	$\nu = 0.88$	$\nu = \frac{1}{2}$
(3) $G \propto \lvert\varepsilon\rvert^{\beta}$	$\beta = 0.45$	$\beta = 1$
(4) $\langle S_x^2 \rangle^{1/2}(\varepsilon \to 0) \propto x^{\rho}$	$\rho = 0.40$	$\rho = \frac{1}{4}$
(5) $D_x(\varepsilon \to 0) \propto R_{zH}^{-1}\,(\varepsilon \to 0) \propto x^{-y}$	$y = 0.28$	$y = \frac{1}{4}$
(6) $n_x(\varepsilon \to 0) \propto x^{-\tau}$	$\tau = 2.20$	$\tau = \frac{5}{2}$
(7) $\langle S^2 \rangle_z^{1/2} \propto DP_w^{\nu/\gamma}$	$\nu/\gamma = 0.506$	$\nu/\gamma = \frac{1}{2}$
(8) $D_z \propto \langle R_H^{-1} \rangle_z \propto DP_w^{-\kappa}$	$\kappa = 0.35$	$\kappa = \frac{1}{2}$
(9) $E \propto \lvert\varepsilon\rvert^{t}$	$t = 2.67$	$t = 3$

$\varepsilon = 1 - \alpha/\alpha_c$; G is the gel fraction; D is the diffusion coefficient; R_n^{-1} is the reciprocal of hydrodynamic radius; and E is the Young's modulus. Subscripts w and z indicate the corresponding average and x the element of the degree of polymerization x.

1.1.2.2   Percolation model

As discussed briefly in the previous section, gelation can generally be discussed within the framework of critical phenomena [7] by having the gel point and critical point correspond. Stauffer applied the percolation theory often used for the general theory of critical phenomena to the crosslinking reaction of polymers [8, 9].

The percolation model of gels starts with setting a lattice and placing monomers in the lattice units. It is assumed that each lattice point is the monomer with the number of the functional group equivalent to the number of the neighboring lattice points. In some cases the neighboring monomers react with a certain probability (connection percolation) and in other cases, the neighboring monomers that are placed under a certain probability always react (position percolation). Either way, the intramolecular cyclic formation that was ignored by the classical model (the FS model) can be readily incorporated into the model (see Fig. 4).

The difference between the FS model and percolation model is in the critical phenomenon. As summarized in Table 1, if the statistical values are normalized by the equivalent distance $\varepsilon(\equiv 1 - \alpha/\alpha_c)$ from the gel point (the critical point), there is a significant difference in critical index for the FS model and percolation model. This difference reflects the difference in size distribution n_x (see Fig. 1 [6]). The difference of the structure in the model is reflected on the fractal dimension D of the fraction that has a certain degree of polymerization x. If the radius of a sphere that corresponds to the volume of the branched polymer fraction with the degree of polymerization x is R, the relationship between x and R is from the fractal dimension D

$$x \propto R^D \qquad (6)$$

If R is replaced by the radius of gyration of the branched polymer fraction with the degree of polymerization x from Table 1, item (4), the fractal dimension $D = 2.5$ for the percolation model and $D = 4$ for the FS model is obtained. The FS model appears to be packed by more than a sphere ($D = 3$) and thus is unrealistic. This contradiction can be understood when a dendrimer is considered. The FS model grows dendritically. When a trifunctional ($f = 3$) monomer grows radially, there will be no space beyond the sixth generation. However, for the FS model, the branches grow theoretically even under such conditions. In the percolation model, the entire space is predistributed to the structural units, thus no such

(a)

Condition prior to gelation ($a = \frac{1}{4} < a_c$).
Each cluster is randomly distributed.

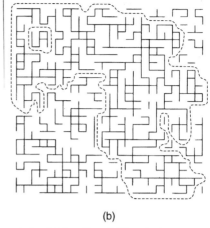

(b)

Gel point ($a = \frac{1}{2} < a_c$). The cluster
surrounded by the dotted line reached
the edge of the lattice. There are also
randomly distributed clusters.

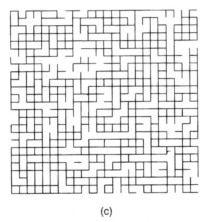

(c)

Condition after gelation ($a = \frac{3}{4} > a_c$).
The entire system is forming a single
cluster.

Fig. 4 Examples of percolation.

unrealistic situation exists. Both the FS model and the percolation model are completely random models that have no relationship among the structural elements and form a homogeneous network structure, but, in the FS model, there is unrealistically dense packing of the structural elements. This shortcoming can be seen in single f-functional condensation polymerization and can be resolved to some extent by considering a branched structure formed by the copolymerization of an f-functional monomer with a bifunctional monomer. This is achieved by considering the gels forming the branched structure by random copolymerization of f-functional and bifunctional monomers. In other words, the homogeneous reaction assumed by the FS model cannot exist.

Thus far, the structure of gels has been discussed from the *fractal* viewpoint by comparing the structure of branched polymers having the degree of polymerization x. In real systems, the molecular weight distribution broadens indefinitely during the process from branching to gelation and becomes infinitely broad at the gel point. In general, the number distribution $n_x(\alpha)$ of the branched polymers with the degree of polymerization that is included in the model is given by [8]

$$n_x(\alpha) = x^{-\tau} f(\varepsilon x^{\sigma}) \tag{7}$$

where τ is the width of distribution and is as narrow as $\tau = 2.20$ in the percolation model and broadens to $\tau = 2.50$ in the FS model (see Table 1, item (6)), and $f(\varepsilon x^{\sigma})$ is the cut-off function that reduces exponentially. The difference in the width of distribution is due to the intramolecular cyclization. The ideal FS model gives the greatest possible width expected theoretically. The weight average degree of polymerization DP_w and z-average radius of gyration $\langle S^2 \rangle^{1/2}$ depend on the shape of the distribution. The experimental fractal dimension is defined by Eq. (6) where x is replaced by DP_w and R by $\langle S^2 \rangle_z^{1/2}$. From Table 1, items (1) and (2), both the percolation model and FS model indicate $D = 2$ and they follow Gaussian distribution as a whole. In other words, the percolation model and FS model both exhibit a completely random gel structure.

1.1.2.3   Aggregation model

The aggregation model can be understood as a more general model of the FS model and percolation model from the fractal point of view, and aggregation phenomena can be found everywhere in nature. Antibody antigen reactions, colloidal suspension, and the aggregation of clouds as well as the Milky Way, all exhibit various aggregation levels and they all grow irreversibly as clusters. Typical examples are inorganic gels. In a sol-gel reaction, particles of a certain size are formed, which then create porous gels by aggregating. However, the aggregation process varies depending on the physicochemical properties. For example, although colloidal suspension usually moves around independently due to the coulombic repulsive forces, aggregation takes place when the electrical charge is blocked. This aggregation process can be divided into two stages: (1) the diffusion of colloidal particles; and (2) the adhesion of colloidal particles by chemical reaction. If the time for each stage of aggregation is significantly different, the slow stage will be the limiting step. The extreme examples of irreversible aggregation phenomena are diffusion-limited aggregation (DLA) and reaction-limited aggregation (RLA); DLA and RLA can be distinguished by the probability of adhesion among clusters. In the case of DLA, the probability of adhesion is roughly equal to the probability of collision, whereas in RLA, it is determined by the reactivity of the clusters. For the polymer gel models, the probability of adhesion can be summarized in Table 2 [10]. For example, in the case of f-functional condensation polymerization (RA_f-type), the cluster (branched polymer) with degree of polymerization i possesses $f(i-2)+2$ functional groups. Therefore, the probability of adhering the cluster with the degrees of polymerization i and j is given by $K_{ij} \propto \{f(i-2)+2\}\{f(i-2)+2\}$. On the other hand, the case where the structural unit (monomer) is one functional A-type and $(f-1)$-functional B-type, and only A reacts with B (ARB_{f-1}-type), the adhesion probability is $K_{ij} \propto 2 + (f-2)(i+j)$. From the viewpoint of aggregation phenomena, RA_f-type is classified as RLA and ARB_{f-1}-type is DLA.

Table 2 Bonding probability of the gelation of polymers during the model aggregation process.

K_{ij}	Physical processes
$\{f(i-2)+2\}\{f(j-2)+2\}$	f-Functional (RA_f-type) condensation polymerized branched polymers
ij	The same as above ($f \gg 1$)
$(ij)^\omega$	The same as above (only the surface of clusters react)
$2+(f-2)(i+j)$	ARB_{f-1}-type condensation polymerized branched polymer
$i+j$	The same as above ($f \gg 1$)
$i^\omega + j^\omega$	The same as above (only the surface of clusters react)
$A + B(i+j) + Cij$	General type ($A_g RB_{f-g}$-type)

In the early DLA simulation [11], that is, with a fractal pattern of dimension approximately 2.5, the Witten–Saners model (WS) was obtained. This model repeats the operation in which multiple small balls that move freely around a small ball fixed at the origin make contact with each other, adhere and form clusters (the particle-cluster aggregation process). If we assume all balls move freely, collide and form clusters irreversibly (cluster–cluster aggregation [12, 13]), more general cases can be handled. In this model, when a model experiment was done under the same conditions as the WS model, the fractal dimension was 1.75. Figure 5 shows the experimental results of a model experiment of cluster–cluster aggregation. As the number of small balls increases, the process of network growth can be seen better.

Classically, the aggregation process is described by the Smoluchowski equation, which is an irreversible chemical reaction kinetic equation used to treat the aggregation of nonfractal particles. Therefore, the collision takes place between two bodies with small fluctuation of concentration, and is assumed to occur randomly.

It has been shown that, if the fractal nature of clusters is carefully considered, the Smoluchowski equation can express the cluster–cluster aggregation process precisely based on the comparison with the simulation by computer [14]. Assuming that the density of the cluster with the degree of polymerization at time t is $C_k(t)$, the following relationship is obtained when there is no spatial interference:

$$\frac{\partial C_k(t)}{\partial t} = \frac{1}{2} \sum_{i+j=k} K_{ij} C_i(t) C_j(t) - C_k(t) \sum_{j=1}^{\infty} K_{kj} C_j(t) \tag{8}$$

The moment $M_n(t)$ of density $C_k(t)$ is defined as follows:

$$M_n(t) \equiv \sum_{k=1}^{\infty} k^n C_k(t) \tag{9}$$

Because the average cluster size is proportional to $M_2(t)/M_1(t)$, the gelation takes place at the time t_c, at which $M_2(t)$ becomes infinite. When the bonding probability K_{ij} is constant, $i+j$, or ij (Table 2), the Smoluchowski equation can be solved analytically [15]. In particular, it will yield results similar to those with the FS model at $K_{ij} = ij$. In the f-functional condensation polymerization system (FS model), the functionality $\{f(i-2)+2\}$ of the cluster with degree of polymerization i is proportional to the volume. However, if the shape of the cluster (branched

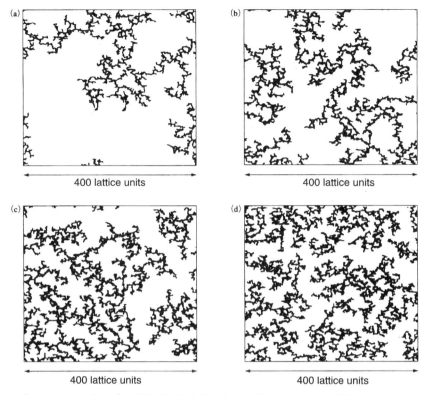

A constant number of particles in the lattice freely diffuse and irreversibly form clusters

(a) Number of particles, 10,000 (particle density 0.0625)
(b) Number of particles, 15,000 (particle density 0.09375)
(c) Number of particles, 20,000 (particle density 0.125)
(d) Number of particles, 25,000 (particle density 0.15625)

The lattice is a 400 × 400 square one

Fig. 5 Simulation of cluster–cluster aggregation.

polymers) and intramolecular reaction are taken into consideration, the effective functionality will be proportional to the apparent surface area of the cluster. Assuming that the apparent surface area of the cluster with degree of polymerization i is proportional to $i^{\omega}(i \to \infty)$ [15, 16], the surface effect factor ω is expected to be within the following range:

$$1 - \frac{1}{d} < \omega < 1 \tag{10}$$

Here, the lower limit corresponds to the surface area of the densely packed body with spatial dimension d and the upper limit corresponds to the FS model where each functional group maintains complete reactivity. If $\omega > \frac{1}{2}$, sol-gel transition takes place within a limited time. In general, $\omega = D/d$ (D is the fractal dimension). The index to define the width of the cluster distribution is (see Eq. (7), τ of Table 1, item (6)),

$$\frac{5d - 2}{2d} < \tau < \frac{5}{2} \tag{11}$$

and for the close packing cluster, from the lower limit of Eq. (10), $\tau = 2.17$.

For ARB_{f-1}-type, when the surface effect is ignored, $\tau = 1.5$ from the Smoluchowski equation, and $\tau = 1.4 \pm 0.15$ from the antibody antigen reaction [17]. The general type of bonding probability [18, 19] (Table 2) is a hybrid of the RA_f-type and ARB_{f-1}-type, and can deal with the aggregation process that is between DLA and RLA.

1.1.2.4 Inhomogeneous structure of gels

The structural models of gels discussed thus far (see preceding subsections 1.2.1, 1.2.2, and 1.2.3) are necessarily random and homogeneous structures because they were considered statistically and theoretically. In such models, the long-range interaction deals with mean-field approximation, and will not be the cause of inhomogeneous structures. On the other hand, gelation is incorporated into the general theory of critical phenomenon in which the gel point is regarded as the critical point. The critical phenomenon is where the intermolecular interaction is extended to infinity. The response can be understood well by considering that the gelation is the state where the bonding of the structural element, whether chemical or physical, extended throughout the system. In this case, the basic intermolecular interaction among structural elements is not considered. Accordingly, gels exhibit fractal structure, and the gel structure is defined by a single fractal dimension.

Consider now the formation process of real gels. In any models of previous sections, gelation takes place when the branched structure spreads to the entire system and occupies the entire space that is given to the system. However, the development of the branched structure does not necessarily start at a single point but rather at multiple locations. Two branched polymers that started growing almost simultaneously from the multiple starting points come into contact and halt their growth. This is shown schematically in Fig. 6. If the clusters formed in such a manner possess branched structures and form domains that are defined by the correlation length ξ when the fractal dimension within the domains is D, the density correlation function of the structural units (monomer) is given by [9]

$$g(r) \approx \left(\frac{\xi}{r}\right)^{3-D} \exp(-r/\xi) \tag{12}$$

When domains are packed closely ($D = 3$), Eq. (12) is the same as the correlation function of the Debye–Busche type [20], and equivalent to the domain that consists of the random aggregates with smooth surface. The structure of the domain can be described by the structural model described in the previous section. In any case, Eq. (12) can be used for approximation. However, the fractal dimension would depend on the model used. For example, for the classical model (FS model), $D = 2$.

As the already formed domains connect with each other, the next larger cluster will be formed. The bonding pattern can be understood by

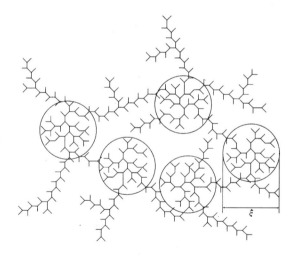

Fig. 6. Schematic diagram of the formation of domains and gelation by the aggregation of the domains (the domain that is characterized by the correlation length ξ forms networks that act as a new unit of a multifunctional structural unit).

looking at models in the preceding subsections 1.2.1, 1.2.2, and 1.2.3. A branched structure is formed hierarchically in such a manner during the gel formation process. It is possible that the bonding pattern that relates to the branch structure formation may not be single. Schematically, inhomogeneous gels can be considered as shown in Fig. 7. On the right-hand side of Fig. 7(a), the portion with high crosslink density functions as a crosslink domain with increased functionality, and suggests that it does not relate to the properties of gels directly. In gels, therefore, it is possible to have different fractal dimensions depending on the scale of observation. Figure 7(b) illustrates a heterogeneous gel model that connects multiple percolation clusters [21]. In this model, new networks are formed by connecting the terminal of the percolation clusters ($D = 2.5$) [22]. Considering the percolation cluster as the new structural unit, the formed networks possess various fractal dimensions depending on the bonding patterns. Figure 7(b) also depicts the schematic diagram of the

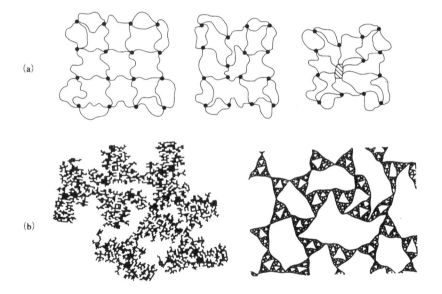

(a) Random heterogeneous gel structure: from the left, homogeneous networks, lightly heterogeneous networks, and very heterogeneous networks.

(b) Fractal heterogeneous gel structure: the left side is the network that is formed by multiple percolation clusters, and the right side is the network that is formed by the random connection of a Sierpinski gasket.

Fig. 7 Heterogenous gel model [21].

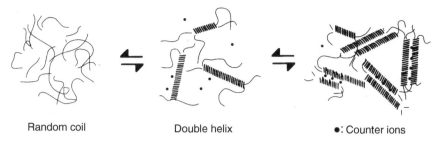

Random coil   Double helix   ●: Counter ions

Fig. 8   Gelation of gellan gum (schematic diagram).

Sierpinski gasket (there can be size distribution) forming networks as a trifunctional structural unit. Therefore, the network as a whole can be described by the FS model ($D = 2$) and the inside structure has the fractal dimension of Sierpinski gasket at $D = 1.585$.

Many physical gels, including biogels, form crosslink domains within which ordered structure exists. In the gelation of gellan gum shown in Fig. 8, double helices are formed at first as the temperature of the gellan gum aqueous solution is lowered. Then a gel is formed by associating those double helices. The structure of the associated double helices is not random but takes a packing pattern that can be expected from a crystal structure [23]. The network as a whole can be described by the model described in the previous section. The crosslink point for the network is no longer a "point" but an ordered structure. In this case, the crosslink domain can also be considered as the phase-separated aggregated phase from the network portion.

Cellulose derivatives gel by increasing temperature and the aggregated phase that is microphase separated forms crosslink domains. In this case, the aggregated phase does not exhibit ordered structure and shows almost a random associated structure at the very early stage of gelation [24]. However, as time passes, crystallization takes place inside the aggregated phase. Although gelation by microphase separation and macrophase separation occurs in a concerted manner, the order of occurrence depends on the thermodynamic state of the system [25].

1.1.3   Dynamic Structure

1.1.3.1   Definition and classification of dynamic structure

One of the characteristic properties of gel is its modulus. Therefore, gels can be defined from their mechanical properties. Stated simply, "gels are viscoelastic solids that possess the plateau of the real part G_i (storage modulus) of complex modulus G^* in a wide frequency range." [26] Phenomenologically, "gels are soft solids that exhibit the abovementioned viscoelasticity, consist of more than two components, and are made of a liquid as one of the components." [27] According to this definition, if G' is flat on the order of seconds and loss modulus G'' is sufficiently smaller than G', it is not necessary for the relaxation time to be infinite to be a gel, and the network, such as the interpenetrating polymer network (IPN), is also a gel. If gels are defined from the dynamic structural point of view, the time scale of observation becomes important.

1.1.3.1.1   Crosslink formation by covalent bonds (chemical gels)
In the case of chemical gels, the gel point corresponds to the point where the longest relaxation time becomes infinite. According to the reptation theory [28], the longest relaxation time of entangled polymers τ_1 is proportional to the third power of the molecular weight M

$$\tau_1 \propto M^3 \tag{13}$$

Therefore, having the longest relaxation time be infinite is equivalent to the molecular weight being infinite, which does not contradict the Flory–Stockmayer definition. The gels formed by covalent bonds (chemical gels) possess infinite relaxation time and plateau modulus. For a small dynamic shear deformation, G' and G'' show little frequency dependence (see Fig. 9). Thus, the following is expected:

$$G' \propto \omega^0; \ \ G'' \propto \omega^0 \tag{14}$$

1.1.3.1.2   Entangled networks
The formation of networks is a result of simple topological interaction of polymer chains. When the product of polymer concentration and molecular weight exceeds a certain value, entangled networks will be formed either in the melt or in the solution. In this case, when the measurement frequency is higher (shorter time scale) than the lifetime of the entanglement, the system behaves as though it were a gel (quasi-gel). Chemical gels and quasi-gels can be readily distinguished by their mechanical properties. If the storage modulus G' and loss modulus G'' are observed as a response to dynamic shear deformation at frequency ω, the mechanical spectra of the chemical and quasi-gels show significant difference at small ω, reflecting the nature of the crosslink points.

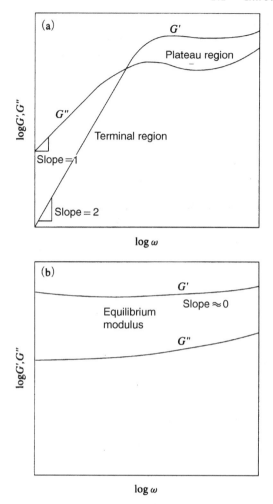

Fig. 9 Dynamic mechanical spectrum of (a) entanglement networks (quasi-gel) and (b) covalent crosslinking networks (chemical gel).

Entangled networks (quasi-gels) show the following relationships:

$$G' \propto \omega^2; \quad G'' \propto \omega^1 \qquad (15)$$

which indicates that, in this region, the quasi-gels behave as, high viscosity liquid.

In the high-frequency region, the chemical and quasi-gels both show similar mechanical behavior. The quasi-gels dissolve by dilution, whereas chemical gels only swell but do not dissolve.

1.1.3.1.3   Physical gels

The gels crosslinked by noncovalent bonds are classified as physical gels. There are many causes of crosslinking, for example, Coulombic force, dipoles, dipole–dipole interaction, van der Waals forces, charge transfer, hydrophobic bonding, hydrogen bonding, etc [29]. Often, the mechanism of crosslinking is not known. In particular, in the case of biopolymers, there is no single source of physical crosslinking. However, for the crosslinking mechanism, there are multiple causes and formed crosslinks often contain higher-order structures, such as multiple helices and the egg box that include ions [30]. These crosslink portions with ordered structure form further crosslink domains.

Accordingly, it is difficult to discuss the generality of physical gels. Here, physical gels are divided into *strong gels* and *weak gels*, both of which behave as a solid in a small deformation regime. While the *strong gels* maintain solid state even under a large deformation as a swollen elastic body, the *weak gels* behave as a liquid having orders under a large deformation. The *weak gels* can be classified further into *thermodynamically weak gels* and *mechanically weak gels*. The *thermodynamically weak gels* have bond energy per crosslink on the order of several kT, and lose and regenerate bonds by thermal fluctuation. The *mechanically weak gels* lose crosslink bonds by mechanical fluctuation. Such crosslink bond energy is also merely several kT and the gels are expected to melt easily at elevated temperature. Therefore, classification between *thermodynamically weak gels* and *mechanically weak gels* are rather subjective and is only a phenomenological distinction.

The *mechanically weak gels* exhibit an extremely narrow linear viscoelastic region, as shown in Fig. 10, in comparison with *strong gels.* The maximum strain γ_{lin} is usually $\gamma_{lin} > 0.2$ for the *strong gels* and, for the *weak gels,* it is approximately $1/1000$ that of the *strong gels*. The *mechanically weak gels* can often be seen in colloids and gels formed by aggregates of small particles. Here, γ_{lin} is determined by the balance of short range attractive and repulsive interactions.

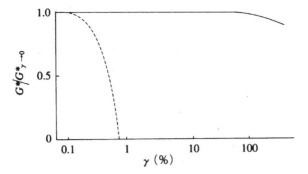

Fig. 10 Mechanical behavior of *weak gels* (dotted line) and *strong gels* (solid line). (The linear region, where the stress is proportional to the strain and is defined by the maximum linear strain γ_{lim}, differs in both cases.)

1.1.3.2 Gels from dynamic structural point of view

As stated previously, the maximum relaxation time becomes infinite approaching the gel point. Specifically evaluating gelation from the viewpoint of dynamic structure requires measurements for *infinite time*. Also, the strain γ must be very close to zero to avoid damage on the crosslinks, because the crosslink density of the networks at the gel point is extremely small. In practice, when the gel point is studied by viscoelastic property measurements, it is assumed that all relaxation modes will appear in the relaxation spectrum [31]. Under this hypothesis, the stress relaxation modulus $G(t)$ can be expressed as $G(t) \approx t^{-s}$. The exponential rule will also apply for the complex modulus. That is,

$$G'(\omega) \approx k_1 \omega^2$$
$$G''(\omega) \approx k_2 \omega^2 \tag{16}$$

The proportionality constants k_1 and k_2 are generally different. That is, the gel point that satisfies Eq. (16) does not necessarily agree with the G', G'' crossover point. Although it is proposed that

$$G'(\omega) = G''(\omega) \tag{17}$$

at the gel point [32], more generally, Eq. (16) should be used. Conceptually, the G', G'' crossover is explained as follows. Assume that a system exists in which a gel is formed in a solution via crosslinking reaction. In the fluid regime (at an early stage of reaction), G' is smaller than G''. As the reaction (crosslinking) proceeds, both G' and G'' will increase.

However, as can be seen from Eq. (15), G' increases faster than G''. At some point, the magnitude of G' and G'' reverses and G' becomes larger than G''. This is the G', G'' crossover and is regarded as the gel point. When further reaction (crosslinking) proceeds, G' increases more and approaches an asymptotic value at a certain time. However, G'' shows a maximum in a parabolic form and eventually reduces to zero.

The index that is defined by Eq. (16) is usually [33]:

$$0.5 \leq s \leq 0.8 \tag{18}$$

Equation (16) applies to the gel point. When gelation proceeds further, $G'(\omega)$ becomes constant. However, it has been confirmed in many systems that $G''(\omega)$ follows the power law. This result indicates that, upon proceeding further with the gelation, the fractal structure at the gel point is more or less preserved.

It is not certain whether or not the gel points ($M_w \rightarrow \infty$) evaluated dynamically and statistically are equivalent. Experimentally, it has been shown that these lie very close to each other [34].

1.1.3.3   Rubber elasticity

One of the most characteristic properties of gel is the rubber elasticity. Rubber elasticity has attracted attention since the early era of polymer science research and has been developed through statistical mechanics [35, 36]. The basis of rubber elasticity is the micro-Brownian motion of the polymer chains [37]. Specifically, the rubber elasticity originates from entropy and is mechanistically different from the energetic elasticity of the crystalline solid. The ideal relationship between stress and strain is given by

$$\sigma = E\left(\lambda - \frac{1}{\lambda^2}\right) \tag{19}$$

where λ is the elongation ratio along the stretching direction and E is Young's modulus. When a single polymer chain with molecular weight M and density ρ is stretched, the following relationship is obtained by calculating the entropy change of deformation:

$$E = \frac{3\rho RT}{M} \tag{20}$$

In this calculation, both ends of the polymer chain are fixed and there is no micro-Brownian motion of the molecule as a whole. The physical cause of fixation of both ends is the crosslinking. When there is room for micro-Brownian motion between the crosslinks, the rubber elasticity appears. That is, in order to discuss the rubber elasticity, the polymer chains must necessarily form a network structure.

For the network structure, let us consider regular networks. When such networks deform, the front factor of Eq. (20) differs by two orders of magnitude. This depends on whether it is assumed that the crosslink point moves proportionately to the deformation of the whole body, although they are fixed spatially (affine model) [38], or whether the networks are fixed to the frame while the internal crosslink points can move freely (ghost model) [39]. In any case, if Eq. (20) is rewritten as

$$E = \frac{3gRTN_e}{V_{\mathrm{mol}}} \qquad (21)$$

Eq. (19) can be used. Here, g is the so-called rubber front factor, V_{mol} is the molar volume of the chain (it is approximately $1/\rho$ in bulk), and N_e denotes the chains in a unit volume that contribute effectively to elasticity and can be calculated from the degree of crosslinking (those chains that are cyclic or the ones that are not participating in the networks will not contribute to elasticity). Because the crosslink points are fixed in the space (they do not contribute to the entropy of the system) in the affine model, the chains that constitute the networks move independently of each other. That is, the topology of the networks, such as the number of the functionality of the crosslink point (the number of chains that come out of the crosslink point), has nothing to do with the elasticity, and the N_e chains that constitute the networks independently contribute to the elasticity. In this case, it is clearly $g = 1$ in Eq. (21). In the ghost model, the crosslink points can move along with the deformation. This degree of freedom of the crosslink points will naturally contribute to the entropy of the system. The movement of the chains that constitute the networks are connected by the crosslink points and are not independent of each other. In this mode, the modulus will naturally vary depending on the network structure. In this case, assuming that the number of functionality of the crosslinks is f, then $g = 1 - 2/f$ [2, 40].

The two models are based on the ideal networks. Here, the word *ideal* means that the chains obey Gaussian statistics, there is no chain-to-chain interaction (they pass through as though they are ghosts), the network itself must have a completely homogeneous structure, and there should not be any crosslink density distribution. Also, fundamentally, the underlying assumption of the mode of network growth is the dendritic structure (Fig. 2) [41].

The variation of the modulus accompanying the sol-gel transition depends on the model as seen in Table 1, item (9). Because the modulus, in a sense, corresponds to the energy density of the system, it can be calculated by evaluating the stored thermal energy in a polymer chain due to the deformation. Only the gel components contribute to the gel modulus. Thus,

$$E \propto RTG/\xi^D \tag{22}$$

where G is the gel fraction given by Table 1, item (3), and ξ is the correlation length and can be considered as a blob of the constituent polymer chains of the networks (there is no correlation among network constituents in the affine model), or as a domain shown in Fig. 6. Furthermore, D is the fractal dimension of the domain that is defined by the correlation length ξ, and is approximated by $D \approx 1/\rho$ (Table 1, item (4)). Eventually, the modulus can be expressed as

$$E \propto |\varepsilon|^t \tag{23}$$

considering $G \propto |\varepsilon|^\beta$, and $\xi \approx \langle S^2 \rangle_z^{1/2} \propto |\varepsilon|^{-\nu}$ (see Table 1, item (9)). From the FS model, $t = 3$ and from the percolation theory $t = 2.67$ (although $t = 1.8$ was predicted previously corresponding to the power law of the conductivity of random networks [9, 42], an index close to $t = 2.67$ was recently obtained [43]).

Equations (21) and (22) indicate that the modulus E is directly proportional to the absolute temperature. This is because the modulus of a gel is assumed to depend only on the entropy rather than enthalpy. However, in reality, the modulus often reduces as the temperature increases. In particular, the changes in enthalpy due to the deformation of physical gels is an important problem. In practice, the modulus of a gel rarely takes into account the contribution of the enthalpy, and is discussed considering only the topology of the networks. In this case, it is assumed that the gels are completely homogeneous for the derivation of Eqs. (21) and (22).

It is easier to understand if the heterogeneous structure of gels is divided into two kinds of heterogeneity [21]. The fractal heterogeneous gel structure model is the case where multiple fractal percolation clusters connect and form networks as shown in Fig. 7(b). In this model, there are crosslinks within the percolation cluster and crosslinks among percolation clusters. In the derivation of Eqs. (21) and (22), these two kinds of crosslinks are not distinguished. The percolation cluster is defined by a certain fractal dimension. In the example of Fig. 7(b), the inside of the percolation cluster is more densely crosslinked than between the clusters. When such gels are deformed, the crosslinks among clusters contribute mainly to the mechanical properties of the system. Thus, the modulus is much smaller than expected from the total degree of crosslinking. In this case, Eq. (22) may apply depending on the bonding type of the percolation clusters.

Although Fig. 7(a) is the model for a random heterogeneous gel structure, there are no self-similarities as with fractal heterogeneous gels and thus it will not follow an exponential law. Lightly heterogeneous gels (Fig. 7(a), center) exhibit mechanical behavior similar to homogeneous gels and Eq. (21) will apply. In other words, homogeneous and lightly heterogeneous gels cannot be distinguished from a dynamic structural point of view [44]. However, static structures, such as the one obtained by small-angle neutron scattering, will show an obvious difference. In highly heterogeneous gels (Fig. 7(a), right), a densely crosslinked portion will function as a crosslink domain and there is no mechanical contribution. Therefore, the modulus will be lower than the homogeneous gels. Accordingly, the modulus of heterogeneous gels reduces in proportion to the degree of heterogeneity.

1.1.4 Conclusions

The structure of gels has been discussed with respect to static and dynamic structural models. Rigorously constructed theoretical models assume homogeneous structure. Real gels are heterogeneous due to various

Table 3 Terminology comparison for percolation and gelation processes.

Percolation	Gelation
Threshold value	Gel point
Finite cluster	Sol molecules
Average cluster size	Average molecular weight
Percolation probability	Gel fraction
Coordination number	Number of functional group
Beté lattice approximation	Flory–Stockmayer model

reasons and their heterogeneity can be understood to some extent by hierarchical construction of homogeneous gel structure models.

Finally, the terminology that is used for the gelation process as well as percolation and aggregation processes are shown in Table 3.

1.2   ANALYSIS OF BIOPOLYMER GELS: HAIR

SACHIO NAITO

1.2.1   Introduction

The human body consists of many polymer gels. The functions of the body (flexible and fast maneuverability and physiological functions) are maintained by gels. The primary conditions of a gel are the necessity of being a polymer, and, further, having local strong interactions among polymer chains that function as chemical or physical crosslink points. Usually, in the body, the networks are swollen by absorbing a solvent (water), and control such life functions as molecular recognition, catalysis, and movement. For example, connective tissues, such as the cornea, vitreous humor, and inner skin, are gels and the surface of body cavities such as the trachea and stomach, are also coated by gels. The human body gels often differ significantly from the gel structures discussed in the previous section 1.1.3.2. In particular, the characteristic higher-order structures of the natural polymers contribute to the formation of crosslink points, and natural gels themselves can also take a hierarchical quasi-ordered structure. According to the classification of Flory's gel structures (see the aforementioned section), the natural gels belong to type (1), the ordered lamellae structure with gel intermediate layers, or type (3), polymer networks that are formed via physical association (local order). The structural analysis of polysaccharides that belong to type (3) is described in Chapter 3, Section 2, "Structure of Gels, Characterization Methods of Properties (Individual Instrumentation)," and Chapter 3, Section 2, Subsection 3, "X-ray, Neutron Scattering." Many of the polysaccharides take the ordered form of a local double helix. Furthermore, by organizing the locally ordered structures to form crosslink domains, thermoreversible gels are formed. Skin and hair form gel intermediate layers during the formation process that are classified as type (1). By taking the form of gel intermediate layers, the constituent protein forms higher-order structures and allows the formation of skin and hair organizations. We will describe here how keratin protein, which is rich in cystine residues (SH group) with approximately 50 kDa average molecular weight, forms strong hair.

1.2.2   System Structure and Morphology of Hair

1.2.2.1   Structure of hair system

Figure 1 shows the schematic diagram of the hierarchical structure of the hair filament [45]. The outermost layers are made up of 5–10 layers of flattened cuticle that protect the corn-shaped cortex. The cells to form hair are aggregated through covalently bonded lipids and proteins (the cell membrane complex (CMC)) and make up the core of the filament [46]. In the cortex are packed more than ten macrofibrils, which consist of microfibrils and a matrix. The properties and organization of hair are mainly due to the structure of this amorphous basic unit, which will be described later.

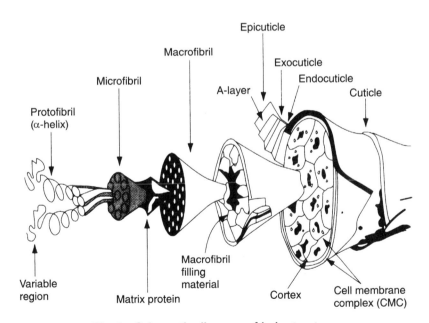

Fig. 1   Schematic diagram of hair structure.

1.2.2.2 Formation of hair

Figure 2 is a diagram of a hair follicle. At the bottom of the hair follicle, where hair is actively synthesized, are found dermal papilla, which include mesenchymal and matrix cells. In the dermal papilla, a capillary enters and provides nutrition and other components necessary for cell division. The outermost layer of matrix cells is covered by a layer of nondivided cells called basal cells. By the cell division of the basal cells, the upper layer cells divide, the keratinization induction mechanism is turned on, and the keratin protein that is necessary to form hair is synthesized. Gradually, the cells are filled with keratin proteins and become corn shaped, resulting in a filament structure that is characteristic to hair (see Fig. 3(a)). In this

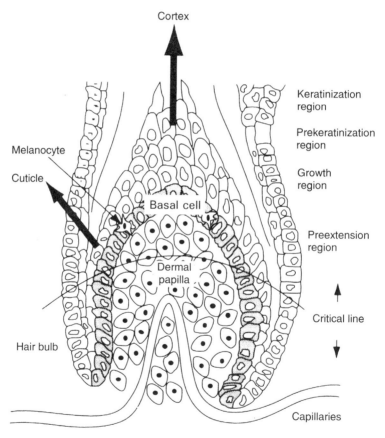

Fig. 2 Diagram of a hair, follicle.

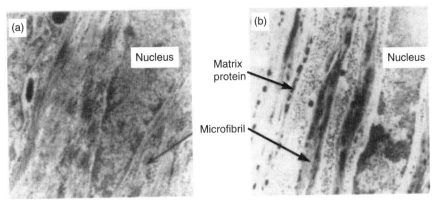

Fig. 3 Transmission electron photomicrographs of (a) prekeratinization region of hair follicle and (b) keratinization region.

region, the life activity of the cells is still maintained. Eventually, in the keratinization region, the matrix protein is synthesized. When it starts interacting with the filament structure, the nucleus of the cell diminishes and the cell dehydrates, a clear keratin pattern starts appearing, and, finally, the cells die (see Fig. 3(b)). To date, from the cell division to the appearance of keratin protein, various observations on the appearance of keratin protein and organized chemical studies have been reported. However, the network formation by SS bonds between (within) proteins and the death of cells are poorly understood.

1.2.3 Hair Proteins and Crosslink Structures

1.2.3.1 Chemical structure and crosslink formation of protein component

Human keratin has two proteins with low (LS) and high (HS) cystine content. According to the latest knowledge, eight hair-specific genes and two other genes that relate to hair are found in the LS protein of hair. Those proteins are type I (5 kinds), which are rich in acidic amino acids with molecular weight of 40–48 kDa, and type II (5 kinds), which are neutral or basic with molecular weight of 58–65 kDa [47–51]. Both type I and type II proteins contain α-helix. Two molecules of type I and type II form a rope and an additional rope pair forms a protofilament. This is the basic unit and the eight basic units then further form a tubular association to form a fibril (filament). This is called an intermediate filament (IF).

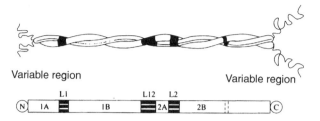

Fig. 4 Schematic diagram of keratin intermediate filament protein structure [8].

Figure 4 illustrates a schematic diagram of an LS molecule [52]. In the molecule, there are nonhelical N- or C-terminal regions and rod regions that contain α-helix. In the rod region, there are four coiled-coiled regions of 1A, 1B, 2A, and 2B, and the segment (linker) L1, L12, and L2 that connect them. Table 1 lists the cystine residues (1/2 Cys) of the IF protein of sheep whose primary structure is known [53]. The average number of crosslink points for the rod region that contains α-helix and nonhelical N- and C-terminal regions is 7 and 15 (mol/molecules), respectively. This corresponds to the Cys content of approximately 220 µmol/g for the IF as a whole, 70 µmol/g for the rod region, and 150 µmol/g for the terminal regions. Accordingly, SS bonds are rich in the amorphous portion of the keratin protein. Although it is not clear for humans, this concentration is reported to be approximately 200 µmol/g.

The HS protein is a globular protein, rich with Cys (approximately 1100 mmol/g) and with molecular weight about 20 kDa, that forms a

Table 1 The number of 1/2 cystine residues in 1F protein.

| Protein types | Domains | | Total | M.W. $\times 10^{-4}$ Molecular weight |
	Rod	N-, C-terminal		
Type-I				4.2–4.6
8c-1	8	17	25	
8a	5	10	15	
Type-II				5.6–6.0
7c	9	21	30	
5	7	18	18	
Average	7	15	22	5×10^{-4}

matrix as an aggregate. The matrix embeds IF and forms a molecular complex. It is called an intermediate filament adhesive protein (IFAP).

In hair, these proteins are formed by crosslinking with SS bonds in the keratin region. Therefore, in order to characterize various crosslinking bonds of protein, a specific protein must be taken out by disrupting the bonds, determining the order, and then rearranging them to reconstruct the original structure. However, if bonds with other molecules are taken into consideration, the number of combinations is enormous and, practically speaking, it is impossible to determine the 3D position.

1.2.3.2   Characterization of crosslinking by physical means

1.2.3.2.1   Preparation of hair swollen like a gel

Here, an attempt is made to characterize the crosslinks based on the rubber elasticity theory of equilibrated swelling of homogeneous polymer networks or a swollen rubber by disrupting the hydrogen bonding of the crystalline region (α-helix) of keratin protein and swelling hair-like gels.

The movement of crosslink point by SH/SS exchange reaction during swelling or stretching can be inhibited if the SH group in hair is chemically treated by N-ethylmaleimide. Furthermore, it has been found that the swollen material shows the thermal expansion coefficient $\beta = 0$ in a wide temperature range (from room temperature to 70°C) in a mixed solvent such as 8 M LiBr and diethylene glycol-mono-n-butyl ether $[HO(CH_2-CH_2)_2-C_4H_9]$ (BC) [54]. Because the temperature coefficient of the equilibrium (eq) stress that is experimentally obtained from the diluted solution $(\partial f/\partial T)_{p,\lambda,\text{eq}}$ is equal to the value of $(\partial f/\partial T)_{V,L}$, the ratio, f_e/f, of the energy component, $f_e = (\partial E/\partial T)_{V,T}$ to the total stress is expressed by

$$\frac{f_e}{f} = 1 - \left(\frac{T}{f}\right)\left(\frac{\partial f}{\partial T}\right)_{p,\lambda,\text{eq}} \tag{1}$$

For a hair, when it is $\lambda < 1.25$, the value of f_e/f is independent of λ and is 0.11. Also, for the hair where the SS bond is decreased by 14% by a reduction reaction, the f_e/f value is 0.08 for $\lambda < 1.55$ and the hair shows gel-like entropy elasticity [54]. The f_e/f value of natural rubber is 0.18–0.25 [55]. In the diluted system using this special mixed solvent, because even a fiber rich with SS bonds such as hair shows almost ideal rubber elasticity at a large deformation, quantitative analysis of the crosslinks is done using this phenomenon.

1.2.3.2.2   Determination of the number of crosslinks and matrix domain volume

In the hair that is swollen by the 8 M LiBr/BC diluted system, there is a globular HS protein that contains a large number of SS bonds. In such a heterogeneously crosslinked system, the ordinary rubber elasticity theory cannot be applied. Hence, a two-phase structure of swollen keratin networks is assumed. This structure consists of the matrix (domain phase) that is a tightly crosslinked and mechanically stable globular HS protein, and continuous networks (rubbery phase) that are made of low crosslink density LS protein chains. The domain phase was hypothesized to provide the filler effect in rubber networks [56]. Equation (2) is the relationship between equilibrium stress F and elongation ratio of rubber phase α:

$$F = G\left(\frac{\sqrt{n}}{3}\right)\left\{L^{-1}\left(\frac{\alpha}{\sqrt{n}}\right) - \alpha^{-3/2}L^{-1}\left(\frac{1}{\sqrt{\alpha n}}\right)\right\} \tag{2}$$

Here, the shear modulus of the swollen hair $G = (\rho RT/M_c)\{(v_2 - \phi_d)/(1 - \phi_d)\}^{1/3}(1 - 2M_c/M)\,\gamma$; n is the number of segments of the network chain; $L^{-1}(x)$ is the anti-Langevin function; ρ is the dry density of the sample; M_c is the number average molecular weight between crosslinks of the rubbery phase; M is the primary molecular weight; R is the gas constant; T is absolute temperature; v_2 is the volume fraction of the polymer within the gel; and γ is the filler effect of the HS domain that exists in the rubbery phase. Equation (3) provides γ:

$$\gamma = 1 - 2.5\kappa\phi_d + 14.1\kappa^2\phi_d^2 \tag{3}$$

where κ is the ratio between the length and the diameter of the rod-like filler. The shape factor is assumed to be spherical fillers. The volume fraction of the domain is expressed by ϕ_d. Using a simplified model, α is correlated to the elongation ratio λ of the swollen hair sample through Eq. (4):

$$\alpha = \frac{(\lambda - \phi_d)}{(1 - \phi_d)} \tag{4}$$

Experimentally obtained F, G, λ, and v_2 are inserted into Eq. (2) and ϕ_d, ρ/M_c, or M_c and κ are obtained as structural parameters. The value of n is obtained by selectively reducing the SS bonds of the HS globular protein with tri-n-phosphine, using the sample with $\phi_d = 0$, inserting into Eq. (2)

the values of $\phi_d = 0$, $\alpha = 1$, $\gamma = 1$, and further using the value 1.250 obtained for M_c/n. The molecular weight of the primary molecule M is assumed to be 5×10^4, and the density ρ is $1.30\,\text{g/cm}^3$.

The domain volume fraction ϕ_d' within the dry sample is given by the following:

$$\phi_d' = \frac{\phi_d}{v_2} \tag{5}$$

Figure 5 shows the plot of the F-λ relationship for various swollen keratin fibers. The solid line indicates a fitted line using Eq. (2). Figure 6 [57] depicts the relationship between the crosslink density, which is defined as the number of crosslink points within $1\,\text{cm}^3$ of the LS protein in the dry keratin and the concentration of SS group [SS] that is determined from ρ/M_c and chemical analysis. Although [SS] varies significantly depending on the type of keratin used, ρ/M_c values are approximately constant at $3.6 \times 10^4\,\text{mol/cm}^3$. The M_c value between crosslink points is about 3600 which corresponds to one crosslink per 31 amino acid residues. This converts to $138\,\mu\text{mol/g}$ $(=10^6/2M_c)$ for the unit weight of the LS protein, which also corresponds to approximately 69% of the total number of crosslink points of the LS protein at $200\,\mu\text{mol/g}$.

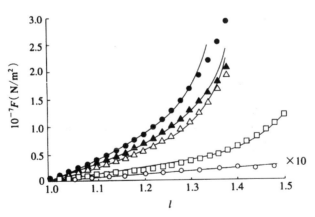

(●) Human hair (○) Reduced human hair (▲) Alpaca hair
(△) Lama hair (□) Opossum hair
(——) line is the analytical curve of the experimental data using Eq. (2)

Fig. 5 The relationship between equilibrium stress F and elongation l of swollen keratin fiber.

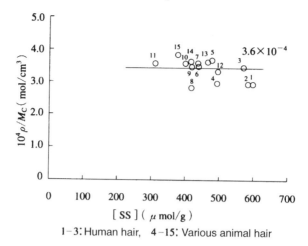

Fig. 6   The relationship between the crosslink density ρ/M_c and disulfide constant [SS] in the region where swollen keratin is rubbery.

Figure 7 illustrates the relationship between the volume fraction ϕ_d of the HS protein that functions as a domain in the swollen gel and [SS]; good linearity is observed. This means that the extrapolated value 148 mmol/g to the [SS] axis is the hypothetical number of intermolecular crosslink points that consist of only the LS protein without containing the HS protein. This value is almost equal to the value obtained from the ρ/M_c–[SS] plot. Because the total [SS] of hair is 627 μmol/g, the [SS] of the HS protein is estimated to be 427 μmol/g. For the detailed distribution of inter- and intramolecular bonding of keratin, the distribution of the SS bonds, which can be disrupted or remain intact with the reducing agent, is evaluated. This is achieved after quantitatively converting the SS bonds in the hydrophilic region to mono-S bonds by lanthionine using KCN aqueous solution or boiling water treatment. In the HS protein, 88.1% of Cys is intramolecularly crosslinked [58–60].

By swelling not only hair but also other keratin fibers, the number of intermolecular bonds of the LS protein can be determined. Also, the volume fraction of the globular domain that consists of the HS protein is calculated. Interestingly, the [SS] of the fiber as a whole varies depending on the species, and the [SS] in the LS protein does not change noticeably in the keratin fiber with high [SS] concentration, indicating that the domain that consists of the HS protein increases.

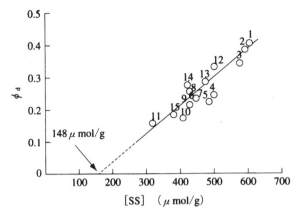

Fig. 7 The relationship between the volume fraction ϕ_d and disulfide content [SS] in swollen keratin (numbers in the figure are the same as the numbers in Fig. 6).

1.2.4 Structure of Hair, Role of [SS] Bonds to Properties

1.2.4.1 *Consecutive, selective reduction of SS bonds and properties of hair*

As described earlier, there are many intermolecular SS bonds in the amorphous region (variable region) at N- and C-terminal regions. It has also been shown that they exist in the environment where they become reactive by KCN and hot water treatment, in other words, in relatively hydrophilic regions. Hence, it is possible to perform consecutive, selective reduction of the SS bonds by controlling the type of reducing agent and reaction conditions [61].

After hair is reduced for 2–300 min by 0.5 M thioglycol acid (TGA) aqueous solution, and blocking the SH group by N-ethylmaleimide, the hair shows rubber-like elasticity in the 8 M LiBr/BC dilute solution. Assuming that the filler effect by the matrix protein is not significant in the early part of the deformation, the crosslink density is calculated by applying Gaussian network theory and obtaining the stress at the early deformation at 50°C.

Equilibrium stress τ is given by

$$\tau = G\left[\alpha - \left(\frac{1}{\alpha^2}\right)\right]$$ (6)

where α is the elongation ratio. The shear modulus G of the gel is given by

$$G = \frac{(\rho RT)}{M_c} \, v_2^{1/3} \left[\left(1 - 2\frac{M_c}{M} \right) \right] \qquad (7)$$

where ρ is the density of the dry sample; R is the gas constant; T is absolute temperature; v_2 is the volume fraction of the polymer in the gel-like swollen hair; M_c is the number average molecular weight between crosslink points in the rubbery phase; and M is the primary molecular weight (50,000). The shear modulus G_w in water at 2% deformation can be obtained by carefully washing with water and removing LiBr/BC.

Three kinds of SS bonds can be found by consecutive, selective reduction of SS bonds by TGA (see Fig. 8). In the following, the characteristic properties by the reduction of various crosslink structures are summarized.

1. In the SS_1 that is reduced early, the crosslink density ρ/M_c decreases linearly. Surprisingly, after removing LiBr/BC, it shows the same G_w as the hair prior to the reduction.
2. For SS_2, the ρ/M_c value decreases linearly while the G_w value also reduces linearly.

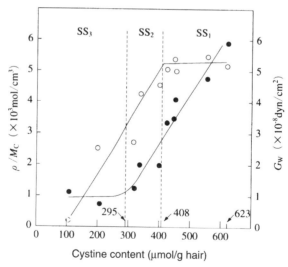

Fig. 8 The relationship between cystine and crosslink density ρ/M_c (●) or modulus in water G_w (○) of TGA-treated hair.

3. In the SS_3 region, the ρ/M_c value shows a constant value whereas G_w decreases linearly.

Therefore, in the hair that is swollen like a gel, the shear modulus decreases following the reduced concentration of crosslink points upon the reduction of SS_1 and SS_2. However, SS_1 in the nonswollen hair does not affect the properties of the molecular chains of the amorphous region where SS_1 resides. The fact that G_w reduces linearly by the reduction of SS_2 and SS_3 indicates the denaturing of keratin protein by the reduction of the SS bonds in this region. In other words, water still remains even after removing LiBr/BC from this region. Interestingly, when hair is reduced by tri-n-butylphosphene, which is difficult to dissolve in water, intramolecular SS bonds that do not contribute to the ρ/M_c value are relatively preferentially reduced as shown in Fig. 9. In the TGAS reduction, it is the reduction of the hydrophobic region that reacts at the end (see Fig. 8).

The SS_1 bond concentration is 215 µmol/g, which corresponds to 35% of the total hair. At the microfibrils of hair, the concentration of the SS bonds is 50% of the whole, of which approximately 60% (20 mmol/g) is said to reside in the amorphous (variable) region [57].

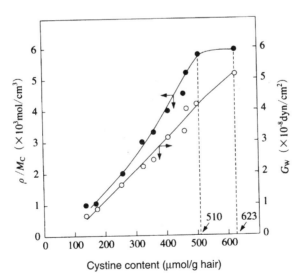

Fig. 9 The relationship between cystine and crosslink density ρ/M_c (●) or modulus in water G_w (○) of TBP-treated hair.

However, because the intermolecular SS bond concentration in the LS protein can be calculated to be 60 µmol/g [62, 63], the existence of the SS bonds between the variable region and matrix protein (IFAP) is predicted. The concentration of the SS_2 bonds is 113 µmol/g and are hypothesized to be the intermolecular SS bonds that exist in the relatively hydrophobic region among rods or the LS protein or IFAP. The SS_3 bonds are considered to reside in the molecule of IFAP.

1.2.4.2 Consecutive, selective reduction of SS bonds and change of hair structure [observation of hair by small angle X-ray scattering (SAXS), and small angle neutron scattering (SANS)]

It is obvious from the reduction of SS bonds in keratin protein that there are SS bonds that do not contribute directly to the properties of hair depending on their environment even though they are the same SS bonds that form intermolecular crosslinking. The role of SS bonds on the actual hair structure and properties are being studied by small angle x-ray scattering (SAXS) and small angle neutron scattering (SANS) in place of traditionally used electron microscopic observation [64, 65]. The SH group of hair can be blocked by N-ethylmaleimide after consecutively reducing SS_1, SS_2 and SS_3 by TGA. As a sample, the hair that is swollen like a gel by 8 M LiBr/BC dilute solution and that washed by water is used. For the neutron scattering heavy water was used for the treatment.

The parameter for the amorphous region can be obtained from Eq. (8) by the Debye plot ($I(q)^{-1/2}$ vs q^2) of the scattering curve:

$$I(q) = 8\pi\phi_A\phi_B(\Delta\rho)\,\frac{2}{c(q^2 + c^{-2})^2} \tag{8}$$

where ϕ_A and ϕ_B are the volume of amorphous portion A and B, respectively; ρ is the electric density; and c is the correlation length.

In the small angle x-ray diffraction, $I_B(0)$ can be obtained by extrapolating the angle to zero in the Ruland plot ($\ln I(q)$ vs q^2) that is from the scattering curve in the nonstructural middle angle region. From $I_B(0)$, the relationship between the structure of the amorphous region and the mechanical properties is given by the following:

$$I_B(q) = I_B(0)\,\exp(Aq^2) \tag{9a}$$

$$I_B(0) = k_B Tr_2 k_T(T_g) \tag{9b}$$

where $I_B(q)$ is the scattering intensity due to the thermal fluctuation of the electron; and $I_B(0)$ correlates with bulk compressive modulus $K_T(T_g)$ at an absolute temperature (T).

To obtain the small angle X-ray scattering pattern, X-rays are irradiated normal to the fiber axis of the hair and the scattering is observed by the Kratky U-slit camera. Figure 10(a) shows the scattering curve of the hair that is swollen like a gel. In the nonreduced hair, the scattering peak can be seen at $q = 5 \times 10^{-3} - 7 \times 10^{-3}$ nm^{-1}, which corresponds to the microfibril distance. Although the hair is apparently in a rubbery state, it can be seen that structures remain that originate from the LS protein in the interior of the hair. In the SS$_1$ reduction, one can see the scattering that is due to the inhomogeneity of the amorphous region. The interior of the hair gradually becomes random in structure and, in the SS$_2$ reduction, it becomes completely amorphous. On the other hand, after washing in water, the SS$_1$ reduction gives the same scattering curve as the nonreduced hair, indicating the structural recovery of microfibrils (see Fig. 10(b)).

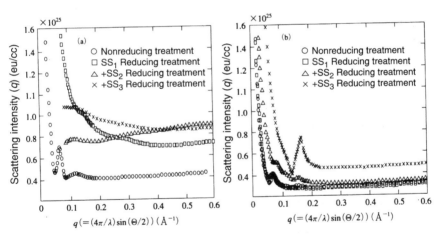

(SS$_1$), (SS$_2$) and (SS$_3$) correspond to the ones in Fig. 8, respectively. The x-ray measurement is done using a Kratky U-slit camera. The entrance slit (incident slit) was set to 70 μm, the height determining slit to 16 mm. The voltage for the x-ray generator was 50 kV with 50 mA. Detection after scattering was done with a linear position sensitive detector and the CuK line ($\lambda = 0.154$ nm) was selectively detected with a waveheight analyzer and nickel foil filter, thickness of 16 μm. Slit correction was made using Glutter's method (repeated insertion method) and scattering intensity was converted to absolute intensity by the Lupolen method.

Fig. 10 Small angle x-ray diffraction curves of reduced and swollen hair and hair washed in water.

The small angle neutron scattering experiments were performed on the pulsed small angle neutron scattering device at the Booster Utilization Facility, High Energy Physics Institute, Ministry of Education (KENS-SAN). The scattering radiation with wavelength $0.4 \leq 1 \leq 1.1$ nm was detected with the 2D position sensitive detector (2D-PSD). The distance between the sample and the detector was 3 m. In order to study the structural anisotropy of the amorphous region with respect to the fiber axis, the hair sample was installed in the direction shown in Fig. 11 to the incoming radiation.

Figure 12 shows the scattering curves of (a) nonreduced, (b) SS_1 reduced, (c) SS_2 reduced (c), and (d) SS_3 reduced hair samples. The characteristics observed on these scattering curves are as follows.

1. Nonreduced hair: The scattering peak that corresponds to microfibril distance as seen in SAXS is not observed. Only the inhomogeneity and structural anisotropy of the amorphous regions are observed.
2. SS_1 reduced hair: The reduction of inhomogeneity and anisotropy proceeds as shown in the reduction of the scattering intensity and

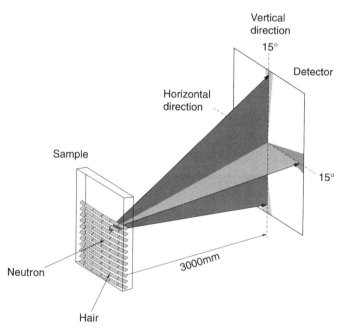

Fig. 11 Neutron scattering measurement of hair sample.

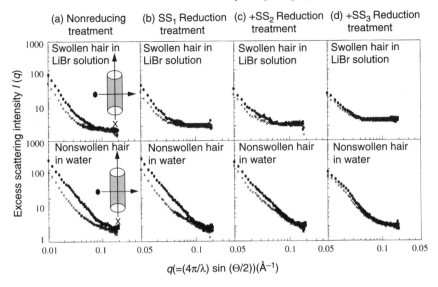

Fig. 12   Small angle neutron scattering curves of TGA reduction-treated hair.

slope of the curve during swelling. After washing with water, the scattering curve becomes identical to the nonreduced hair and there is no effect of SS bond reduction.

3. SS$_2$ reduced hair: During swelling, significant reduction of scattering intensity and slope of the curve is seen. The anisotropy diminishes. After washing, inhomogeneity and anisotropy of the amorphous structure no longer recover.

4. SS$_3$ reduced hair: Although the hair exhibits a superficial fibrous appearance, there is complete loss of anisotropy within the fiber, which approaches a homogeneous suspension system.

The change of structural parameter following the reduction of each SS bond is listed in Table 2. According to SAXS, there are no changes in the correlation length after washing with water, while according to SANS, it increases as the SS bonds reduce and, in water, it decreases markedly. This discrepancy can be understood by knowing that SAXS detects mainly the scattering caused by the electron density change by the localization of Cys (S) in the amorphous region. On the other hand, the diffusivity of D$_2$O caused by the structural changes is being detected in SANS. The changes of the $I_B(0)$ value due to the scattering caused by the thermal fluctuation of electrons correlates well with the equilibrium stress (G, G_w)

Table 2 Physical changes of amorphous structure by the SS bonds of consecutive reduction treatment of hair.

| | Correlation length (nm) | | | | $I_B(0) \times 10^{-24}$ (eu/cm^3) | |
| | Small angle x-ray scattering | | Small angle neutron scattering | | Small angle x-ray scattering | |
	Swollen	Nonswollen	Swollen	Nonswollen	Swollen	Nonswollen
Nonreduced	6.0	5.3	4.2	4.4	2.0	0.7
SS$_1$ reduction	n.d.	4.2	3.8	4.6	4.7	0.6
SS$_2$ reduction	n.d.	5.4	2.7	3.8	6.0	1.1
SS$_3$ reduction	n.d.	5.1	2.6	3.3	6.4	2.4

obtained by the consecutive, selective reduction of SS bonds. The reduction of SS$_1$ will not influence the thermal fluctuation of electrons in water.

Based on these results, it is suggested that there are SS bonds whose breakage will not affect the structure and properties of hair, and also that the disruption of the SS bonds in the more hydrophobic environments increases water absorption and the structure is irreversibly changed. In particular, the SS$_2$ that is regarded as the intermolecular bonds in the relatively hydrophobic region between IF$_{rod}$ and IF$_{rod}$ or IFAP and IFAP is essential for the formation of microfibrils because even after the reduction or LiBr elimination, the orientation of the LS protein is diminished (Fig. 10(b)).

1.2.5   Change of the Structure and Properties of Hair that is Independent of the Chemical Changes of SS Bonds

Hair is a shape memory fiber and it is difficult to permanently change its shape. Often, as a method to change the shape permanently, there is a generally used method to reduce the SS bonds. After the deformation, the SS bonds are reformed by oxidizing the SH group. The result is called a permanent wave. As discussed so far, the structure and property of hair by reduction is related to the structural changes in the amorphous region by the diffusion of water in the hydrophobic region. Accordingly, even by the nonaqueous treatment such as with 120°C propylene glycol (PG), the structure and properties of hair can be changed irreversibly if an aggressive treatment to interrupt hydrogen bonding is adopted (see Fig.

13). A hair was treated at 90 or 120°C for 20 min in deuterated PG and cooled to room temperature; this sample was used for SAN. At 90°C, only when there is a large amount of PG in the hair is deformation possible. For the sample treated at 120°C, it was again washed with water and then dried. The result is shown in Fig. 14. Although macroscopic behavior is similar to the reduction/oxidation of SS_2 the PG/120°C treatment shows that the scattering intensity is proportional to the -4th power at the small angle region ($q < 0.03$). The Porod law is thought to be applicable in this region.

Also, even after removing the PG with excessive water washing, the structure in the amorphous region does not change [64]. Specifically, permanent deformation of the hair shape is possible during the disruption and rearrangement of hydrogen bonding in the hydrophobic region and the accompanying ideal phase separation of the amorphous region without resorting to the reduction of SS bonds. On the other hand, when SS_2 bonds are reduction/oxidation treated, the scattering intensity is stronger than the nonreduction treated hair and the angular dependency of the slope becomes small. If SS bonds are cleaved, the hair is irreversibly denatured and the D_2O in the hydrated amorphous region is distributed evenly even after crosslinking again. These results are shown schematically in Fig. 15.

1.2.6   Structure and Properties of Keratin Protein Model Gel

It has already been mentioned that the properties of hair depend on the higher-order structure of the amorphous region of protein. Also, it was

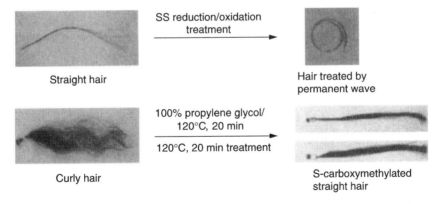

Straight hair

SS reduction/oxidation treatment
$\longrightarrow$

Hair treated by permanent wave

Curly hair

100% propylene glycol/ 120°C, 20 min
$\longrightarrow$
120°C, 20 min treatment

S-carboxymethylated straight hair

Fig. 13 Permanent deformation of hair shape without SS reduction treatment.

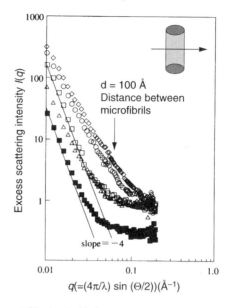

○ Nontreated hair
◇ Hair in which SS₂ is reduced with
   TGA, then further oxidized.
□ Hair treated with deuterated propylene
   glycol at 120°C
△ Hair treated with deuterated propylene
   glycol at 90°C
■ Hair treated with deuterated propylene
   glycol at 120°C, and then washed in
   water and dried.

Fig. 14 Small angle neutron scattering curves of hair treated in various ways.

suggested that the hydrogen bonding in the interior of the amorphous region plays an important role in the formation of the higher-order structure. However, hair is a complex aggregate that consists of various keratin proteins and thus its analysis is limited. Finally, a model gel for the simplified hair structure will be described. The HS protein contains a high concentration of Cys and is difficult to dissolve. On the other hand, the LS protein will be soluble after reducing the SS bond and chemically treating the SH group by N-ethylmaleimide or monoiodide acetic acid. However, to recombine the SS bonds from these water soluble proteins is naturally impossible.

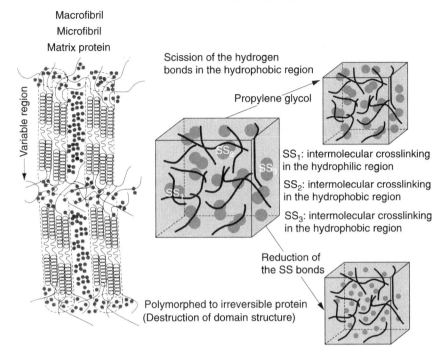

Fig. 15 Schematic diagram of a macrofibril and structural change in the amorphous hydrophobic region caused by propylene glycol and reduction treatment of the SS bonds.

Wool is reduced by 0.3 M TGA aqueous solution (pH 11) that contains 8.0 M urea, neutralized to pH 7 by acetic acid, and oxidized by 1.5 M NaBrO$_3$ aqueous solution in the presence of a sufficient amount of unreacted TGA. After filtration of the insoluble component, a water soluble protein is obtained [65]. This protein is: (1) from amino acid analysis, the SH group that is chemically treated with S-carboxymethyl-alanyl disulfide (CMAD; $-SSCH_2COOH$); (2) from SDS-polyacrylamide electrophoresis, the mixture of type-I and type-II LS keratin proteins; and (3) from the CD spectrum, a polymer with α-helix (Fig. 16). This is called carboxymethylalanyl disulfide keratin (CMADK).

When 4% CMADK aqueous solution was heated at 100°C, opaque aggregates were formed as shown in Fig. 17. These aggregates quickly dissolved into a transparent solution in a dilute TGA aqueous solution. Therefore, in these aggregates, a part of CMAD is thought to return to the

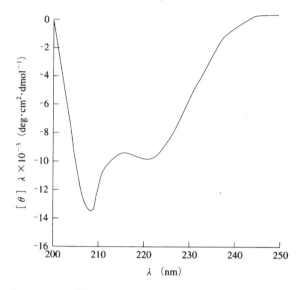

Fig. 16 Local property of the circular dichroism of CMADK (S-carboxymethyl-alanyl disulfide keratin) in tris-hydrochloric acid buffer solution (pH 7).

intermolecular SS bonds and forms the gel. When CMADK aqueous solution and the gel are compared using SAXS (Fig. 18) and SAN (Fig. 19), the scattering intensity is proportional to the -2nd power of the angle in the small angle region ($q < 0.03$) and CMADK is dispersed randomly in an aqueous solution. On the other hand, in the gel, the scattering intensity is proportional to the -4th power of the angle and the Porod law is applicable in this region [66]. Also, from the fact that the SAXS and SAN results agree with each other, it is thought that crosslinking occurs among the LS keratin proteins, electron density difference is generated due to the aggregation formation of rod domains that contains α-helix with low Cys content, and ideal phase separation takes place because heavy water cannot penetrate into this region. This gel shrinks in 100% PG at 120°C or 11 M LiBr aqueous solution at 90°C, and swells in 8 M urea aqueous solution. The former is the reduction of volume by the globuliza-tion of peptide chains due to the disruption of the hydrogen bonding of the α-helix; the latter is the increase in volume by the reduction of the cohesive force among peptides by the reduced hydrophobic interaction among rod domains that contain the α-helix (see Fig. 17). In other words,

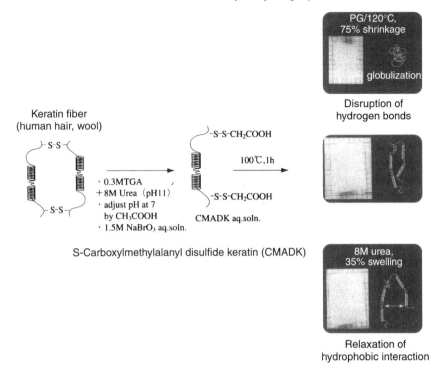

Fig. 17 Schematic diagram of CMADK (S-carboxymethylalanyl disulfide keratin) gel formation and structural changes.

for the higher-order structure formation of LS protein, the crosslinking among IF or IF_{rod} is essential, which agrees well with the aforementioned results on hair structure changes by treatment. Accordingly, it is possible that the basic structure of hair is understood *in vitro*. It is interesting that this model can predict the crosslinking by SS bonds and higher-order structure formation during keratinization of hair follicles.

1.2.7   Conclusions

For the higher-order structure formation and properties of hair, the results on the crosslinking and property development by SS bond obtained by both chemical and physical analysis have been described. Fibrous proteins such as hair are found to show nearly ideal rubber elasticity in 8 M LiBr/BC dilute solution. Starting from the elucidation of the number of crosslink points and crosslink pattern from high elongational curves, the

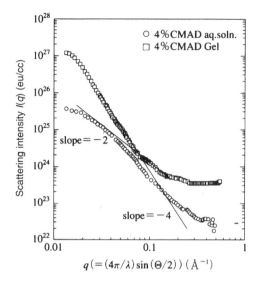

Fig. 18 Small angle x-ray scattering curves of CMADK (S-carboxymethyl-alanyl disulfide keratin) aqueous solution and gel.

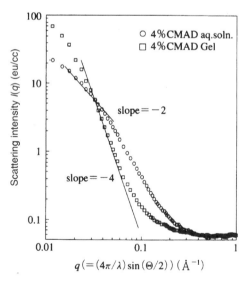

Fig. 19 Small angle x-ray scattering curves of deuterated CMADK (S-carboxy-methylalanyl disulfide keratin) aqueous solution and gel.

relationship between the crosslinking and property development is observed by nondestructive polymer structure analytical techniques, such as small angle x-ray scattering and small angle neutron scattering. As a result, although SS bonds have a role in the formation of the higher-order structure of the constituent protein, it has become gradually clear that the macroscopic shape and properties of hair are characterized by the structure of the amorphous region itself.

In the future, it will be necessary to pay attention to the relationship between the higher-order structure formation of the amorphous region and the hydrogen bonding among amino acid side chains that constitute protein.

REFERENCES

1 Hermans, P.H. (1949). *Colloid Science* II, H.R. Pruyt, ed., Amsterdam: Elsevier.
2 Flory, P.J. (1956). *Principles of Polymer Chemistry*, Ithaca, New York: Cornell University Press.
3 Flory, P.J. (1974). *Dis. Faraday Soc.*, **57**: 7.
4 Gordon, M. (1962). *Proc. Roy. Soc., London*, **A268**: 240.
5 Good, I.J. (1955). *Proc. Camb. Phil. Soc.*, **51**: 240.
6 Kajiwara, K. and Gordon, M. (1973). *J. Chem. Phys.*, **59**: 3623.
7 Stanley, H.E., (1971). *Introduction to Phase Transitions and Critical Phenomena*, Oxford: Clarendon.
8 Stauffer, D. (1976). *J. Chem. Soc., Faraday Trans., II*, **72**: 1354.
9 de Gennes, P.G. (1979). *Scaling Concepts in Polymer Physics*, Ithaca, New York: Cornell University Press.
10 Hayakawa, A. (1994). *Theory of Aggregation: Fluctuation, Chaos, and Fractal*, Japan: Science.
11 Witten, T.A., and Sanders, L.M. (1981). *Phys. Rev. Lett.*, **47**: 1400.
12 Meakin, P. (1983). *Phys. Rev. Lett.*, **51**: 1119.
13 Kolb, M., Botet, R., and Jullien, R. (1983). *Phys. Rev. Lett.*, **51**: 1123.
14 Ziff, R.M., McGrady, E.D., and Meakin, P. (1985). *J. Chem. Phys.*, **82**: 5269.
15 Ziff, R.M. (1980). *J. Stat. Phys.*, **23**: 241.
16 Botet, B., Jullien, R., and Kolb, M. (1984). *Phys. Rev.*, **A30**: 2150.
17 von Schulthess, G.K., Benedek, G.B., and De Blois, R.W. (1980). *Macromolecules*, **13**: 939.
18 Drake, R.L. (1972). *Topics in Current Aerosol Research* 3: Part 2, G.M. Hidy and J.R. Brock, eds., New York: Pergamon.
19 Spouge, J.L. (1983). *Macromolecules*, **16**: 121.
20 Debye, P. and Bueche, A.M. (1949). *J. Appl. Phys.*, **20**: 518.
21 Vilgis, T.A. and Heinrich, G. (1992). *Angew. Makromol. Chem.*, **202/203**: 243.
22 Adolf, A. and Martin, J.E. (1991). *Macromolecules*, **24**: 6721.
23 Yuguchi, Y., Urakawa, H., and Kajiwara, K. *Macromol. Symp.* (in press).
24 Yuguchi, Y., Urakawa, H., Kitamura, S., Ohno, S., and Kajiwara, K. (1995). *Food Hydrocoll.*, **9**: 173.
25 Tanaka, F. and Ishida, M. (1995). *J. Chem. Soc. Farday Trans.*, **91**: 2663.

26 Burchard, W. and Ross-Murphy, S.B. (eds.) (1990). *Physical Networks: Polymers and Gels*, London: Elsevier Applied Science.
27 Almdal, K., Dyre, J., Hvidt, S., and Kramer, O. (1993). *Polym. Gels Networks*, **1**: 5.
28 Doi, M. and Edwards, S.F. (1986). *The Theory of Polymer Dynamics*, Oxford: Oxford University Press.
29 Tsuchida, E. and Abe, K. (1982). *Adv. Polym. Sci.*, **45**: 1.
30 Clark, A.H. and Ross-Murphy, S.B. (1987). *Adv. Polym. Sci.*, **83**: 57.
31 Scanlan, J.C. and Winter, H.H. (1991). *Makromol. Chem. Macromol. Symp.*, **45**: 11.
32 Winter, H.H. and Chambon, F. (1986). *J. Rheol.*, **30**: 367.
33 Martin, J.E., Adolf, D., and Wilcoxon, J.P. (1989). *Phys. Rev.*, **A39**: 1325.
34 Trappe, V., Richtering, W., and Burchard, W. (1992). *J. Phys. II (France)*, **2**: 1453.
35 Treloar, L. (1958). *The Physics of Rubber Elasticity*, Oxford: Clarendon.
36 Kubo, K. (1996). *Rubber Elasticity* (reprint of the first edition), Tokyo: Shokabo.
37 Kuhn, W. (1939). *Kolloid-Z.*, **87**: 3.
38 Kuhn, W. and Grun, F. (1942). *Kolloid-Z.*, **101**: 248.
39 James, H.M. and Guth, E. (1943). *J. Chem. Phys.*, **11**: 455.
40 Saito, N. (1967). *Polymer Physics*, Tokyo: Shokabo.
41 Dobson, G.R. and Gordon, M. (1965). *J. Chem. Phys.*, **43**: 705.
42 Stauffer, D. (1985). *Introduction to Percolation Theory*, London: Taylor & Francis.
43 Vilgis, T.A. (1988). *Makromol. Chem., Rapid Commun.*, **9**: 513.
44 Vilgis, T.A. (1992). *Macromolecules*, **25**: 399.
45 Fraser, R.D. (1981). *Inter. Report of CSIRO Division of Protein Chemistry*, Parkville, Australia.
46 Naito, S., Ooshika, M., Yorimoto, N., and Kuroda, Y. (1995). *Proc. 9th Int. Wool Text. Res. Conf.*, Biella, II, 367.
47 Heid, M.H., Werner, E., and Franke, W.W. (1986). *Differentiation*, **32**: 101.
48 Lynch, M.H., O'Guin, A.M., Hardy, C., Mak, L., and Sun, T.T. (1986). *J. Cell Biol.*, **103**: 2593.
49 Stark, H.J., Breitreutz, D., Limat, A., Bowden, P.E., and Fusenig, N.E. (1987). *Differentiation*, **35**: 236.
50 Bowden, P.E., Stark, H.J., Breitreutz, D., and Fusenig, N.E. (1987). *Curr. Top. Dev. Biol.*, **22**: 35.
51 Tobiasch, E., Winter, H.H., and Schweizer, J. (1992). *J. Invest. Dermatol.*, **94**: 512.
52 Bowden, P.E., Hainey, S., Parker, G., and Hodgins, M.B. (1994). *J. Dermatol. Sci.*, **7**: 152.
53 Fraser, R.D.B., MacRea, T.P., Sparrow, L.G., and Perry, D.A.D. (1988). *J. Bio. Macromol.*, **10**: 106.
54 Arai, K., Sasaki, N., Naito, S., and Takahashi, T. (1989). *J. Appl. Polym. Sci.*, **38**: 1159.
55 Treloar, L.R.G. (1975). *The Physics of Rubber Elasticity*, Oxford: Clarendon Press, p. 281.
56 Arai, K., Ma, G., and Hirata, T. (1991). *J. Appl. Polym. Sci.*, **42**: 1125.
57 Arai, K., Hirata, T., Nishimura, S., and Naito, S. (1993). *Appl. Polym. Sci.*, **47**: 1973.
58 Arai, K., Sakamoto, M., Naito, S., and Takahashi, T. (1989). *J. Appl. Polym. Sci.*, **38**: 29.
59 Naito, S., Arai, K., Mitsushige, H., Naotugu, N., and Sakamoto, M. (1996). *Appl. Polym. Sci.*, **61**: 1913.
60 Naito, S. and Arai, K. (1996). *Appl. Polym. Sci.*, **61**: 2113.
61 Gillespie, J.M. (1987). *J. Polym. Sci. Part C*, **20**: 201.
62 Parry, D.A.D. and Frasser, R.D.B. (1985). *Int. J. Biol. Macromol.*, **7**: 203.

63 Naito, S., Inimura, T., Imokawa, G., and Kurita, K. (1993). *3rd Pacific Polym. Conf., Preprint*, Gold-Coast, p. 255.

64 Naito, S., Tchimura, T., Schimizu, S., Kurita, K., and Furusaka, M. (1995). *KENS Report-X*, 156.

65 Yoneyama, M., Kawada, T., Yoneda, K., Arai, K., Naito, S., and Hojiri, H. (1995). *Proc. 9th Inter. Wool Text. Res. Conf.*, Biella, II, 450.

66 Miyauchi, Y., Naito, S., Shimizu, S., Kurita, K., and Furusaka, M. (1997). *KENS Report-X*, 107.

Section 2
Structure of Gels, Characterization Techniques

MASAMICHI KOBAYASHI[*]

2.1 Infrared Spectroscopy

The most fundamental problem in evaluating the structure of polymer gels is the structural analysis on the molecular level, in particular, the determination of molecular conformation and quantitative analysis.

[*]Contribution from the late Masamichi Kobayashi, Honorary Professor, Osaka University, Japan.
Graduated from the Division of Applied Chemistry in the First Department of Science and Engineering, Waseda University in 1951. Research Associate in the Division of Physical Science, Osaka University in April 1962 by way of the Osaka Technology Testing Center at the Industrial Technology Institute. Became Professor in April 1983 and retired in March 1996. Passed away on April 1, 1997 at the age of 64.
Professor Kobayashi's research focused on vibrational spectroscopy, diffraction crystallography, and structural chemistry fundamentals that include dynamic grating. He studied the many different difficult structural shapes existing in long chain molecules such as gels, polymers, oligomers and fatty acids from their molecular structure to the higher-order composition, which have been at the core of today's advances in this field. He published many books, among them *The Dynamics of Fatty Structures* (Kyoritsu Publ., 1992), edited the report of the Polymer Testing Seminar, was a member of the editorial board of such publications as the *Polymer Journal*, and had membership in many associations. Professor Kobayashi received the Polymer Association Award as well as the Polymer Science Merit Award for his contributions to associations and for the significance and the quality of his research. He was awarded the Fourth Rank and Order of Merit, Third-Level Rising Sun Award posthumously (Takaji Tashiro, Professor of Physical Science Research Division, Osaka University Graduate School).

Various physical gels from semicrystalline polymer solutions are known to form higher-order placement of the polymer chain. Also, in the ordering process of polymers such as gelation or crystallization, ordering of the molecular conformation is fundamental. Infrared spectroscopy is one of the most useful methods to obtain knowledge regarding these processes. The interaction between the polymer and solvent in the gel formation process is also an important problem, and information on this interaction can also be obtained. This section examines the fundamental principles, the measurement procedure and analysis methods using representative systems.

2.1.1 Vibration of Polymer Chains and Infrared Spectra

Many semicrystalline chain polymers possess chemical structures in which a simple structural unit repeats along the chain direction. In the polymer melt or solution, polymer chains take a statistically random structure by the distribution of the rotational angle (molecular conformation) of the C$-$C bond of the main chain. On the other hand, in the crystalline structure, an ordered structure appears with a specific pair of molecular conformations along the polymer chain. For example, polyethylene, $-(CH_2CH_2)_n-$, takes a planar zigzag structure with repeated trans-type (T) conformation, whereas isotactic polypropylene, $-[CH_2CH(CCH)_3)]_n-$, forms 3/1 helix with alternating trans and gauche (G) to form a repeating TG structure. Accordingly, ordered linear polymer corresponds to a 1D crystal. The polymer will possess a line symmetry, where a specific structure is a characteristic period (fiber period) along the extended polymer chain. This line symmetry corresponds to the unit cell of a 3D crystal. In the case of polyethylene, two CH_2 groups, and for isotactic polypropylene, three monomer units, form this unit cell. The molecular vibration of such ordered polymer chains is characterized by the phase angle (expressed by the angle δ) of the vibration among the line symmetry units and the vibrational mode (expressed by the number j) within the line symmetry unit. When the line symmetry unit includes N atoms, $3N$ normal modes exist ($j = 1N$ to $3N$). The normal vibration is expressed by $v_j(\delta)$ as a function of δ for each j. Also, δ takes the values between $-\pi$ and $+\pi$. At $\delta \neq 0$, the relationship $v_j(-\delta) = v_j(+\delta)$ holds. The positive and negative values of δ correspond to the positive and negative propagating waves along the polymer chain.

These waves possess the same vibrational energy (degeneracy). On the other hand, $\delta = 0$ expresses the standing wave and is called in-phase vibration.

When infrared and Raman spectra are considered, the most important point is that only the in-phase vibration at $\delta = 0$ is observed in the aforementioned ideally ordered chain. In this sense, this vibration is called an optically active vibration. Among $3N$ in-phase vibrations, four have zero vibration frequency. They correspond to the molecular motion as a whole by translation along two normal directions of the chain direction and by rotation. The curve of v_j (δ) plotted as a function of δ is the dispersion curve. The dispersion curve separates into $3N$ curves called the vibrational branch depending on j. The branch $v_j = 0$ at $\delta = 0$ is called the acoustic branch, and $v_j > 0$ at $\delta = 0$ is called the optic branch. As a representative example, the dispersion curves of polyethylene are shown in Fig. 1 [1]. In this case, since $N = 6$, 18 vibrational branches can be obtained. However, for ease of observation, nine branches $(v_1 – v_9)$ are displayed by separating into zero to π and π to 0 around $\delta = \pi$ using the situation where each corresponding branch is degenerate at $\delta = \pi$. In this figure, $v_1 – v_5$ are in-plane vibration and $v_6 – v_9$ are out-of-plane vibration with respect to the plane of the zigzag chain.

When the line symmetry is broken down by the defects in the polymer chain, in principle, all vibrations will be optically active; however, the intensity of the band depends strongly on the local phase of the vibration. When the conformational order accompanying the ordering process of gelation and crystallization is studied by infrared and Raman spectroscopy, the relationship between the local phase lag and intensity is important.

Thus, we will consider how ordered segments influence the band intensity when such segments are formed among the disordered polymer conformations. As the simplest model, we will consider the system where the same atoms are aligned at the same interval in the x direction while these atoms are vibrating normal to the chain direction. When the number of atoms is infinite, it is an ideal 1D crystal. The frequency of the vibration with phase angle $\delta = 0$ corresponds to the translational motion along the y direction being zero. Its dispersion curve is expressed by one acoustic branch as shown in Fig. 2(b). When the number of atoms is finite (M), M normal modes will be allowed corresponding to the phase angles of $\delta = n\pi/M$ $(n = 1, 2, \ldots, M)$ (see Fig. 2(a)). When M is an even number, only the mode whose n is an odd number will be infrared active and its

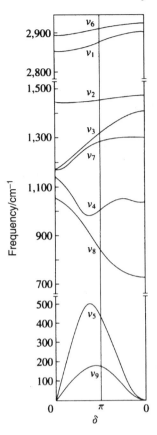

Fig. 1 Dispersion curves of the molecular vibration of planar zigzag polyethylene chains.

absorption intensity at $n = 1$ will be much more intense than the other
modes (see Fig. 2(c)). For the acoustic modes, as M increases, the $n = 1$
mode approaches the translational mode with vibration frequency of zero.
When one translational unit consists of multiple atoms, optic branches
exist, all of which will have zero slope at $\delta = 0$. Therefore, the value of M,
that is, the length of the ordered segment, becomes greater than a certain
value, and the frequency of the $n = 1$ mode approaches very close to the
optically active mode of the infinite chain. Also, the same thing can be
said about the band intensity of the monomer residue. Accordingly, the
ordered segments longer than a certain length behave the same as the

infinitely long chains. This length is called the critical chain length and the number of monomer units m will be used to express this length.

The value m depends on the mode and its value will be greater for the stronger interaction among the monomer units.

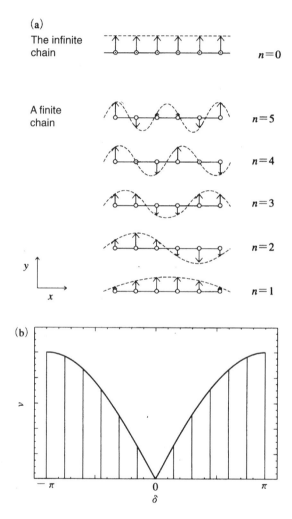

Fig. 2 Molecular vibration of the infinite chain ($M = \infty$) and finite chain ($M = 6$), which contain one atom in a unit cell.

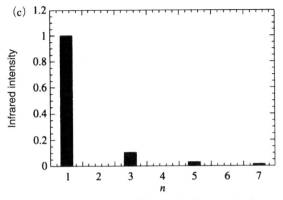

(a) Phase angle, $\delta = n\pi/M$, and vibration mode of neighboring atoms.

(b) Dispersion curves of the infinite chain and the phase angle and frequency (the position of the vertical line and the height) allowed for a finite chain.

(c) Phase angle (expressed by the n value) and infrared absorption intensity of a finite chain.

Fig. 2   (*continued*).

2.1.2 Determination of Critical Chain Length

It is necessary to experimentally determine the critical chain length for the characteristic band of the ordered conformation when the conformational order that accompanies gelation or crystallization is quantitatively evaluated from an infrared spectrum. In this section, stereoregular polystyrene, that is, isotactic polystyrene (*i*-PS) and syndiotactic polystyrene (*s*-PS), will be used; *i*-PS takes 3/1 helix, which consists of TG-type conformational chains in the crystalline phase (triclinic). On the other hand, the crystalline polymorphism of *s*-PS is much more complex than *i*-PS. In terms of molecular conformation, it can be classified into the planar zigzag structure consisting of a TT-type chain and 2/1 helix structure, which consists of a TTGG-type chain (see Fig. 3). The infrared spectrum of highly crystalline material gives rise to characteristic bands unique to the conformation of the regularity. Figure 4 illustrates the polarized infrared spectra unidirectionally oriented in three samples where the solid and broken lines indicate the transition moments perpendicular and parallel to the chain axis respectively. The characteristic bands to the specific

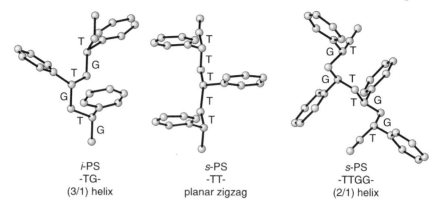

Fig. 3   Energetically favorable stereoregular conformations of isotactic (*i*-PS) and syndiotactic (*s*-PS) polystyrene.

conformations are indicated by the wavenumbers. These bands either disappear or change into broad bands upon melting or in solution. The response of the band intensity to the regularity of the conformation differs in each band reflecting the critical chain length.

As an experimental method of determining the critical chain length, the author developed the intramolecular isotope dilution method using the copolymer made of an ordinary monomer (H-monomer) and a deuterated monomer (D-monomer) [2, 3]. By changing the H/D copolymer composition, the statistical chain length of the H-monomer (or D-monomer) can be controlled freely. It is possible to generate the arbitrary statistical chain length by introducing the mass defects while maintaining the conformational regularity. Molecular vibration is influenced not only by conformation but also by mass defects. By analyzing the relationship between the copolymer composition and band intensity, the *m* value for each specific band can be obtained by the following procedures.

Let the mole fraction of the total H-monomer in the H/D copolymer (random copolymer) be X, the mole fraction F of the H-monomer in the chain that consists of m consecutive monomers is

$$F(m) = X^{m-1}[m - (m - 1)X] \qquad (1)$$

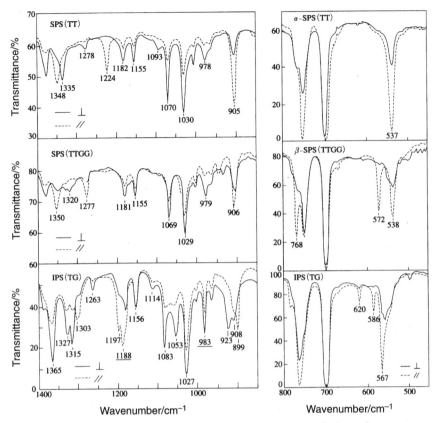

The solid and broken lines indicate infrared bands perpendicular and parallel to the chain axis, respectively.

Fig. 4 Polarized infrared spectra of uniaxially oriented *i*-PS and *s*-PS, which take two conformations.

The integrated intensity of the band $I(X)$ from the critical chain length m, which is measured under high conformational order is

$$\frac{I(X)}{I(1.0)} = F(m)X \qquad (2)$$

Thus, by contrasting the $I(X)/I(1.0)$ vs X relationship measured on each band with the $F(m)$ vs X curves using Eq. (1), the m value can be obtained. Figure 5 shows the results on the three infrared bands of *i*-PS (TG-type) [2, 3], whereas Fig. 6 depicts the results of the 572 cm^{-1} band for TTGG-type *s*-PS and the 1124 cm^{-1} band for TT-type *s*-PS [4].

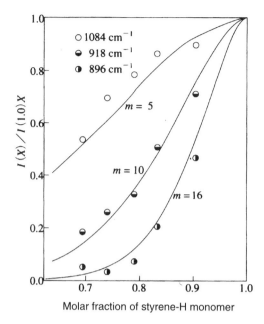

Fig. 5 Determination of the critical chain lengths for three infrared bands, which are characteristic of the TG-type stereoregular chains of *i*-PS. (The intensity of the bands was determined for the *i*-PS/CS$_2$ gel cooled to $-104°$C.)

2.1.3   Gelation of *s*-PS and Conformational Order

The *s*-PS dissolves in various solvents at high temperature. In many instances, these solutions gel when they are left at room temperature. When the infrared spectrum of this gel is compared with polystyrene with differing stereoregularity and aggregation state, characteristic bands in the TTGG-type regular chain are observed. This indicates that TTGG conformation exists in the gel (see Fig. 7) [5–7]. The regularity conformation that is formed in a gel is always TTGG-type regardless of the solvent. However, the rate of gelation or conformational order depends strongly on the solvent. Such changes occur in several minutes in benzene or CCl$_4$ with several percent polymer concentration at room temperature. On the other hand, such processes proceed slowly with several tens of hours in CHCl$_3$. Figure 8(a) shows time-resolved infrared spectra of CCl$_4$ solution at 30°C. The bands at 572 and 549 cm^{-1} unique to the TTGG regularity increase as a function of time, whereas the broad band around 540 cm^{-1} which is due to random conformation, decreases. Judging from the

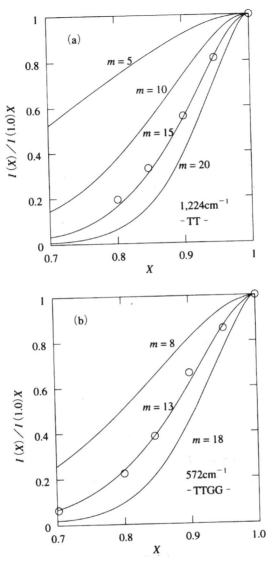

(a) The 1224 cm⁻¹ band, which is characteristic of the TT-type chain.

(b) The 572 cm⁻¹ band, which is characteristic of the TTGG-type chain (obtained from the s-PS/o-C₆H₄Cl₂ gel cooled to – 60°C).

Fig. 6 Determination of the critical chain length for the characteristic infrared band of the stereoregular chains of s-PS.

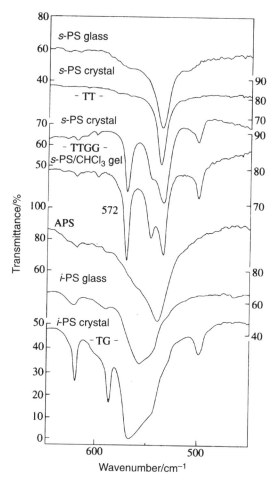

Fig. 7 Infrared spectra of polystyrene with different stereoregularity and state of molecular aggregation.

observation of equal absorption points in a series of spectra, it is possible to treat the spectral changes as being caused by the compositional changes of the two-component system, which consists of the regular (A) and random (B) structures [8].

Let the mole fraction of the monomer units in each component be x_A and x_B, the corresponding extinction coefficient of the band be ε_A and ε_B,

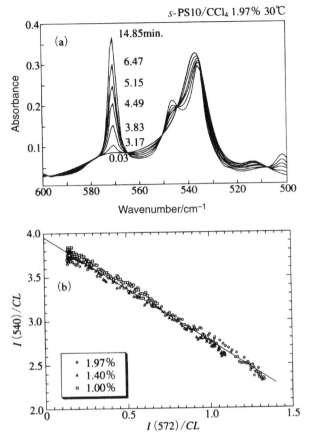

Fig. 8 Gelation process and molecular conformation of the *s*-PS/CCl₄ system.

the integrated intensity be I_A and I_B, the path length of the optic cell be L, and polymer concentration be C. Then,

$$I_A = \varepsilon_A x_A CL \tag{3}$$

$$I_B = \varepsilon_B (1 - x_A)CL = \varepsilon_B CL - \left(\frac{\varepsilon_B}{\varepsilon_A}\right) I_A \tag{4}$$

When I_B/CL vs I_A/CL is plotted from the spectra observed with various concentrations and times, a linear relation is obtained from Eq. (4). Figure 8(b) shows the results of such plots for A and B components using 572

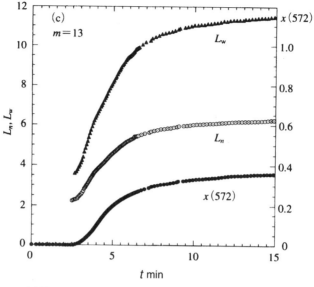

(a) Time resolved infrared spectra
($M_w = 30 \times 10^4$, $C = 7.57\%$, 10°C)

(b) $I(540)/CL$ vs $I(572)/CL$ plot

(c) Time-dependent variation of $x(572)$, L_n and L_w

Fig. 8 (*continued*).

and 540 cm^{-1} bands as characteristic bands, respectively. From the slope and intercept of the straight line, ($\varepsilon_B/\varepsilon_A$) and ε_B values can be obtained, respectively. Also from both values, the ε_A value will be obtained. Using Eq. (3), the measured intensities, $I_A/CL = I(572)/CL$, can be converted to x_A. Here, x_A is the weight fraction $W(m)$ of the TTGG conformation with the critical chain length m greater than 13, which contributes to the band at 572 cm^{-1} in the total monomer units. This quantity is one of the absolute measures of conformational order.

When the molecular conformation order follows Bernoulli statistics, $W(m)$ can be expressed by using the probability z of a monomer unit existing in the ordered conformation as

$$W(m) = z^{m-1}[m - (m - 1)z] \tag{5}$$

Also, the number average length L_n and weight average length L_w of the conformational order are expressed as $1/(1-z)$ and $(1+z)/(1-z)$, respectively. Figure 8(c) depicts $x_A = x(572)$, L_n, and L_w calculated from the measured values of the $572\,cm^{-1}$ of s-PS as a function of time.

As previously mentioned, TTGG-type conformational order is formed upon gelation. It is necessary to examine how this conformation relates to the crosslinking by molecular aggregation, which is the fundamental structure of this gel formation. Thus, the formation of molecular aggregation as a function of time was done using small angle neutron scattering (SANS) [8, 9]. Because SANS measurement requires at least 20 min for each measurement, a slow gelation system s-PS/CDCl$_3$ was chosen and then compared with the results of time-resolved infrared spectroscopy. Figure 9 shows the results of time-resolved SANS measurement. Here, the abscissa is wavenumber vector $q = (4\pi/\lambda)\sin\theta$ (θ: Bragg angle, λ: wavelength). The ordinate is the product of scattering function $I(q)$, which is corrected for the inelastic scattering due to the H atoms of the solvent and s-PS, and q^2. This is the so-called Kratky plot. Around the starting point, the flat profile (Kratky plateau), which is characteristic for the Gaussian chain conformation polymer solution, is seen. As the time passes, the scattering around $q = 0.3\,nm^{-1}$ increases thus indicating molecular aggregation formation. The total integrated intensity,

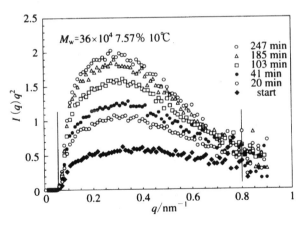

Fig. 9 The time-dependent variation of small angle neutron scattering of s-PS/CHCl$_3$ system during gelation process ($M_w = 36 \times 10^4$, $C = 7.57\%$, $10°C$).

$Q = 4\pi \int I(q)\, q^2 dq$, is proportional to the extent of gelation. If the time-dependent change and the degree of conformational order $x(572)$ are compared (see Fig. 10), it can be found that both are increasing.

The rate of gel formation and the conformational order depend strongly on the polymer concentration. This indicates that the conformational order proceeds with the aggregation of molecular segments. In order to elucidate its mechanism, the dynamic mechanical behavior of conformational order is analyzed. As a model, we will consider the process of

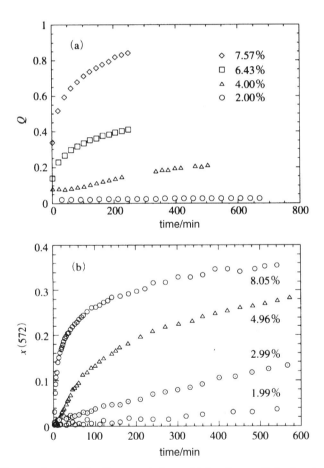

Fig. 10   Comparison of the time variation of (a) gel crosslink structure and (b) molecular conformation of s-PS/CHCl$_3$ during gelation process (10°C).

cluster (Cr) formation by gathering p random segments, Sr, which consists of r monomer units. This rate equation is

$$\frac{d[\mathrm{Cr}]}{dt} = k[\mathrm{Sr}]^p \tag{6}$$

where [Cr] and [Sr] are the molar concentration of the ordered chain cluster and random segments, respectively, and k is the rate constant. The integrated intensity of the $572\,\mathrm{cm}^{-1}$ band, $I = I(572)/L$, is

$$I = Ap \sum_{r=m}^{\infty} r[\mathrm{Cr}] \tag{7}$$

Thus, the initial slope with respect to time t is

$$\left(\frac{dI}{dt}\right)_0 = Ap \sum_{r=m}^{\infty} \frac{rk[\mathrm{M}]_o^p}{r^p} \tag{8}$$

where A is the proportionality constant, m is the critical chain length of the $572\,\mathrm{cm}^{-1}$ band, and $[\mathrm{M}]_0$ is the initial concentration of the monomer unit in Sr, which is proportional to the polymer concentration. If we assume only $r = m$ contributes to I at the beginning of gelation, the following equation is obtained:

$$\ln\left(\frac{dI}{dt}\right)_0 = \ln(Apm^{1-p}k) + p\ln C \tag{9}$$

Hence, the log–log plot of the initial rate of increase in absorption intensity $(dI/dt)_0$ and C will be a straight line. From its slope, the value of p should be obtained.

Figure 11 indicates that Eq. (9) holds at any temperature. The value of p, which is obtained from the slope of the straight lines, depends on temperature. For example, $p = 5.2$ at $10°\mathrm{C}$. This indicates that approximately five segments are gathering to stabilize the TTGG conformational chain order. As the temperature reduces, the value of p, that is, the number of the segments necessary to stabilize the TTGG conformational chain order, decreases. This number becomes 1 below $-15°\mathrm{C}$ (see Fig. 12).

In other words, the mechanism of conformational order changes from segment aggregation to intramolecular self-ordering as the temperature decreases. This crossover temperature depends on the solvent, and it is around $-15°\mathrm{C}$ for $\mathrm{CHCl_3}$ whereas for o-dichlorobenzene it moves to room temperature. Such solvent dependency is considered to be due to the difference in the interaction between s-PS and solvent molecules. In

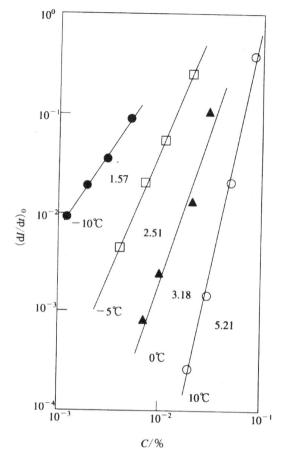

Fig. 11 The p value, which is obtained from the linear relationship of the log–log plot of $(dI/dt)_0$ vs C and its slope during the gelation process of s-PS/CHCl$_3$.

general, it has been confirmed by SANS that the solvent with faster gelation rate forms higher molecular aggregate density.

2.1.4   Gelation Behavior and Conformational Order in *i*-PS/Solvent System

When an *i*-PS/CS$_2$ solution is cooled from room temperature, infrared bands, which are due to the TG regularity, appear around $-50°$C. The

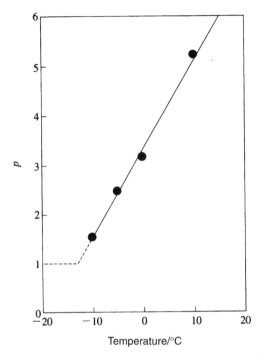

Fig. 12 Temperature dependence of the p value ($C = 20\%$).

intensity gradually increases and around the melting point of the solvent ($-110°C$) the intensity corresponds to the highly crystallized solid (see Fig. 13). From the fact that the bandwidth of the ^{13}C NMR spectrum suddenly increases around this temperature, gelation and conformational order are also related in this system. Unlike the previous case of s-PS, the spectral changes occur in reverse order. Thus, disordered conformation and ordered conformation are in the thermal equilibrium state at each temperature. If the probability of any monomer unit is incorporated into the ordered chain at an absolute temperature T, we obtain

$$\frac{1-z}{z} = \exp\left(\frac{\Delta S}{R}\right) \exp\left(\frac{\Delta H}{RT}\right) \qquad (10)$$

where ΔS and ΔH are the entropy and enthalpy difference between the monomer unit in the disordered conformation and ordered conformation, respectively. Experimentally, $\Delta S = 41.6\,\text{J/K mol}$ and $\Delta H = 13.0\,\text{kJ/mol}$ have been obtained. The value z can also be calculated as a function of T.

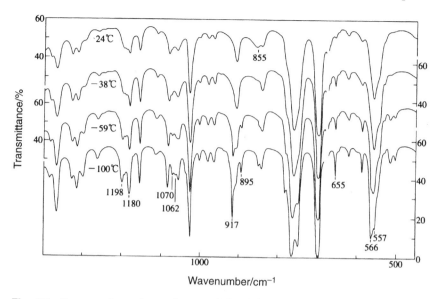

Fig. 13 Temperature dependence of the infrared spectra of *i*-PS/CS$_2$ (the solution gels below $-50°$C and the structure is reversible by reversing the temperature).

The absorption intensity of the band (expressed as the fraction I/I_∞ against the maximum value I_∞) having the critical chain length m equals $W(m)$ in Eq. (5). Figure 14 shows the comparison between the temperature dependence of measured and calculated $W(m)$ for several characteristic bands. It can been seen that a highly ordered TG-type helical structure is formed at low temperatures.

In the infrared spectrum of *i*-PS/CS$_2$ gel, several bands unique only to gel can be observed in addition to the characteristic bands for the solid-state TG-type 3/1 helix. On the other hand, those bands at 1187 and 983 cm^{-1}, which increase their intensity with the increased degree of crystallinity, will not change upon gelation. Judging from the fact that most of the 3/1 helix bands show the same frequency, relative intensity and band shape as the crystalline solid, the main chain structure in the gel can be regarded as close to the TG-type 3/1 helix.

On the other hand, the unique bands for the gel and the two aforementioned crystalline bands are both assigned to the benzene ring modes. Hence, it is hypothesized that the side chain stereoscopic conformation is different from the crystal due to the interaction with the solvent.

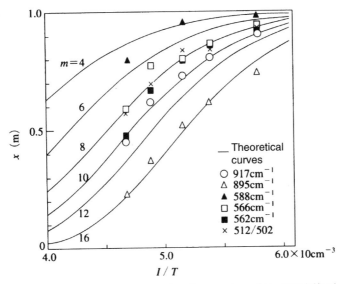

Solid line indicates the calculated curves from ΔH = 13.0 kJ/mol and ΔS = 41.6 J/molK

Fig. 14 Temperature dependence of the weight fraction of the monomer residue, which is in the TG-type stereoregular chain with longer than the various critical chain lengths of *i*-PS/CS$_2$ gel.

2.1.5 Gelation of Amorphous Polystyrene

In addition to the already-stated stereoregular polystyrene, an amorphous polystyrene that has random stereoregularity also gels. A commercial polystyrene (*a*-PS) that is synthesized via free radical polymerization has the regularity with respect to the racemic dyadic notation $[r] = 0.65$, which indicates slight richness in syndiotacticity [10]. When *a*-PS/CS$_2$ solution ($C = 20\%$) is cooled from room temperature, it gels around $-50°$C and the characteristic 572 cm^{-1} band for the TTGG chain appears and increases its intensity. On the other hand, the band at 560 cm^{-1}, which originates from the TG chains, decreases in relative intensity. Accordingly, the gelation in *a*-PS/CS$_2$ is caused by the formation of a TTGG chain in the syndiotactic portion of the chain. When an *i*-PS sample is treated by hexamethylphosphoramide, a partially syndiotactic sample whose tacticity depends on the treatment time is obtained. The racemic dyadic values $[r]$ obtained by ^{1}H NMR will be indicated as EPS ($[r]$) in Fig. 15. Figure 15

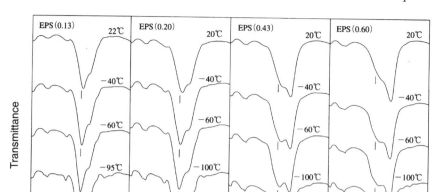

Fig. 15 Temperature dependence of the infrared spectra of *i*-PS (EPS)/CS$_2$ solution (gel) ($C = 20\%$) (the symbol | indicates the characteristic band of TG-type stereoregularity at 560 cm^{-1}).

also illustrates the temperature-dependent infrared spectra of the CS$_2$ solution of each sample. The band that appears at low temperature decreases as the [r] value increases. Also, the bands at 917 and 894 cm^{-1} of the long critical chain length sample are barely observed. These bands quickly diminish as the [r] value increases.

2.1.6   Gelation of Poly(ethylene oxide)

Poly(ethylene oxide) (PEO) $(-CH_2CH_2-)_n$ has two crystalline polymorphisms. In the ordinary triclinic structure, seven monomer units form a helical structure with two turns within one fiber period.

There is no special symmetric operation in the molecule itself. Those seven monomer units are mutually nonequivalent and form a skewed 7/2 helix. The fundamental vibrations of a homogeneous 7/2 helix are classified into A$_1$, A$_2$, E$_1$, E$_2$ and E$_3$, of which A$_2$ (9 vibrations) and E$_1$ (20 vibrations) are infrared active and A$_1$ (10 vibrations), E$_1$ (20 vibrations), and E$_2$ (21 vibrations) are Raman active. The polarization characteristics of the infrared bands are that A$_2$ modes are parallel to the fiber axis and E$_1$ is perpendicular to it. The infrared spectrum of a solid PEO sample can be approximated by the homogeneous 7/2 helical model. However, a group of bands corresponds to the progression bands ($n \neq 1$) due to the seven connecting monomers at low temperatures (see Fig. 16 (a)).

The infrared spectrum of PEO that is dissolved in CS_2-CHCl_3 ($1:1$ mole) mixed solvent does not exhibit the bands that correspond to ordered conformations, indicating PEO is in a random structure. Upon cooling this

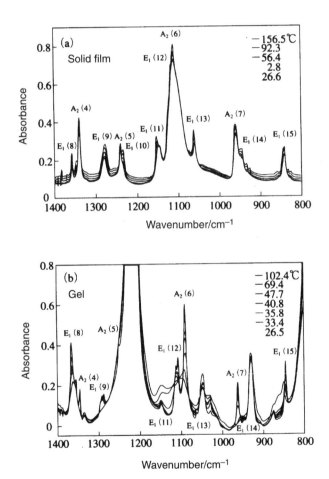

Fig. 16 Temperature variation of the infrared spectra of crystalline solid of poly(ethylene oxide) (PEO) and PEO/(CS_2-$CHCl_3$ equimolar mixed solvent) solution (gel).

solution to $-35°C$, this solution gels. At the same time, the band due to the TTG-type ordered conformation appears and the intensity gradually increases as the temperature is lowered (Fig. 16(b)). Also, this change is reversible. Upon comparison of the low-temperature gel spectrum with that of the crystalline solid, several characteristic differences can be found. The band position shifts significantly (up to $15\,cm^{-1}$) and the relative intensity is also different. Furthermore, when cooled, the doubly degenerate E_1 bands split into two components around $-70°C$. Hence, the PEO molecules in the gel takes a TTG-type homogeneous helix. However, below $-70°C$, the nonsymmetric force field around the molecule breaks down the double degeneracy resulting in the splitting of the E_1 modes (site splitting). It is presumed that the nonsymmetric force field formation is due to strong interactions such as the complex formation between PEO and solvent molecules. The gelation behavior and its accompanying infrared spectrum of PEO solution depends strongly on solvent, indicating the existence of unique polymer-solvent interactions.

2.2 LIGHT SCATTERING

HIDETOSHI OIKAWA

2.2.1 Introduction

When visible incident laser light irradiates a molecule, it polarizes vibrationally due to the photoalternating current field, and vibrational dipoles are induced. As a result, the molecule becomes a scattering body of the secondary light and scattered light will be generated. This is called Rayleigh scattering.

The intensity of scattered radiation I_s is given by the statistical average of the multiple of the conjugate of scattered radiation field (Rayleigh's 4th power law):

$$\frac{I_s}{I_0} = \left(\frac{8\pi^4}{\lambda_0^4 L^2}\right)(1 + \cos^2\theta)\sum N_i\alpha_i^2 \tag{1}$$

where I_0, λ_0, L, θ, N_i, and α_i are the intensity of incident light, wavelength of incident light, distance between the scattering body and the detector, scattering angle, number of the scattering body i, and polarizability of the scattering body i. For the light scattering of a solution, polarizability fluctuation $\delta\alpha$ based on the concentration fluctuation δC through the dielectric constant fluctuation is considered and the α in Eq. (1) is replaced with $\delta\alpha$; $\delta\alpha$ is a function of local pressure, temperature, and change of concentration. In a dilute solution, $\delta\alpha$ will be approximated by δC. The fluctuations of pressure and concentration are independent of each other. Thus, the fluctuations of the solvent density and the concentration are all additive.

   In the static light scattering (SLS) technique, a time-averaged intensity of scattered light is measured, whereas in the dynamic light scattering (DLS) technique, the autocorrelation function (ACF), which is the intensity fluctuation of the scattered light, is determined. Although light scattering is an extremely important analytical technique in the field of polymer science, many examples cannot be discussed here due to limited space. Thus, important reviews and monographs [11–20] are listed and readers are referred to the references cited therein.

2.2.2   Theoretical Background of Light Scattering of Gels

Aside from the experimental parameters that are unique to gels, in SLS measurements, in principle, the scattering vector q dependence on I_s is of interest [11, 13, 16]:

$$q = \left(\frac{4\pi n_0}{\lambda_0}\right) \sin\left(\frac{\theta}{2}\right) \tag{2}$$

where n_0 and θ are the refractive index of the solvent and the scattering angle. The reciprocal of q possesses the dimension of real space, and it follows that the smaller the q, the longer the influence of concentration fluctuation. In gels, the relationship between I_s and q is often expressed by the Ornstein–Zernike equation (OZ) (Eq.(3)) or the Debye–Buiche equation (DB) (Eq. (4)):

$$I_s(q) = \frac{I_s(0)}{(1 + \xi^2 q^2)} \tag{3}$$

$$I_s(q) = \frac{I_s(0)}{(1 + \alpha^2 q^2)^2} \tag{4}$$

where $I_s(0)$ indicates the likelihood of the occurrence of concentration fluctuation and denotes the functions of concentration, temperature, and bulk modulus [16], ξ and α are static correlation lengths, and ξ is the characteristic length of the concentration fluctuation in the gel.

Due to the heterogeneous structure represented by α, when the spatial scale of refractive index fluctuation is larger than the concentration fluctuation, the DB equation is used. However, the OZ and DB equations are not necessarily unambiguous. It is desired to measure broad q-dependency using techniques such as SAN [16, 20]. On the other hand, for the clusters that are formed during the gelation process such as the sol-gel transition and the branched structure of microgels, the Kratky plot ($q^2 I_s$ vs q), which is the result of the calculation based on the hard-sphere and soft-sphere models, is proposed [18]. Furthermore, microgels and aggregation of the domain in the gel are discussed with respect to fractal dimension D from the approach of I_s to an asymptotic value in the high q region:

$$I_s(q) \sim q^{-D} \tag{5}$$

In relation to aggregate structures, it is classified as a mass fractal when $D < 3$ and a surface fractal when $D > 3$ [21]. In particular, when $D = 4$, it is termed the Porod region and is regarded as a smooth interface. The D value during the gelation has been predicted [22]. For the q-dependency of I_s, a combination of the the OZ and DB equations, the equation that incorporates elongation function, and OZ and DB equations that contain fractal dimension have been proposed [16].

If ACF of DLS is assumed to be simple exponential damping, the damping constant Γ is given by the following equation [12, 14]:

$$\frac{\Gamma}{q^2} = (1 - \phi)^2 D_{\text{coop}} \tag{6}$$

where ϕ is the gel concentration; D_{coop} is the cooperative diffusion constant and corresponds to the concentration dependence of gel osmotic pressure. This D_{coop} differs from the parallel diffusion coefficient, and increases with ϕ. This takes the same form as the Einstein–Stokes equation [12, 23] against the blob, which indicates the range of the excluded volume of polymer chains:

$$D_{\text{coop}} = \frac{k_B T}{6\pi\eta_0\xi_H} \tag{7}$$

where k_B, T, and η_0 are the Boltzmann constant, absolute temperature, and the viscosity of solvent, respectively, and ξ_H is the screening distance (dynamic correlation length) that corresponds to the blob size. Within the range where the C* theorem [23] applies, the ξ_H can be regarded as the mesh size of the gel networks. However, if there are entanglements in the gel, the ξ_H will be smaller than the distance between crosslinks.

The ξ_H of the gel that is in the equilibrium swelling state will be scaled against ϕ as shown in the following equation [23]:

$$\xi_H \sim D_{coop}^{-1} \sim \phi^{-0.75} \tag{8}$$

DLS is also useful for the analysis of the gelation process, and fractal analysis can be done from the ACF of the long correlation function [14]:

$$\mathrm{ACF} \sim \tau^T \tag{9}$$

2.2.3   Measurement Examples of SLS

Figure 1(a) [24] is the OZ plot of the quasi-gel (transitional gel) that is formed in gelatin semidilute solutions at elevated temperature. The ξ value decreased from 5.1 nm (2%) to 3.5 nm (5%) as the concentration increased. The data deviates from the straight line in the plot because of the heterogeneous structure of the quasi-gel. Figure 1(b) shows the DB

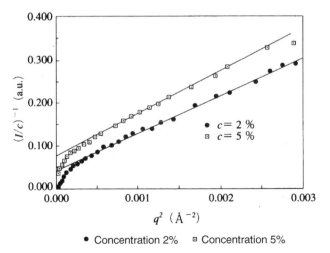

• Concentration 2%   ▫ Concentration 5%

Fig. 1(a)   OZ plot of quasi-dilute solution (quasi-gel) of gelatin [24].

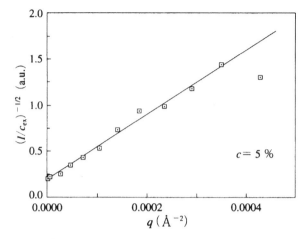

Fig. 1(b)   DB plot of the excess scattering intensity at low q region of gelatin quasi-dilute solution (quasi-gel) (concentration 5%) [24].

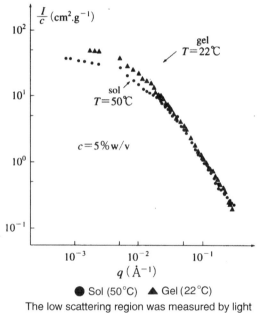

● Sol (50°C)   ▲ Gel (22°C)

The low scattering region was measured by light scattering, and high scattering region by small angle neutron scattering

Fig. 2   Scattering vector dependence of scattering intensity of gelatin gel [25].

plot of the excess intensity component of scattered light in the q region where $a = 13.5$ nm at 5%. Figure 2 illustrates the q dependence of scattered light intensity of the gel that is formed from the gelatin semi-dilute solution at room temperature [25]. At the high q region, there is almost no difference in the intensity of scattered light and ξ of the gel is 3.8 nm (5%). Figure 3 depicts a typical DB plot of bimodal networks that consist of a long chain ($M = 22{,}500$) and a short chain ($M = 770$) of polydimethylsiloxane (PDMS) [26].

When the short chain component is 50%, a is approximately 40 nm. The relationship between the crosslinking reaction of both terminal groups and heterogeneous structure is discussed.

Figure 4 shows the measurement results on a poly(vinyl acetate) microgel and the Kratky plot of the calculated branching number dependence assuming the soft-sphere model [27]. A characteristic of this plot is the appearance of a maximum as the amount of branching increases.

Figure 5 shows the results of the fractal analysis of the microphase-separated structure of poly(acryl amide) gel formed under the presence of poly(ethylene glycol) [28]. While the solution of the same composition shows $D = 4$, the gelled system shows approximately $D = 3$, suggesting a highly porous structure [29]. Also, upon quenching and gelling the

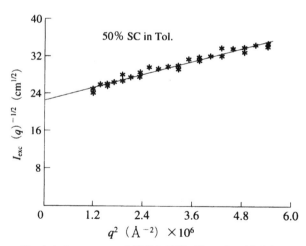

Short chain component (SC) is 50%. The solvent is toluene

Fig. 3 A DB plot of bimodal polydimethylsiloxane gel with both terminals crosslinked. The gel was prepared from a long chain prepolymer ($M = 22{,}500$) and short chain prepolymer ($M = 770$).

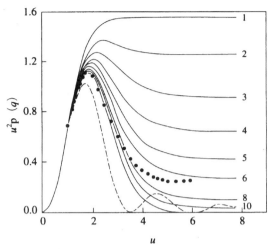

Here $p(q)$ is the particle scattering factor and u is the product of scattering vector and radius of gyration; the number in the graph is the number of branches

Fig. 4   A Kratky plot of the calculated results by soft-sphere model (solid line) and hard-sphere model (broken line) and the measurement results of poly-(vinyl acetate) microgel [27].

polymer liquid crystal synthetic polypeptide (PBLG), phase separation with spinodal decomposition-type phase separation is suggested. The later process is analyzed fractally using the general Zimm equation and the results are shown in Fig. 6 [30]. In this case, $D = 2.7$ and $x = 250\,\text{nm}$ and it is found to be reduced from the characteristic distance of concentration fluctuation at the early SD process. The time-resolved scattered light intensity change is shown in Fig. 7 when the poly(vinyl alcohol) solution of DMSO/water mixed solvents is gelled by quenching [31].

A typical SD is shown that follows Cahn's theory, and the size of the heterogeneous structure due to phase separation is estimated to be approximately $1\,\mu\text{m}$.

2.2.4   Examples of DLS Measurement

Figure 8 shows the gel concentration dependence of D_{coop} of a polystyrene (PS) ($M = 700,000$) semidilute solution and PS gel. According to the results, the applicability of Eq. (8) in the semidilute region and the C^*

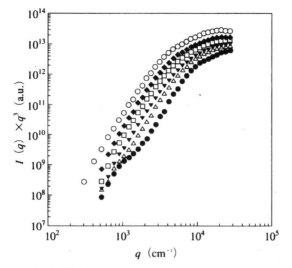

In the high scattering region, the scattering intensity
approximates the 3rd power of the scattering vector

Gelation time					
• 49s	▵100s	▾151s	▫250s	◆450s	○2152s

Fig. 5   Scattering vector dependence of the scattering intensity of poly(acryl
amide) gel prepared in the presence of poly(ethylene glycol) (4 wt%) [28].

theorem as the first approximation has been confirmed. Also, from the
measurement of PDMS model networks, it was confirmed that the distance
between crosslink points obtained from the average molecular weight of
prepolymer agrees well with ξ (see Fig. 9).

Typical analytical methods regarding the damping constant of a
nonexponential damping function include the cumulant method that
expands the exponential terms about the average value $\overline{\Gamma}$ [32], the
multiexponential method [32], and the elongational exponential function
method (the KWW equation) [33]. On the other hand, the histogram
method [32] and the CONTIN method [33] are usually employed in order
to obtain the distribution of Γ. Figure 10 [34] displays the Fast and Slow
mode D_{coop} when ACF, which is measured on the aforementioned PDMS
bimodal networks, is analyzed by double exponential; D_{fast} of the fast
mode is the cooperative diffusion mode that almost obeys Eq. (8); D_{slow} of

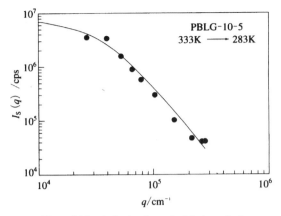

The solid line indicates the calculated results from
the generalized Zimm equation; the volume fraction
is 7.3×10^{-3}

Fig. 6 Scattering vector dependence of the scattering intensity during the late stage spinodal decomposition of poly(γ-benzyl-L-glutamate) liquid crystal gel when quench gelled from 333 K to 283 K [30].

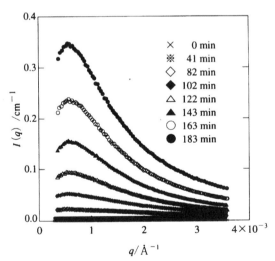

The time in the graph indicates elapsed time after the
quenching; the solvent was DMSO/water (60/40, v/v)
mixed solvent, and the polymer concentration was 5 g/dl

Fig. 7 Time resolved scattering intensity change of quench gelled poly(vinyl alcohol) gel from 100°C to 25°C.

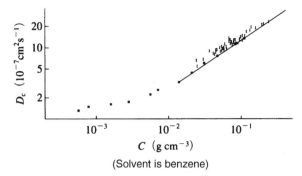

(Solvent is benzene)

Fig. 8 Concentration dependence of the cooperative diffusion coefficient of dilute and semi-dilute polystyrene ($M = 700{,}000$) solutions and polystyrene gel [32].

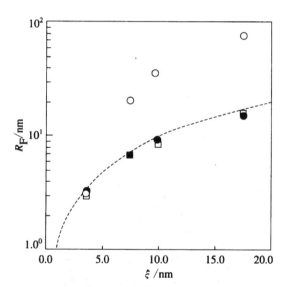

The distances between crosslink points were obtained from the average molecular weight of the polydimethylsiloxane prepolymer (□), Flory–Lehrer's equation using equilibrium degree of swelling (○), and from C* theory (●)

Fig. 9 Comparison between the persistence length and distance between crosslinks of uni-modal polydimethylsiloxane, crosslinked at both ends of the chain [33].

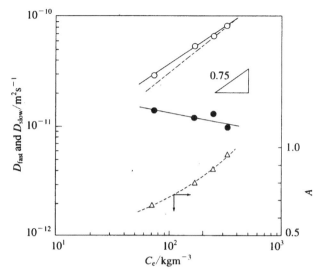

The average molecular weight of the long chain and short chain is 41,000 and 2500, respectively. When the short chain concentration changed from 20 to 80 wt%, gel concentration increased. The broken line near the solid line is the cooperative diffusion constant obtained by the cumulant method

Fig. 10 The diffusion constant of the fast mode (○) and slow mode (●) of bimodal polydimethylsiloxane, crosslinked at both ends of the chain. The gel concentration dependence of intensity A of the fast mode is also shown.

the slow mode does not depend so much on the gel concentration, and is regarded as the cooperative self-diffusion mode of the domain that is formed by short chains (Fig. 11) [35].

Similar analysis is also applied to a chemically crosslinked gelatin gel and the results indicate that the magnitude of the ξ_{slow} of the slow mode agrees almost with the size of the gelatin microcrystals that takes a triple helix structure. These microcrystals become physical crosslink sites in the gel. Actually, there has been on-going discussion on the existence and assignment of the slow mode [14, 36]. Figure 12 shows the autocorrelation of PVA-sodium borate gel. In this case, the Γ of the slow mode does not depend on q and is called the relaxation mode [37]. Its relaxation time agrees almost with that measured by the viscoelasticity measurement of the gel, and the correlation between the concentration fluctuation and viscoelasticity of networks has been suggested.

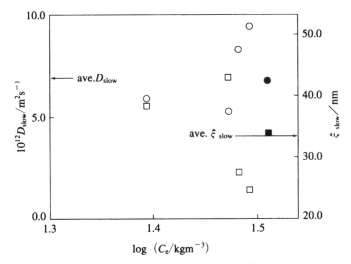

Fig. 11 Gel concentration dependence of the diffusion constant (○) and persistent length (□) of the slow mode of chemically crosslinked gelatin gel [35] (● and ■ are the diffusion coefficient and persistent length when the concentration of the crosslinking agent is zero).

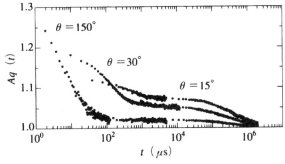

Measurement temperature was 25°C and gel concentration 2 wt%; θ is the scattering angle

Fig. 12 Self-correlation function of poly(vinyl alcohol)/sodium borate gel [37].

Figure 13 illustrates the changes of ACF during the gelation process of silica gel [38]. The ACF follows an exponential function type, elongational exponential function prior to the gel point, and Eq. (9) at the gel point. The exponent γ is 0.27 and is lower than the theoretical

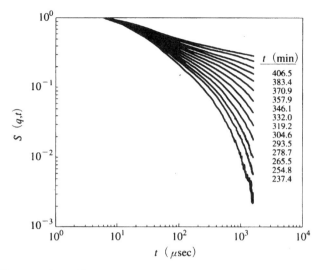

Fig. 13 Changes in self-correlation function of silica gel (silicon dioxide) at sol-gel transition (gelation time is 406 min) [38].

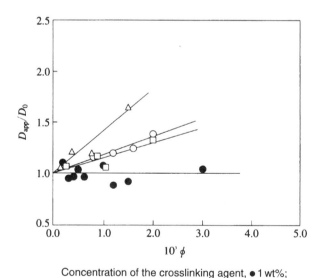

Concentration of the crosslinking agent, ● 1 wt%;
○ 5 wt%; □ 8 wt%; and △ 12 wt%

Fig. 14 The relationship between translational diffusion coefficient and gel concentration of polystyrene microgel at the swelling state [39].

value. Figure 14 shows the relationship between the parallel diffusion constant of swollen PS microgel and f. The slope depends on the crosslinking agent concentration and thus core-shell type networks are expected. Readers are referred to the literature concerning the non-ergodicity problems due to the heterogeneous structure of gels [40, 41].

2.3 X-RAY—NEUTRON SCATTERING

KANJI KAJIWARA

2.3.1 Fundamentals of Electromagnetic Wave Scattering

X-ray scattering is caused by the electromagnetic wave interaction of the electromagnetic x-ray and the electrons within matter and can provide information on the spatial distribution of the electron density of the material. The neutron is a particle wave and is scattered by nuclei. However, due to its particle and wavelike nature, it can be described similarly to the x-ray scattering by the general theory of electromagnetic wave scattering [42, 43]. Let us consider the cases where the x-ray and neutron are scattered by particles of a certain size (diameter D). Both the x-ray and neutron are scattered by the scattering materials within the particles (in the case of the x-ray it is electrons, and for the neutron it is nuclei). Both the scattered x-ray and neutron interfere with each other. Consequently, because the scattered electromagnetic waves increase the phase lag as the scattering angle increases, the scattering intensity is the maximum at scattering angle zero (the direction of incident light) and reduces as the scattering angle increases. If the wavelength of the electromagnetic wave is λ, the scattering intensity becomes zero at the scattering angle on the order of λ/D. The scattering angle θ and spatial distance r have a reciprocal relationship. Considering that the wavelengths of the x-ray and neutron are around 0.1–1.0 nm, from the scattering profile at the small angle portion within 1°, the information on spatial correlation can be obtained up to about 100 nm. This situation can be understood well by examining the simulated scattering profile shown in Fig. 1. Here, the intensity of the scattered electromagnetic waves can be obtained by the product of the amplitude of the scattered electromagnetic waves and their conjugate electromagnetic waves. For statistically isotropic materials, the

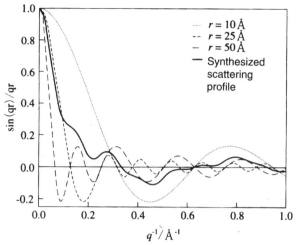

The distance between the two points is indicated in the graph

Fig. 1   Scattering angle dependence of the scattering intensity by the interference of the scattering from two scattering points.

observed scattering intensity $I(q)$ can be obtained by summing scattering intensity changes that are caused by the interference of the scattered light from the two points with distance r for all scattering point pairs [44]:

$$I(q) = V \int_0^\infty 4\pi r^2 \gamma \, (r) \frac{\sin qr}{qr} \, \mathrm{d}r \tag{1}$$

where V is the volume of scattering matter. The magnitude of the scattering vector is given by

$$q = \left(\frac{4\pi}{\lambda}\right) \sin\left(\frac{\theta}{2}\right) \tag{2}$$

The correlation function $g(r)$ is the average of the product of the density fluctuation of the two points with distance r and has the following relationship with dynamic radius distribution function (Fourier transform of the scattering intensity) $p(r)$:

$$p \, (r) = \frac{1}{2\pi^2} \int_0^\infty I(q) qr \sin(qr) \mathrm{d}q = Vr^2 \gamma(r) \tag{3}$$

That is, if the measured scattering intensity is inverse Fourier transformed with respect to the magnitude q of the scattering vector, a dynamic radius distribution function can be obtained. This distribution expresses spatial correlation of electron density of two points with relative distance r in the statistically isotropic scattering media. When $r = 0$, Eq. (3) gives invariance Q that is independent of the structure of the scattering media:

$$Q \equiv V\gamma(0) = \frac{1}{2\pi^2} \int_0^\infty q^2 I(q) dq \qquad (4)$$

As can be seen from Eq. (4), the invariance Q is proportional to the square average of the electron density fluctuation of the scattering media. When the electron density difference $\Delta\rho$ is constant, it is proportional to the volume of the scattering media. In this case, the correlation function can be rewritten using the function $\gamma_0(r)$ that depends on the shape of the scattering media:

$$\gamma(r) = (\Delta\rho)^2 \gamma_0(r); \quad \gamma_0(0) = 1; \quad \gamma_0(r \geq D) = 0 \qquad (5)$$

Beyond the maximum distance D of the scattering media, $\gamma_0(r)$ diminishes.

As is obvious from Eq. (3), if the scattering intensity that is observed from the gel is inverse Fourier transformed, the dynamic radius distribution function can be obtained. The fluctuation of the density within the gel, that is, information concerning the gel structure, can be gained to some extent. However, Eqs. (3) and (4) should be applied with caution for various reasons; that is, because the actual measurement range of the scattering vector is limited, the measured scattering intensity cannot be extrapolated from $q \to 0$ and $q \to \infty$.

2.3.2   Gel Models and Scattering

Readers are referred to Chapter 3, Section 1, for detailed gel models. Here, we will consider the scattering from a simple model gel prior to dealing with actual gel formation.

As the simplest model, we will start with the Flory–Stockmayer dendritic model (the FS model; see Chapter 3, Section 1, Fig. 2) [45]. If the individual structural unit of the FS model is the scattering point, Debye's particle scattering can be applied [46]. This corresponds to the case where the scattering points distribute uniformly in the solvent, and the scattering intensity is given by

$$I(q) = \sum_{i=1}^{n} \sum_{j=1}^{n} A_i A_j \frac{\sin(qr_{ij})}{qr_{ij}} \tag{6}$$

where A_i is the scattering amplitude of i-th and j-th scattering body, respectively, and r_{ij} is the distance between the scattering bodies i and j. Accordingly, the scattering from an f-functional random polycondensation polymerization system is given as a function of reaction probability α by extending Eq. (1) of Chapter 3, Section 1 [47]:

$$I(q) \approx \frac{A^2(1 + \alpha\phi)}{[1 - (f - 1)\alpha\phi]} \tag{7}$$

$$\phi = \exp\left(\frac{-b^2 q^2}{6}\right) \tag{8}$$

where it is assumed that the scattering amplitude from each scattering point is the same, the mean square distance from the neighboring scattering point is b^2, and the spatial correlation of the scattering points follows the Gaussian distribution.

This model is fundamentally dilute (gaseous state), and the interactions among scattering points are ignored. The weight average degree of polymerization approaches infinity passing the gel point ($\alpha_c = 1/(f - 1)$), the scattering intensity $I(q)$ is finite as long as $\alpha\phi < 1/(f - 1)$, and the scattering can be observed. If the scattering from the f-functional random condensation polymerization is displayed by the Kratky plot ($q^2 I(q)$ vs q) and log–log plot (log $I(q)$ vs log q), the scattering profile at the small angle region ($q \to 0$) before and after the gel point shows a significant difference (see Fig. 2). Therefore, after the gel point, $q \to \{(6/b^2) \times \ln[(f - 1)/\alpha]\}^{1/2} > 0$ and $I(q)$ approaches infinite value. If the formation of gels start with the formation of clusters in the solution, those clusters grow and touch each other, and then the structure spreads to the entire system. The scattering intensity from such systems can be given by the product of the sum of scattering from the individual cluster and the interference term that reflects the spatial distribution of the cluster. Therefore, if the change in solute composition by the presence of clusters formed through bonding is known (in practice, they are the size distribution and spatial distribution of the clusters), the structural parameters of

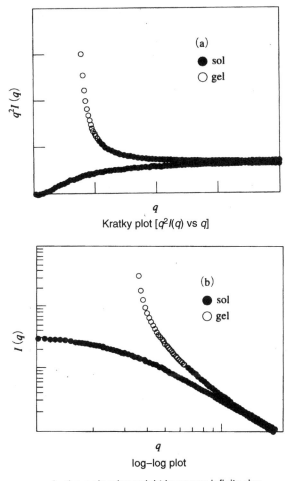

Kratky plot [$q^2I(q)$ vs q]

log–log plot

As the molecular weight becomes infinite due to gelation, the characteristic divergence of the Kratky plot occurs at $q \rightarrow 0$; for detailed divergence conditions, refer to the text

Fig. 2 Scattering profile of sol and gel as predicted from the FS model.

the scattering intensity can be calculated. Here, the individual cluster is considered to form a domain that is defined by the correlation length ξ. The clusters are also thought to be independent and have no correlation at the distance beyond ξ.

Assuming that the domain formed early reflects the isolated chain and is formed by the Gaussian chains, the density fluctuation within the domain is given by the density correlation function that is defined by the correlation length ξ [48]:

$$\gamma(r) \approx \left(\frac{\xi}{r}\right) \exp\left(-\frac{r}{\xi}\right) \tag{9}$$

The scattering function that is obtained from the Fourier transform of this correlation function (the Ornstein–Zernike type) is the Lorenz type and is given by the following equation:

$$I(q) \approx \frac{\xi^3}{1 + \xi^2 q^2} \tag{10}$$

Assuming that the gelation is the infinite network structure that is formed by the crosslinking among domains, the molecular movement of the polymer chains in the vicinity will be restricted due to crosslinking and locally frozen heterogeneous structures. If the density fluctuation of such a heterogeneous structure is expressed by the density correlation function that is defined by the breadth of the inhomogeneity a,

$$\gamma(r) \approx \exp\left(-\frac{r}{a}\right) \tag{11}$$

the scattering function that originates from such a structure is

$$I(q) \approx \frac{a^3}{(1 + a^2 q^2)^2} \tag{12}$$

According to Eq. (12), it can be seen that the scattering function follows the Porod law $(I(q) \propto q^{-4})$ at the large q, reflecting the two-phase heterogeneous structure with relatively smooth interface. As the crosslinking proceeds, the density fluctuation is frozen, and a solid heterogeneous structure is created within the liquid networks. Using such a model, a phenomenological study on the polymer network formation based on statistical crosslinking has been reported [49].

As the domain continues further crosslinking, it becomes dense with almost uniform particles, corresponding to the Gaussian-type density correlation function that is defined by the radius of gyration R_G

$$\gamma(r) \approx \exp\left(-\frac{3r^2}{4R_G^2}\right) \tag{13}$$

the scattering from such domains can be approximated by the Guinier-type scattering function [50]:

$$I\left(q\right) \approx \exp\left(-\frac{R_G^2 q^2}{3}\right) \tag{14}$$

In addition, various density correlation functions can be obtained depending on the mechanism of domain formation [51].

The gelation of globular protein can be approximated by the aggregation process of microparticles (see Chapter 3, Section 1). Globular protein gels usually take the form of fractal structure and the scattering profile can be organized in the following form:

$$I(q) \propto q^{-D} \tag{15}$$

where D is the fractal dimension and in the range $1.5 \leq D \leq 3.0$, although it is in the range of 1.75–2.5 according to the simulation of aggregation process. In the case of other biopolymer gels, in particular, polysaccharides, and physical gels of synthetic polymers with high stereoregularity, the crosslink domains are often formed by a hierarchical higher-order structure of polymer chains. The crosslink domains have a semicrystalline structure. Assuming that the crosslink domain is oriented randomly in the gel, the scattering from the domains can be accurately calculated according to Eq. (6) by Debye using a molecular model.

Among experimentally determined scattering functions, the scattering that originates from the domain structure can be obtained from a linear sum of Eqs. (10), (12) or (14), or the direct calculation from a molecular model. However, it is necessary to consider interference when there is correlation among domains. For example, when the domain acts as f-functional structural units and forms a branched structure after the domain formation, the scattering function, Eq. (7), of the FS model can be applied. Here, it is necessary to replace the square of scattering amplitude A_2 with Eqs. (10), (12) or (14) that directly originate from the domains.

When there are strong repulsive forces among domains, the interaction influences the spatial distribution of the domain. Along with the sum of the scattering from the individual scattering of the domain, the interference among domains will be observed. If the domain is assumed to be approximately spherical and undergoes rigid sphere interaction, the observed scattering intensity is given by the product of the scattering function $P(q)$ and the interference term $S(q)$ that originates from the interactions among domains:

$$I(q) \approx P(q)S(q) \tag{16}$$

$P(q)$ is the scattering when there are no repulsive interactions and it is necessary to take into consideration the scattering from each domain and position correlation; $S(q)$ is the interference due to the interaction among domains, and if the interaction is spherically symmetric, it is given by

$$S(q) = \frac{1}{1 - (2\pi)^{2/3}\left(\dfrac{\varepsilon}{v_1}\right)\beta(q)} \tag{17}$$

where ε is a constant of approximately 1, v_1 is the volume fraction of the domain, and $\beta(q)$ is the interaction potential function among the domains in the Fourier space.

When the domains (including interactions) can be approximated by rigid spheres, the following equation is obtained:

$$S(q) = \frac{1}{1 + 8\left(\dfrac{v_0}{v_1}\right)\varepsilon\Phi(2qR)} \tag{18}$$

Here, v_0 is the volume of the sphere that represents the interaction potential, and potential function $\Phi(r)$ is given by the scattering amplitude of the sphere with radius r:

$$\Phi(r) = \frac{\sin qr - qr \cos qr}{3(qr)^3} \tag{19}$$

Equation (18) indicates that the interaction is represented by a homogeneous sphere with radius $2R$. However, R is not necessarily the radius of the domain. More generally, the shape of the domains is not spherical, their size has distribution, and the interaction is not isotropic. The domains themselves are also fluctuating in solutions and gels due to thermal movement. This effect can be expressed by multiplying the Debye–Waller factor $\exp(-\alpha^2 q^2)$ by the potential function of the denominator of Eq. (18), where α can be considered as the range of distance of thermal fluctuation. If the domain distribution is broad and the thermal movement is active, the molecular movement is much faster than the measurement time, and the anisotropic interaction can be regarded as directionally averaged and isotropic. Furthermore, assuming that the domain size and its mode of movement distribution can be expressed by a random process, the mutual potential function $\beta(q)$ is regarded as a Gaussian type [52]. Accordingly,

$$\beta(q) \approx -\exp\left(-\xi^2 q^2\right) \tag{20}$$

where ξ is the correlation length of interaction that is a measure of the interaction distance (the distance where interaction reduces to $1/e$). In this case, Eq. (17) becomes

$$S(q) \approx \frac{1}{1 + 2A_2 M_w c \, \exp\left(-\xi^2 q^2\right)} \tag{21}$$

The constant c is determined to give a proper expression of the first term of the virial expansion with respect to c (concentration), while A_2 is the second virial coefficient.

2.3.3   Application Examples of Analysis

As a practical application example, the gel structure analysis of gellan gum aqueous solution will be discussed here [53]. Gellan gum is a polysaccharide (Fig. 3) with four sugar repeat units. Due to its carboxyl group on the repeat unit, it functions as a polyelectrolyte. Its aqueous solution gels at low temperature in the presence of counter ions. The gelation of gellan gum is attributed to the double helix formation and the association of the double helices (see Fig. 4). From the x-ray fiber diffraction pattern of the stretched gellan gum gel sample, the crystal structure formed from the packing of the gellan gum double helices has been analyzed [54]. The scattering function is calculated by Eq. (6) (the Debye equation) from the molecular models of a single helix, double helix, and association of a double helix (see Fig. 5). In this case, each scattering point is replaced by the sphere with the same van der Waals radius of the constituent atoms of the models.

4-Methyl ammonium salt (TMA) of gellan gum does not gel and turn viscous when cooled. Figure 6 shows the small angle x-ray scattering

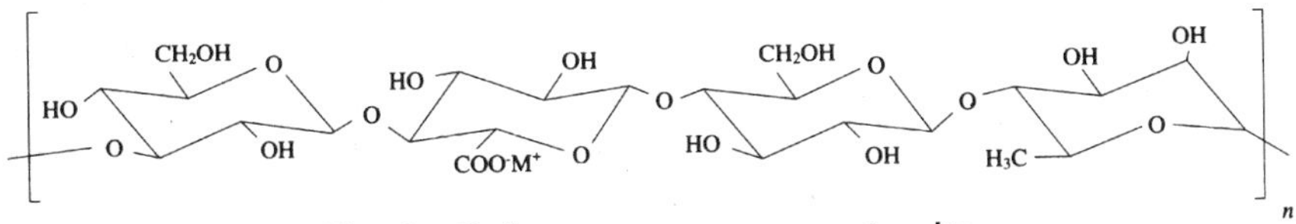

Fig. 3   Gellan gum sugar repeat units.

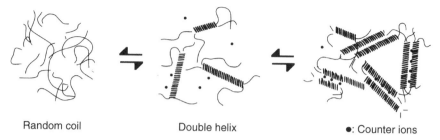

Random coil   Double helix   •: Counter ions

Fig. 4   Schematic diagram of the gelation of gellan gum.

profile of a gellan gum-TMA aqueous solution at 10°C and 60°C. It can be seen that the 10°C results calculated for a double helix model, and the results obtained at 60°C agree well with the calculated scattering profiles for a single helix model. In this calculation, the static repulsive interaction is taken into account using Eq. (20). It is apparent that a double helix is formed at 10°C and the static repulsive interaction intensified. When the counter ion is changed to potassium, the gellan gum gels upon cooling. Figure 7 illustrates the small angle x-ray scattering profile of gellan gum-K aqueous solution at 10°C (gel). The hatched line indicates the calculated

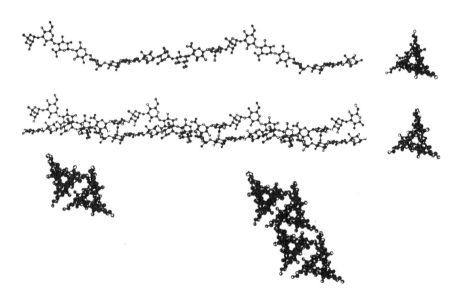

Fig. 5(a)   Molecular models for single helix, double helix, two associated double helices, and four associated double helices of gellan gum.

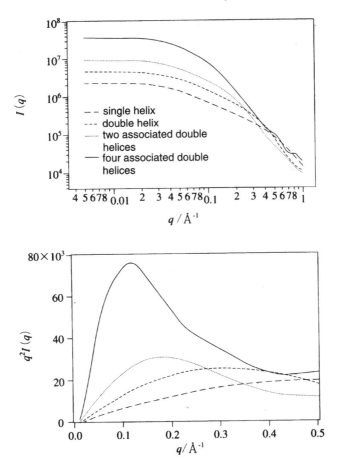

Fig. 5(b)   Scattering function calculated from the corresponding molecular models (shown in log–log plot (upper curves) and Kratky plot [$q^2 I(q)$ vs q] (lower curves)).

scattering profile of two double helices that do not take into consideration the static repulsive interaction. The dotted line is the case where the static repulsive interaction is approximated by Eq. (18). Finally, the solid line shows the results of additional consideration of thermal fluctuation of crosslink domains by incorporating the Debye–Waller factor. The model calculation and the actually measured small angle x-ray scattering profile agree well. Accordingly, the correctness of the schematic diagram of gelation shown in Fig. 4 can be seen.

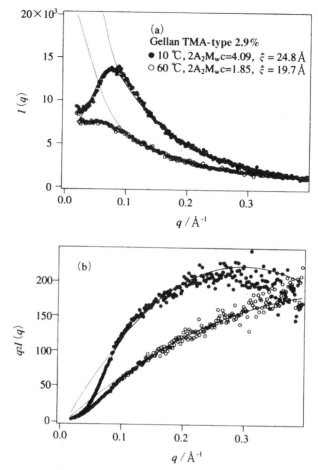

The dotted line indicates the calculated profile where static repulsive interaction is ignored (see Fig. 5) and the solid line shows the calculated profile where the static repulsive interaction is approximated by Eq. (17). The results are in agreement with the double helix structure at 10°C and single helix structure at 60°C

Fig. 6 Small angle x-ray scattering profile of gellan gum-TMA aqueous solution and the calculated scattering profile using a molecular model.

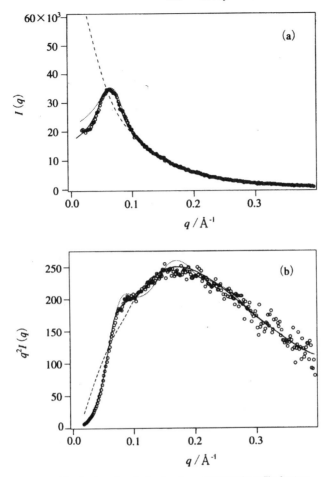

The broken line indicates the calculated profile for two associated double helices where static repulsive interaction is ignored (see Fig. 5) and the dotted line shows the calculated profile where the static repulsive interaction is approximated by Eq. (15). The solid line indicates the case where thermal fluctuation is taken into consideration by the Debye–Waller factor ($a \cong 1\,\text{nm}$).

Fig. 7 Small angle x-ray scattering profile of gellan gum-K aqueous solution and the calculated scattering profile using a molecular model.

2.3.4 Conclusions

Small angle x-ray scattering and small angle neutron scattering are suitable for the analysis of an amorphous structure 0.5–100 nm in size. Small angle x-ray scattering is electron scattering and its intensity is proportional to the square of the atomic number. Therefore, by replacing a certain portion of a gel by a heavy atom, measurement can be made with emphasis on the structure of that specific portion. The x-ray from synchrotron radiation allows time-resolved measurements at 0.5 s intervals. Simultaneous mechanical, thermomechanical, and molecular spectroscopic measurements will allow a dynamic mechanical study of the system being gelled, real-time observation of the structure responding to the external stimuli, and the analysis of a wide range of gel structures. Small angle neutron scattering occurs when the neutron (particle wave) is scattered by nuclei. Because the scattering depends on the neutron that has nuclei and spin number, there is no simple correlation, as in the case of x-rays, between the scattering intensity and atomic number. Fortunately, as ^1H has negative scattering amplitude, and ^2H (D) has positive scattering amplitude, partial labeling and change of contrast (for example, a specific portion of polymer networks is labeled and the structure of this portion is analyzed) can be made without changing the chemical and physical properties of the materials of interest. At the present time, the measurement time of several hours is a weakness of the method.

2.4 MICROSCOPY

2.4.1 Confocal Laser Scanning Microscopy

YOSHITSUGU HIROKAWA

2.4.1.1 Introduction

Various analytical instruments as shown in this section are used for the analysis of gels. Generally speaking, analysis can be made from a chemical and physical point of view. For example, nuclear magnetic resonance spectroscopy (NMR), infrared spectroscopy (IR), and elemental analysis can be used for the chemical analyses of polymers that constitute gel networks, crosslink point, and solvents. On the other hand, detailed analysis of physical structures that are important to elucidate the gel structures, such as the molecular weight between crosslink points and its distribution, the concentration of the crosslink points and their positions, and the number of branches, is difficult at the present time.

Attempts to study gel structures have been made by imaginary space observation of x-ray scattering and real space observation of microscopy. As scattering techniques, light scattering, x-ray scattering, and neutron scattering are used. However, when scattering techniques are used, they usually do not require pretreatment of samples as with electron microscopy, and the *in situ*, real-time observation of the gel is possible. These methods are extremely useful and provide information on the structure on the order of nanometers, which is close to both the molecular and the micrometer level. However, these scattering methods are for observation of structures in the imaginary space and thus they reflect only the average structures of the sample. In contrast, the study by microscopy is the real space observation and, as such, it has the advantage of making it easy to study a concrete structure, though only the local structure may be studied.

For the microscopic real space observation, scanning and transmission electron microscopies are often used. However, due to the nature of the instrument, electron microscopy requires that samples be dried to observe the structure. Therefore, although the image might reflect the structure of the original material, it is not an *in situ* method and thus it is somewhat doubtful whether the real structure is observed. Also, even if the sample is studied by electron microscopy using a cryostage, the sample is frozen and, thus, there is no guarantee that the real structure is being studied. Furthermore, it is difficult to obtain information on the internal structure of gels in a 3D manner.

In recent years, the progress in the development of analytical instruments has been remarkable, thus making it possible now to study what was difficult to observe in the past. For example, using atomic force microscopy, observation of the gel surface has been made possible [55]. This method allows observation of sizes at the atomic and molecular level. Thus further application in this area is desired. Atomic force microscopy is used to study the roughness of a surface by detecting the interatomic force between the detector with a single atom at the tip of the needle and the sample. Accordingly, although this method is fundamentally useful in observing surface structures, it is not suitable for internal structure observation.

To observe the internal structures of gels, ordinary optical microscopy may be considered. However, even if the focus is on a particular location, the images before and after the focal point are observed to overlap due to the depth of focus, making it difficult to observe positional correlation in a 3D manner. Confocal laser scanning microscopy (CLSM) is a kind of optical microscopy that utilizes a special optical system and the cross section of the sample can be observed nondestructively. Hence, it is extremely useful in observing the internal structures of gels [56, 57].

2.4.1.2 Characteristics of CLSM

Confocal laser scanning microscopy (CLSM) has been applied mainly in biology [58]. For example, introducing a fluorescence probe in a cell, a specific portion of the cell is stereoscopically observed. Today, this instrument is used not only in biology but also has begun to see widespread use in the materials research field [59].

Figure 1 is a diagram of the confocal laser scanning microscope (CLSM). This microscope utilizes laser light, and various lasers can be used depending on the objects. For example, a UV laser is used for fluorescence imaging (Ar ion laser at 365 nm) and a visible laser for reflection imaging (Ar ion laser at 488 nm). For a detector, normally a

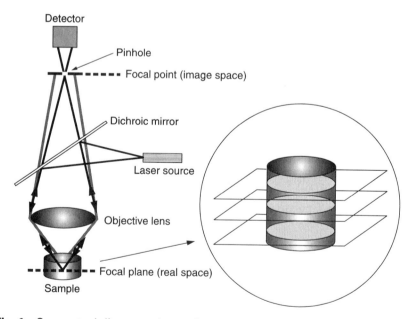

Fig. 1 Conceptual diagram of a confocal laser scanning microscope (CLSM).

photomultiplier or CCD camera is used. Here, a digital signal is recorded and image analysis is possible if necessary.

This instrument is characterized by its optical system. The light from the laser source passes through the dichroic mirror, objective lens, and finally the sample.

The reflected light from the sample again passes through the objective lens and dichroic mirror, and reaches a pinhole. The pinhole is positioned precisely on the image space focal point of the objective lens. As a result, the reflected light from the material space of the sample (sample side) passes through the pinhole and finally reaches the detector. On the other hand, although the light reflected from the nonfocal points also reaches the pinhole, it is blocked by the pinhole that functions also as the spatial filter and cannot reach the detector. Hence, by positioning the pinhole on the image space focal point of the lens and by strongly limiting the focal depth of the lens, only the image that is on the focal point of the image space can be observed as the cross-sectional image of the sample.

Because the observed image is only the image that is on the focal point of the lens, a new cross-sectional image can be observed by changing the distance between the lens and sample. Accordingly, by systematically changing the distance between the lens and sample, a series of cross-sectional images can be observed. This series of cross-sectional images can be reconstructed into a 3D image using a computer, which can also be used to elucidate the structure inside the sample.

The resolution depends on the instrument and the condition of the sample. It is 0.1–0.2 μm at best. The range of observable scale by this method is compared with other methods in Fig. 2.

2.4.1.3   *Observation of polymer gels by CLSM*

Figure 3 indicates various levels of observation from the molecular level to the level of observation by the human eye. When a gel is prepared, the form of the gel that can be observed by the human eye depends on the shape of the container that is used for gelation. On the other hand, if the molecular level observation of the structure is possible, one might be able to see network chains and crosslink points of gels and their spatial position relationship. Unfortunately, this is difficult to achieve at the present time. It is hoped that such observation will be possible by the development of such instruments as the atomic force microscope in the future. Between the macroscale and nanoscale, which is on the order of the molecular

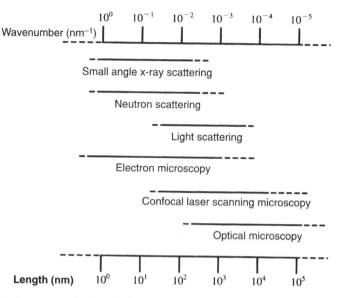

Fig. 2 Various analytical techniques for gel structures and their appropriate structural scales.

level, it is possible to expect a microlevel phase structure due to the concentration variation of the gel network.

Figure 4(a) shows the homogeneous network chains that possess equal molecular weight between crosslink points with no distribution and whose crosslink points spatially distribute homogeneously.

This is the "ideal network" that is often used as a theoretical model. On the other hand, the network shown in Fig. 4(b) is a heterogeneous gel network with different molecular weight between crosslinks having a distribution. The crosslink points spread spatially in a nonuniform manner. Actual gels are believed to take more or less such a structure. Figure 4(c) shows the heterogeneous gel where, similar to Fig. 4(a), the gel has equal molecular weight between crosslink points with no distribution; however, the spatial placement of the crosslink points is inhomogeneous. The ideal networks such as those in Fig. 4(a) show no difference in refractive index within the gel and thus are difficult to study with CLSM. In contrast, the cases of Fig. 4(b) and (c) have different refractive indices in various positions within the gels and thus the reflected images can be observed.

Figure 5 shows the image of the interior of a gel observed using a CLSM. Reflection images were obtained by systematically changing the

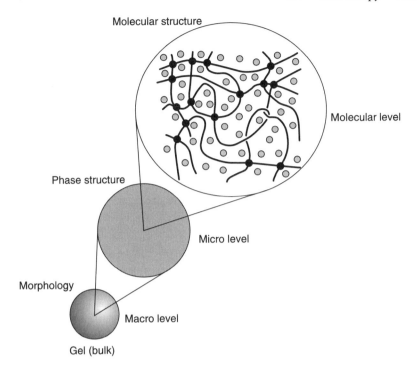

Fig. 3 Structure of gels (there are various structures in gels from macro level to molecular level).

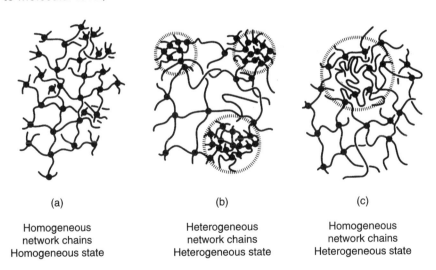

(a)	(b)	(c)
Homogeneous network chains Homogeneous state	Heterogeneous network chains Heterogeneous state	Homogeneous network chains Heterogeneous state

Fig. 4 Schematic diagrams of various gel network chains.

distance between the sample stage of CLSM and the objective lens and the cross-sectional images were observed at every 1 μm. The gel used for this study is a milky-white N-isopropylacrylamide (NIPAAm) gel prepared at 27°C. After the NIPAAm gel is prepared from the NIPAAm monomer and a crosslinking agent, and if it is prepared at a low temperature, it will be colorless and transparent; but at an elevated temperature it will be turbid. The turbid gels will not recover transparency even when the temperature is lowered due possibly to the freezing of the heterogeneity by crosslinking [60, 61].

As shown in Fig. 5(a), the cross-sectional images by reflection show bright domains with a black background. Also, the vertical cross-sectional view shown in Fig. 5(b) and (c) gives similar images. These images are thought to be due to the heterogeneity within the gel. Based on the

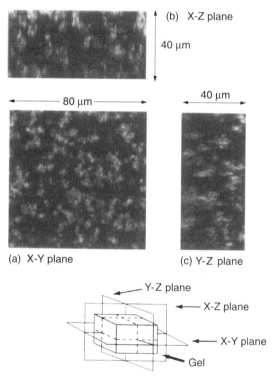

Fig. 5 Images of the internal structure of NIPAAm gel observed with a CLSM.

observation from the fluorescent probe that fluoresce only in a hydro-phobic environment, the white portions in Fig. 5 are assigned to be hydrophobic regions. Hence, the white portions in Fig. 5 can be regarded as the area with a high concentration of gel networks.

Figure 6 depicts a reconstructed 3D computer-made view of a series of observed images. The white portions in Fig. 6 are where the concen-tration of gel network chains is high. By analyzing this image, it can be found that the domains with high and low network chains are forming a 3D co-continuous structure. By analyzing this image further, it is possible to learn the detailed characteristics of the structure.

It is difficult to use CLSM to study the interior structure of a transparent gel that is prepared at a low temperature. However, there are occasions where even in transparent gels the internal structure can be studied if the network chain is made of different polymers. Figure 7

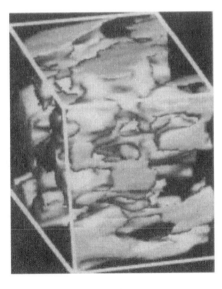

40 μ m   40 μ m

30 μ m

Fig. 6   Reconstructed 3D image of NIPAAm gel from CLSM images.

Fig. 7 Cross-sectional image of the cornea of pig using CLSM (sample provided by Dr. T. Matsuura at Department of Ophthalmology, Nara Medical School, Japan).

illustrates the cross-sectional view of the cornea of a pig as an example of a biological, transparent gel. This cornea is considered to have a lamellar structure made of different polymers whose morphology is clearly shown here. Accordingly, CLSM is useful in observing the internal structure of some gels.

2.4.1.4   Future development
It has been difficult to observe *in situ* the internal structure of gels in a real space, but CLSM has made it possible. The observation of the internal structure of gels by CLSM has just begun. The knowledge of the internal structure of gels obtained by CLSM is believed to become essential along with the information obtained by various scattering methods, NMR, and thermal analysis.

It is hoped that many new insights on the gels will be obtained in the future through the improvement of instrumentation as well as the observation conditions.

2.4.2   Scanning Electron Microscopy (SEM)
MAKOTO SUZUKI

One of the useful methods to study gel structures is scanning electron microscopy (SEM). For ordinary observation in SEM, samples are dried, given electrical conductivity to the surface by evaporation of gold and other materials, and observed in a vacuum (approximately 10^{-3} Pa). However, it is fundamental for a majority of gels to be wet. Methods to study structures under wet conditions will be described in the following sections.

2.4.2.1   Critical point drying method

The structure of biomaterials deforms when dried. This is due to the surface tension of water during drying. In order to avoid this phenomenon, evaporation can be done at a critical point where there is no surface tension observed. Unfortunately, it is not practical because the critical point of water is $374°C$ at 22.06 MPa. In contrast, the critical point of carbon dioxide is $31.4°C$ at 7.375 MPa.

The method involving the use of carbon dioxide was developed by Anderson and the procedure is as follows. First, the sample is immersed in ethanol and dehydrated. After the ethanol is replaced by isoamyl acetate, as this dissolves in both ethanol and liquid carbon dioxide, the sample is placed in a pressure vessel into which liquid carbon dioxide is introduced. Upon increasing the temperature to approximately $40°C$ while the container is closed, the pressure increases to about 12–13 MPa and exceeds the critical point. While maintaining the temperature and leaking carbon dioxide slowly, a dry sample is obtained [62]. By this method, the understanding of the microphysiology of microbes has advanced dramatically [63]. Currently available instruments include the HCPD-2 by Hitachi, the JCPD-5 by JASCO, the CPDO-30 from Balzers, and the 4770 by Parr.

When critical point drying is performed to obtain dry polymer gels, the first treatment of ethanol itself changes the structure of gels. An appropriate combination of solvent that is suitable for critical point drying has not been established and further study is needed.

2.4.2.2   Low-vacuum SEM

Ordinary SEM is used at a pressure less than 10^{-3} Pa. However, samples with high water content, such as biomaterials and polymer gels, largely deviate from the original structure during the drying process under such vacuum. To solve this problem, the SEM that can be used at a low vacuum level has been developed [64]. In this SEM, a pressure differential is maintained by graded evacuation between the sample chamber and the electron beam system. Under this low vacuum, the reflected electron is used to detect and obtain images rather than the ordinary secondary electron from the sample. This is because the secondary electron detector uses a high voltage, and it sparks at a low vacuum level. However, high resolution cannot be expected from this method because the reflected electrons escape from the depth of 0.1–1 µm. In practice, the magnification used is below 10,000×.

Danilatos proposed a method to improve the resolution of low-vacuum SEM [65]. It is the secondary electron detection method that involves placement of an electrode on the sample surface and recovery of the secondary electrons by applying the several hundred volt positive bias. Due to the high voltage, the secondary electrons are accelerated and ionize the gaseous molecules resulting in amplification of the number of electrons. As a result, sufficiently intense secondary electron signals can be obtained. On the other hand, the reflected electron has a similar energy level and almost no amplification will occur because the probability of collision with the residual gaseous molecules is extremely small compared with the secondary electrons. With regard to the resolution, when the electron beam enters the sample chamber, it scatters by colliding with the gaseous molecules thereby increasing the background noise. However, as a spot can be observed relatively clearly even around 2000 Pa, a high-resolution observation is possible under the saturated water vapor pressure at room temperature (2700 Pa). Also, because the ionization of the gaseous molecules generates positive ions, part of which showers the sample, even the samples with high resistivity will not exhibit a charge-up phenomenon resulting in a good quality image. This is a suitable method to observe untreated samples, and an instrument that uses this method is the ESEM from Nikon. As expected, the interior of the sample cannot be observed and, if it is desired, the rapid freeze etching method described in the next section must be used.

2.4.2.3   *Rapid freeze etching method*

If another solvent is used to dehydrate the polymer hydrogel to observe the internal structures, it often causes shrinkage and other structural changes. Thus, if possible, it is desirable when observing internal structures to use water in the sample. One of the methods is the rapid freezing method [66]. Due to the rapid cooling, the formation of microcrystals of ice can be suppressed and amorphous vitrification takes place. The cooling rate necessary for amorphous vitrification varies widely at 3000 K/min for red blood cells, 10 K/min for yeast, and 6 K/min for a colon bacillus. In order to achieve a cooling rate higher than 104 K/min, the sample is dropped into the liquid nitrogen slush and frozen rapidly. When liquid nitrogen is evacuated by a rotary pump, the accelerated evaporation of nitrogen decreases the temperature by the heat of vaporization, resulting in a sorbet-like slush ($-208°$C).

When the sample is dropped into liquid nitrogen under this condition, the formation of gas bubbles is minimized and a good frozen sample is obtained. The rate of cooling is estimated to be greater than 104 K/min [66]. An alternative is to use liquid propane whose thermal conductivity is regarded as being higher and which can be prepared by cooling with liquid nitrogen. Another approach is the method of colliding a sample into a block of copper that is cooled by liquid helium. This method allows even faster cooling, and the depth of amorphous ice formation is on the order of 2 µm. However, the values cited here are only approximations and the actual values must be verified by analyzing the state of ice and the system from various directions.

The sample rapidly frozen in such a manner is transferred into the cryostage that is precooled to approximately $-120°$C to $-130°$C in the vacuum evaporation apparatus and evacuated. The sample is then cut by a knife to expose the cross section. If the temperature of the stage with the freeze-sectioned sample is raised to $-80°$C, the ice will sublime without passing through the liquid state, that is, without causing structural changes by surface tension, the level of ice can be lowered leaving behind the polymer networks. This is called freeze etching and the procedure is shown in Fig. 1.

If etched for a long time, a dry sample can be obtained. Under such deep etching conditions, depending on the residual stress or the way the sample was fixed to the stage, the sample may deform or fracture, and, thus, care is needed. Figure 2 illustrates a cross-sectional photomicrograph of a poly(vinyl alcohol) hydrogel membrane (thickness approximately 50 µm) that was prepared by a repeated freeze drying method. Only the polymer networks are seen due to the deep etching, and it is presumed that the original structure is well maintained. However, an even more reliable method is to use shorter etching time. Because the sample is cooled to $-120°$ to $-130°$C after subliming a very small amount of ice from the cross-sectional surface, the structure of the networks rarely changes.

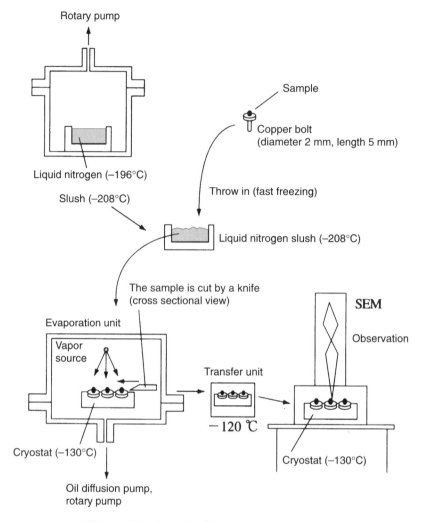

Fig. 1   Fast freezing freeze etching method.

Under this condition, SEM samples are prepared by evaporating platinum or carbon. The photograph taken by this method is shown in Fig. 3(a). The amorphous ice is seen above and below the membrane. In Fig. 3(b), the internal networks of the gel and the amorphous ice that fill the space can be observed clearly [67].

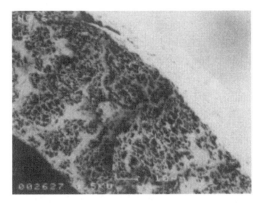

Fig. 2   Cross section of a deep-etched polymer gel.

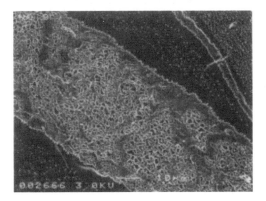

Fig. 3(a)   A cryo SEM image of a properly etched polymer gel cross section.

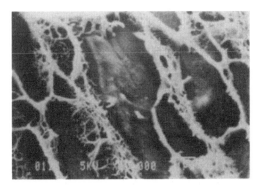

Fig. 3(b)   The amorphous ice between polymer gel networks.

Here, a brief explanation on the coating will be given. Coating is necessary in order to avoid the charge-up that results in defocusing when the electron beam is impinged upon a sample. The easiest method is the evaporation of gold, but the particle size is slightly rougher than desired at about 10–60 nm. However, if evaporation is done on a cryostage (at approximately $-120°C$), fine gold particles can be obtained due to the restriction of gold particle migration on the surface. Gold sputtering produces a more homogeneous coating than evaporation due to the excellent coverage of the hidden surface. On the other hand, it is not suitable for shadowing, which is commonly used for TEM sample preparation. Platinum is considered to form smaller particles than gold. Furthermore, Au-Pt gives a particle size of 3–6 nm, which is an order of magnitude smaller than a gold particle. Because carbon coating is almost transparent to the electron beam, it is used to enhance intensity and contrast when a replica for TEM is prepared. Even for SEM, it is convenient to improve electrical conductivity.

Among evaporation methods, the beam evaporation method has the advantage of avoiding contamination from the beam source because it can coat away from the heated source. It is also suitable for shadowing due to the excellent directionality of the beam. In the cryo SEM, if the magnification is 20,000–30,000×, gold evaporation will suffice. For high-resolution analysis, observation of the replica by transmission electron microscopy (TEM) or in lens-type SEM (example: DS-130 by ABT) is suitable.

For the observation, the SEM stage needs to be also cooled to less than $-120°C$ to maintain the sample structure. This is why it is called cryo-SEM. To ensure observation of the surface, caution must be exercised not to let moisture freeze on the sample when the sample is transferred from the evaporation apparatus to SEM after the evaporation through air. Therefore, in order to prevent the sample from contacting the air, a transfer unit is necessary to maintain low temperature and vacuum for a short transfer period. An example of an experimentally manufactured unit is depicted in Fig. 4. The photograph is the scene when the unit is fitted to the evaporation apparatus and is cooled by liquid nitrogen. After this, the evaporated sample is carried to the SEM maintaining the vacuum and cold temperature. Commercial instruments that can perform cryoetching and evaporation include the JFD7000 from JASCO and the CT 1500 from Oxford.

Fig. 4 A transfer unit attached to a beam evaporation unit.

2.4.3 AFM, STM

KEN NAKAJIMA AND TOSHIO NISHI

Due to such advantages as high resolution that can approach the real atomic and molecular scale, and the ability to perform real-time measurement that cannot be matched by traditional microscopy, scanning tunneling microscopy (STM) and atomic force microscopy (AFM) have attracted considerable attention since their introduction from researchers in various fields. The operational procedure of these microscopes is to position an atomically sharp detector needle to less than several nanometers from the surface of a sample, probe the interaction between the detector needle and the sample, scan the sample surface two-dimensionally, and obtain the surface image (an unprecedented method). If the interaction that is probed is the tunneling of the electron that is well known in quantum mechanics, the technique is called STM (T indicates tunneling). If, on the other hand, atomic force (van der Waals force) is used, it is called AFM.

In recent years, various microscopes that have operational mechanisms similar to STM and AFM, and probe different interactions, have been introduced as members of the scanning probe microscopy (SPM: P indicates probe method employed) family. In particular, friction force microscopy (FFM) is a new method used to probe the frictional force between the sample and detector needle in studying microscopic mechanical information, and many researchers have begun using this method.

In addition to the clear superiority of resolution of STM and AFM in comparison to other types of microscopy, there is another advantage. Unlike electron microscopy that can only be used under vacuum, they do not require a special environment for the observation. This originates from the STM and AFM adaptation of a universally existing interaction in any environments of the probe. This is very important. For example, it is possible to observe biorelated polymers *in vitro* in a saline solution, or various reaction process *in situ* under a special gas environment. Furthermore, the microprocessing on the order of molecules using STM and AFM has been a popular subject recently. At present, techniques to extract or add individual atoms from a clean semiconductor or metal surface are being researched. The manufacture of new devices using these techniques will generate a technological revolution in the near future, and further progress is strongly desired.

Today, these new microscopic techniques are being applied to organic molecules, polymers, and biopolymers throughout the world. However, STM and AFM that have exhibited usefulness on flat surfaces such as semiconductors and metals face many difficulties when used on soft materials and only a few application examples have been reported. In particular, it is safe to say that almost no reports on the application of STM on polymer gels have appeared in technical journals. Here, the discussion will be made with particular emphasis on the technical limitations and their solutions in order to assist those who wish to apply these techniques to polymer gels in the future.

2.4.3.1 Fundamental principles of STM

Scanning tunneling microscopy (STM) was invented in 1982 by Binnig and Rohrer and the 7×7 reconstructed structure of the (111) face of a Si single crystal was observed in real space for the first time [68]. Also, the fact that they received the Nobel Prize in Physics in 1986 for development of STM is extremely well known. As shown in Fig. 1, its fundamental principle is to move a metallic detector needle such as tungsten or platinum-iridium to the sample surface at around 1 nm, apply an appropriate bias V_T between the detector needle and surface, and measure the tunneling electric current, I_T; I_T can be expressed as follows by the semiclassical WKB approximation:

$$I_T \propto \exp(-2\kappa s) \tag{1}$$

where s is the distance between the detector needle and sample surface. Using the local work function ϕ that varies at various locations on the surface, κ can be expressed as follows:

$$\kappa = \frac{2\pi(2m\phi)^{1/2}}{h} \tag{2}$$

where m is the mass of the electron and h is Planck's constant. Assuming that f in Eqs. (1) and (2) is for ordinary metals at several electronvolts, the variation of s by 0.1 nm leads to the change of I_T by an order of magnitude. It is this sensitivity upon which the high resolution of STM is based. If the detector needle position is scanned in the x, y plane using a piezoelectric element, maintaining the I_T to be a constant by controlling the feedback voltage on the z-direction piezoelectric element, and displaying the voltage on the various points on the 2D surface, the surface roughness of the sample on the order of atoms will be reflected. This measurement method is called constant current mode and is the most popular method.

Also, without moving the piezoelectric element in the z-direction, the variable current mode that monitors the tunneling current I_T, which changes as a function of the distance to the sample surface, is often used for special purposes. It requires caution because the change of I_T is also affected by the local work function ϕ as shown in Eq. (2). For example, in the vicinity of the steps or dislocation of a solid surface, the local work function rather than the distance to the sample surface strongly influences the image. Furthermore, on the other atoms or molecules adsorbed onto the sample surface, the situation is even more complex and the interpretation of the image is not straightforward. In such cases, as it is common to observe charge transfer, a detailed theory to accommodate this phenomenon is necessary, but at the present time, no such established theory exists. However, various possibilities have been proposed, including the variation of the work function due to the adsorbent [69], the resonance tunneling effect [70], and the formation of adsorbent potential due to the multibody interactions [71]. In the future, correlation between experiment and theory will be the most important subject in this field.

Although the fundamental principle is important, even more important is the technical problem regarding the STM operation. The examples are infinite and include vibration damping of the measurement system, sound proofing, the elimination of the thermal drift due to the temperature

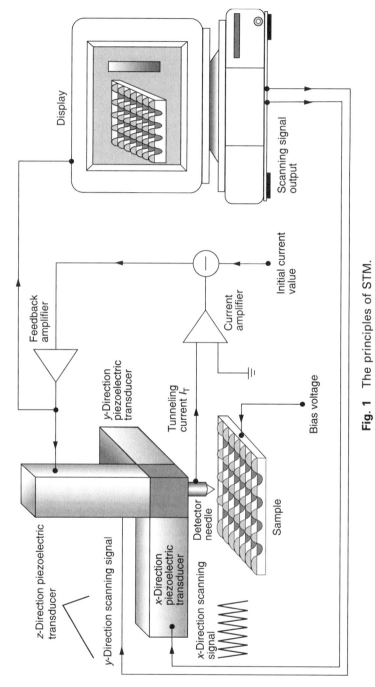

Fig. 1   The principles of STM.

change, the shape of the detector needle, the contamination and oxidation of the sample surface, a detection system for a minute signal, hysteresis of the piezoelectric element, etc. In particular, in many cases, it is impossible to control the shape of the detector needle. It is important to realize that, quite often, we are forced to rely upon luck. In this case, due to the protruded object on the sample surface, the detector needle structure rather than the sample surface itself may appear in the measurement. There are several examples reporting this as a special surface structure of the sample. In order to solve this problem, the FI-STM, an instrument that is a combination of field ion microscope and STM, has been proposed [72]. A superior method has already been established in this instrument; first the atomic organization of the detector needle is controlled by the field ion effect, a single atom needle is produced, and it is used as the detector needle of STM. Unfortunately, however, this method can only be operated under ultrahigh vacuum due to its nature and, thus, has many limitations. At present, specialized commercial instruments that can be operated in air or solution are available. However, without the effort to optimize the system for the sample of interest, it is rarely possible to obtain good images. Due to space limitation, it is not possible to discuss these problems further. However, readers are referred to excellent monographs on these subjects [73–75].

The application to organic molecules and polymers is now recognized as an even more difficult problem. Until several years ago, many systems had been observed by STM, and many researchers were amazed by the broad application ranges of STM. However, since then, many observations have been found to be artifacts. The first reason for difficulties originates from the absolute requirement of electrical conductivity of STM on the sample surface. To solve this problem a thin film is placed on a conductive substrate. Although there are coating methods such as simply coating a film and spin coating, the Langmuir-Blodgett technique is probably the most frequently used method [76]. This is an excellent method because the close packing of the organic molecules naturally avoids the difficult problem of fixing the substrate during the STM observation. At any rate, STM observation seems possible if the film thickness is less than 2–3 nm although there is no theoretical support.

As caution is required for the substrate used, highly oriented pyrolytic graphite is often employed as a substrate. The substrate can provide an easy to clean and smooth surface by peeling and readily gives an image as shown in Fig. 2. Thus it tends to be regarded as the best substrate to study in air and solution. However, as reported in the literature [77], due to the microscopic flakes of HOPG as well as the appearance of the detector needle convolution image, its use requires some caution. The way to separate the real image and the false image is to read many research papers and to patiently repeat the experiments many times. Perhaps the number of researchers who use HOPG as substrate will decrease in the future; however, a recent trend is to use an Au (111) single crystal surface or evaporated Au as the substrate.

Accordingly, there are many items that require caution when STM is used, which may be why the number of researchers who use STM to study organic materials has decreased. In the next section, examples of STM observation of polymer gels in our laboratory will be shown, and it is hoped that many more researchers will make the effort to start using STM again.

2.4.3.2   Application of STM on gels

As described in the previous section, the observation of a polymer gel surface by STM is very difficult. In general, many polymer gels are macroscopical insulators. Furthermore, due to its heterogeneous structures, ordinary measurement will not provide a clear image even if they are electrically conductive. Nonetheless, it is not necessary to prepare gels in

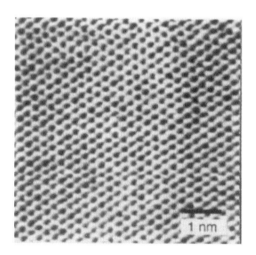

Fig. 2   A STM image of HOPG ($I_T = 9\,nA$, $V_T = 13\,mV$).

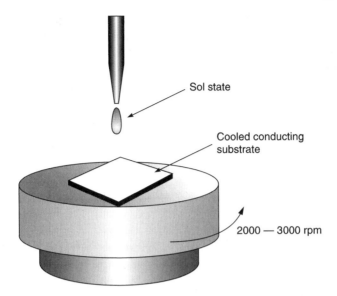

Fig. 3 Preparation method of a gel sample.

an ordinary manner if it is desired to obtain information on the microscopic information of the crosslink region. In this section, the STM
samples are prepared under this approach and the examples will be
introduced where the crosslink structures are visualized on the order of
nanometers [78]. As shown in Fig. 3, in order to prepare the sample, the
sol-state solution that is kept at several degrees above gelation is placed on
the cooled substrate that is spinning rapidly. The samples obtained by this
method, though dependent on the concentration of the solution, are very
thin gel films. The samples used are polysaccharides from a microbe with
the chemical repeat unit structure shown in Fig. 4.

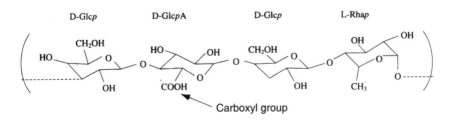

Fig. 4 The structure of a gellan gum chemical repeat unit.

The substrate used is HOPG. As shown in Fig. 5, this polysaccharide is a random coil in solution at elevated temperature, but changes to double helix structure as the temperature is reduced (the transition temperature is approximately 80°C). Interestingly, if the temperature is lowered further to approximately 35°C, the double helices begin associating with each other, leading to a crosslink domain, and eventually gels. If an appropriate cation exists in the solution, the mechanical properties of the gel are strongly influenced. This is probably because of cation interaction with the dissociated carboxyl group in the chemical repeat unit and change in the structure of aggregates. However, there has never been a report in which such microscopic structures were observed directly, and the clarification by STM has been desired. Figure 6 depicts the STM image of potassium-ion-added gellan gum obtained with a scanning range of 60 nm, tunneling current of 0.5 nA, and bias voltage of -1.0V. Several double helices can be seen in the middle of the figure and they are forming a crosslink domain. The pitch of each double helix agrees with the earlier reported value by x-ray diffraction method [79], which supports that the obtained image is real. Furthermore, as a result of the direct observation by STM, it was found that the length of the crosslink domain changes freely as the type of added salt changes. This can be explained knowing that, at the first approximation, the statistic mechanical length of the domain is proportional to the bonding strength among ions. If the length of this crosslink domain has a correlation with the mechanical strength, this observation may have bridged the macroscopic and microscopic properties of the gel.

Accordingly, in the past few years, STM has changed from a simple observation tool to a tool "to investigate" the system. Certainly, such research will increase in the future.

2.4.3.3   Fundamental principles of AFM

In some polymer samples, electrical conductivity is nearly zero. In such a case, atomic force microscopy (AFM) is very useful as it makes use of the van der Waals forces that always exist between the sample surface and probe; thus it is fundamentally applicable to any sample. It also provides more detailed understanding of the surface and interface of polymers due to the different interactions detected by both STM and AFM.

Atomic force microscopy was invented in 1986 by Binnig, Quate and others [80]. It is surprising to see such a remarkable technique developed only a few years after the development of STM. Its principle,

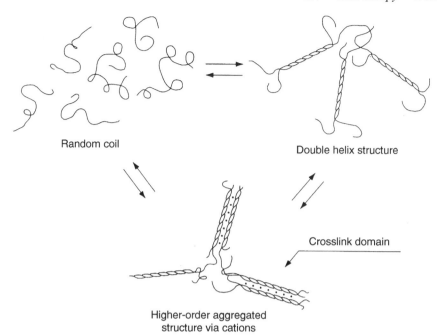

Fig. 5 The crosslinking mechanism of gellan gum.

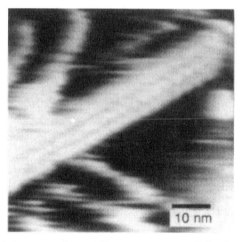

Fig. 6 An STM image of a gellan gum crosslink domain ($I_T = 0.5\,nA$, $V_T = 1.0\,V$).

as shown in Fig. 7, is to scan the surface of an insulating material by a flexible cantilever, detect the minute movement of the lever by a laser or interference of light, and obtain the information on the surface roughness of the sample surface. The progress of AFM in just the past year is remarkable. The ultrahigh-vacuum AFM that is operated in the attractive force region has comparable resolution to STM for semiconductor surface analyses [81]. However, although there are reports of several angstrom resolution for organic crystals, the majority of cases are on the order of several nanometers for organic material applications [82]. Especially in the case of AFM, there are commercial instruments that can provide very stable images of the order of microns. Already in conferences, there have been several interesting reports on polymer gels focused at this level of resolution [83].

However, in this section, AFM will be defined as the technique to measure surface force. We will discuss our study on the viscoelastic properties of polymer surfaces that are especially important for adhesion and friction areas.

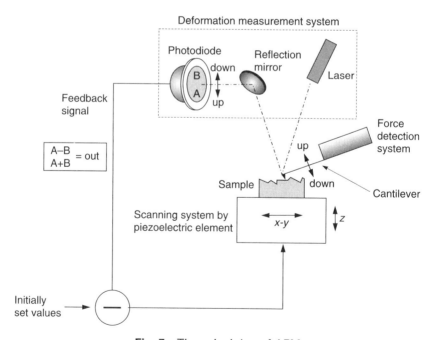

Fig. 7   The principles of AFM.

2.4.3.4 Application of AFM to gels

The AFM first measures the force curve to determine the point of detector needle movement prior to its function as a microscope. This procedure is explained in the following. As shown in Fig. 8, the force curve measurement is similar to the ordinary surface force apparatus (SFA). The difference is that the AFM uses a cantilever that has relatively lower modulus than the sample surface and thus the deformation of the lever is measured. As a reference for later discussion, Fig. 8 shows the ideal force curve that does not accompany the deformation of the sample. First, the sample approaches the very small needle that is fixed on the cantilever. Subsequently, the cantilever jumps from point A in the figure to point B that is near the sample (adhesive force) and it establishes contact with the sample. Afterwards, the cantilever feels a repulsion force and tilts backwards. Normally, the repulsion force measurement by AFM is done in the vicinity of the change from the attractive force to repulsion force. In this case, if the sample does not deform, the force slope in the region (1) gives the modulus of the cantilever itself. At this time, the sample gradually parts from the detector. Then, the cantilever again bends down and jumps backward (point C) after passing through the maximum slope.

As a result, the ordinary force curve will not reflect the mechanical properties of the sample itself. However, if the modulus of the cantilever is larger than the sample surface, the sample can be deformed. The example introduced here is measured in this manner [84]. The AFM used has been specially manufactured with optimized observation of polymers in mind.

The samples used are the polystyrene (PS) and poly(vinyl methyl-ether) (PVME) blends whose mechanical properties can be freely changed by controlling the blend ratio. In this case, in order to make the sample the measurement standard, the molecular weight of the polystyrene used is relatively low to avoid phase separation at room temperature. Ordinarily, the modulus of PS is on the order of GPa and PVME is on the order of MPa; the moduli of blends can vary continuously in the range of these values as a function of the blend ratio. This is why this blend system was chosen as the standard of mechanical properties. The cantilever used is a V-shaped lever with thickness of $0.8\,\mu m$ and length of $100\,\mu m$. At the tip, a very small Si_3N_4 scanning needle is fixed. The modulus of the cantilever is $0.68\,N/m$, which is stiffer than the one for ordinary contact measurement. Two different measurements were conducted. The first is the

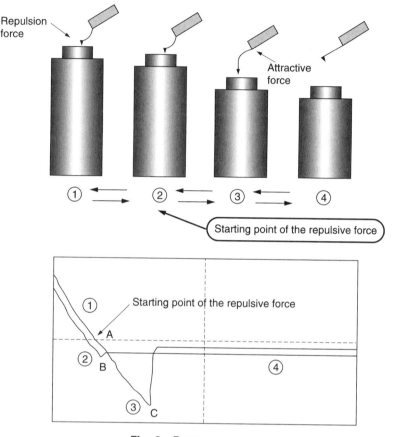

Fig. 8   Force curves.

measurement of a semistatic force curve to determine the elastic properties of the polymer. That is, the rate of change of distance with the sample is extremely slow. Unfortunately, such a measurement cannot be done with the commercial instrument without modification. Here, as shown in Fig. 9, we have succeeded in distinguishing the two forces with different origins (perhaps, the thermodynamic cohesive force between two surfaces, and the capillary force that is caused by the water that exists at the interface or low molecular weight materials). It is also possible to calculate the modulus of the sample through its fitting with the force slope obtained during the contact process. Today, as a standard, a Hertz contact is often assumed for the calculation [85].

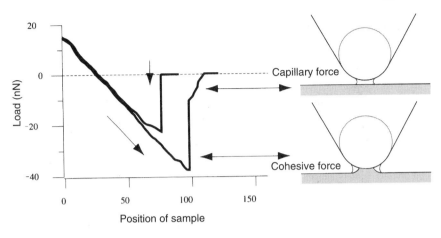

Fig. 9 Force curves of 100% PS sample and the corresponding cohesive force and capillary force.

In the second experiment, the rate dependency of the force curve was investigated in order to highlight the viscous contribution of the polymer blend. As a result, polymer behavior such as time-temperature super-position, which is often observed in such a dynamic method, is clearly observed. Although the sample with 40% PS behaves as a rubbery material, the same blend behaves as a glassy material at a faster stimuli. This behavior is shown in Fig. 10. It is seen that the mechanical behavior changes significantly as the order of movement rate changes. Surprisingly, this behavior is observed on the order of nanometer size. It is probably this research that discovered that the time-temperature superposition principle can be applied to such a microscopic size. The two aforementioned examples clearly demonstrate that AFM is useful to detect viscoelastic properties of polymers. In the future, this might be called nanorheology. It is hoped that many researchers, especially those who are working on polymer gels where mechanical properties are important, make use of AFM from this viewpoint.

2.4.4   Conclusions

Thus far, the application of STM and AFM to polymer gels has been discussed. This field is new and the accumulation of knowledge and technology is scarce. However, it is certain that these techniques, such as STM, AFM, and FFM, can be used to study polymers, in particular, a

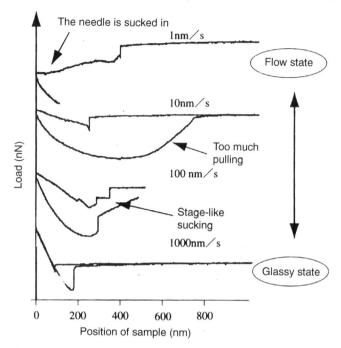

Fig. 10 Behavior similar to that of the temperature-time superposition principle of the rate dependence of the force curves of 40% PS sample.

series of materials (including polymer gels) that are called functional polymers, on the atomic and molecular level in a real space, analyze the microscopic properties, and resolve various problems that are not yet understood today. For this purpose, it is also true that the challenge remains to correlate the results of STM and AFM, which can provide information not obtainable with ordinary methods, with the observation of polymers.

It is also possible to use STM and AFM to manufacture futuristic functional materials through ultramicroscopic processing of polymers. This is because STM and AFM possess the unique capability of causing atomic and molecular level changes through the interaction between the detector needle and the sample and observing it.

By developing instrumentation utilizing these properties actively, ultramicroprocessing, such as scissoring or bonding polymer chains through pulsed current such as controlled tunneling current, atomic force, or frictional force, will become feasible. In the future organic electronics elements may be produced upon successful development of such a capability.

Finally, other than the techniques discussed in the foregoing, the probe to highlight polymer properties includes viscoelastic properties and NMR. These probes are also combined with STM and AFM and integrated instruments are being developed to elucidate the polymer properties. One of the inventors of STM, Rohrer, had proposed the possibility of such a direction several years ago [86].

2.5 NUCLEAR MAGNETIC RESONANCE SPECTROSCOPY (NMR)

2.5.1 Synthetic Polymers

ISAO ANDO, HIROMICHI KUROKO, SHUKEI YASUNAGA, AND MASATOSHI KOBAYASHI

2.5.1.1 Introduction

The recent development of nuclear magnetic resonance spectroscopy (NMR) has been remarkable. It is now essential for physics, polymer science, materials science, organic and inorganic chemistry, pharmaceutical research, and medicine. It is well known that NMR has contributed significantly to the analysis of the structure and dynamics of polymers [87–89]. The NMR studies on the structure and dynamics of polymer gels have begun to show progress.

As in Fig. 1, when classifying NMR into pulsed NMR, solution NMR and solid-state NMR, it is convenient to know which NMR technique should be used for a liquid, solution, solid, liquid crystal or gel sample. For example, using the solution NMR spectrum of a polymer gel, that is, a high-resolution solution NMR spectrum, there is a combination of rapid movement and slow moving parts that prevents signal yield. Therefore, to obtain complete information on the polymer gel, various NMR techniques need to be combined. In this section, examples in which NMR is used for studies on microscopic and macroscopic structures and dynamics will be introduced.

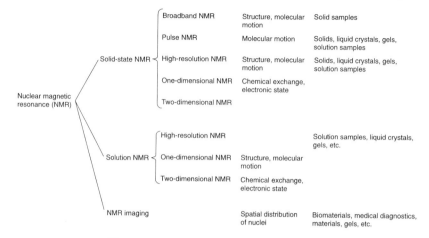

Fig. 1   Classification of NMR methods.

2.5.1.2 Pulsed NMR technique

Pulsed NMR irradiates a high-power pulse onto a sample, and detects its response—free induction decay (FID)—as a function of time. From the rate of damping of FID, relaxation time can be obtained, which allows molecular mobility information to be obtained. In general, pulsed NMR is a broadband NMR technique that cannot provide knowledge on an individual nucleus that has been separated by chemical shift as in high-resolution NMR. However, if the sensitive ^{1}H nucleus is the subject of investigation, the relaxation time of the entire system can be obtained accurately in a short time and when components with differing relaxation times coexist, their composition can be determined.

By combining several pulses and performing pulsed NMR measurements, spin–lattice relaxation time (T_1) and spin–spin relaxation time (T_2) can be obtained. The NMR parameters can provide information on gel solvents and the microscopic molecular movement of polymers. When the relaxation is dominated by dipole–dipole interaction, the correlation time τ_c of molecular motion from the relaxation time is based on the Bloembergen–Purcell–Pound (BPP) theory [90]. Examples of the structure and dynamics of polymer gels are described in what follows.

2.5.1.2.1   Sol-gel transition of poly(N-isopropylacrylamide) solution from the behavior of ^{1}H T_1 and T_2

Poly(N-isopropylacrylamide) (PNIPAAm) solution shows a sol-gel transition around 31°C. The ^{1}H T_2 of pure water increases linearly up to 50°C.

However, the ^{1}H T_2 of the water in a PNIPAAm solution first increases up to 30°C, then rapidly decreases at the gel point (31°C), and subsequently increases again when the temperature is further increased [91]. This indicates that water mobility increases as temperature increases, and then water mobility is severely restricted as the system gelled.

Figure 2 shows the change in molecular mobility for the main and side chains of PNIPAAm molecules at the sol-gel transition of the PNIPAAm gel using ^{1}H T_2 measurements. The T_2 value of the main chain is almost constant in the range of 10–30°C, then suddenly decreases at the gel point (31°C), and resumes constancy in the range of 40–50°C. This indicates that the main chain movement is strongly restricted by gelation. In contrast, the T_2 value of the side chain increases suddenly at the gel point, suggesting that the side chain motion increases drastically by sol-gel transition; T_1 also behaves similarly. As shown in this section, ^{1}H pulsed NMR provides detailed information on the molecular motion of the solvent and polymer main chains and side chains in the polymer gel at the sol-gel transition.

2.5.1.2.2 Phase transition of poly(N-isopropylacrylamide) gel from the behavior of ^{1}H T_2

The PNIPAAm gel made by free radical polymerization of N-isopropylacrylamide (NIPAAm) and a crosslinking agent, methylene bisacrylamide (MBAAm), undergoes volumetric phase transition that expels water by shrinking. As shown schematically in Fig. 3, the gel (gel A) with a swelling ratio of $q = 30$ ($q = M_{swollen}/M_{dry}$: $M_{swollen}$ is the mass of swollen gel and M_{dry} is the mass of dried gel) exhibits mostly surface shrinking and exhibits heterogeneity. On the other hand, the gel with a swelling ratio of $q = 13$ (gel B) shrinks homogeneously. The time variation of ^{1}H T_2 of water in the gel during heating is shown in Fig. 3. The T_2 relaxation time decreases as a single component, indicating overall reduction of water mobility. In comparison, gel A shows two components, long and short T_2 components, after 20 min.

This indicates that there are two components of water with different molecular mobilities. The short T_2 component corresponds to the water whose mobility is restricted in the shrunken phase (skin layer) at the surface, whereas the long T_2 component corresponds to water that possesses good mobility within the interior of the gel [92]. Accordingly, the ^{1}H pulsed NMR technique can determine relaxation time accurately

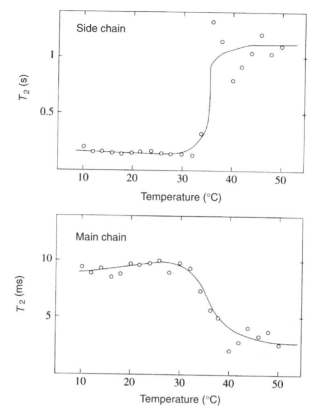

Fig. 2   The temperature dependence of T_2 of the main chain and side chain of PNIPAAm molecules in PNIPAAm deuterium oxide solution (10 wt%).

over a short measurement time, and is therefore suitable for the study of structural changes as a function of time.

2.5.1.3   Measurement of self-diffusion coefficient by pulsed NMR technique

In polymer gels that consist of 3D polymer networks and a solvent, it is essential to analyze the translational mobility of the molecules, including network polymers, solvent, counter ions, and other molecules that exist in the gel, in order to evaluate the fundamental properties of gels and the development mechanism of macroscopic properties. The self-diffusion coefficient (D) is an important physical quantity for the evaluation and

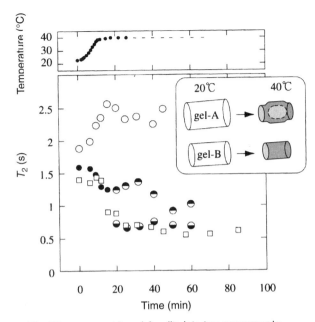

The T2 component in gel A splits into two components
after heating; schematic diagram showing the shape
changes of gels A and B is also shown.

Fig. 3 Time-dependent changes of T_2 of water in gel A (●) and gel B (○)
during heating.

proof of the translational movement of molecules or ions, and it becomes
necessary to determine the D of various parts (network polymers, solvent,
counter ion, etc.) that constitute the gel.

There are various methods to determine D, including the pulsed
NMR method, the tracer method, and chromatography. Readers are
referred to a detailed description, including characteristics, of measure-
ment, in Part 2, Chapter 2, Section 4, Transport and Permeability (Material
Diffusion). The measurement of D is done in what follows here by pulsed
NMR.

The characteristics of pulsed NMR for determining D of gels can be
summarized as follows.

1. The D of each part of the gel, such as the network polymer, solvent,
    counter ions and others, can be measured separately and simulta-
    neously.

2. A broad range of D $(10^{-5}–10^{-10}\,\text{cm}^2\,\text{s}^{-1})$ can be measured under various conditions (temperature, pressure).
3. Synthesis of a special probe material and introduction of the probe into the gel are unnecessary.
4. In the absence of probe introduction, change of state and failure will not occur.
5. The measurement time is in the range of several tens of seconds to several minutes. It is much shorter than in other methods.
6. There are no special requirements for the shape and state of samples. Measurements on opaque samples and microparticles are also possible.

A nucleus with magnetic moment can be, in principle, a subject of measurement for the evaluation of a diffusion coefficient. Various observations are being made and assorted methods are being developed. However, due to its sensitivity and widespread use, the ^{1}H NMR measurement is by far the most popular technique. Regarding measurement techniques, pulse sequence and an application method for magnetic gradients have been devised and many newly developed methods are in use. We will describe the most generally used method, the pulsed-field-gradient spin echo (PGSE) technique, which is useful for determining D in gels. Other methods and application examples can be found in reviews [93–95].

2.5.1.3.1 Principle of PGSE technique

When many spin moments are included in the sample under magnetic field, macroscopic magnetization M is generated. This M is the function, $M(r, t)$, that depends on space and time [96]:

$$\frac{\partial M(r, t)}{\partial t} = \gamma M \times H(r, t) - \frac{(M_x \mathbf{i}' + M_y \mathbf{j}')}{T_2} - \frac{(M_z - M_0)\mathbf{k}'}{T_1} + D\nabla^2 M \quad (1)$$

where γ is the magnetogyric ratio, $H(r, t)$ is the magnetic field, M_x, M_y, and M_z are the x, y and z components of magnetization, M, $\mathbf{i}'$, $\mathbf{j}'$ and $\mathbf{k}'$ are the unit vector in the x, y and z directions, and M_0 is the equilibrium value of M.

In Eq. (1) $D\nabla^2 M$ expresses the contribution of the diffusion on the rate of change of magnetization when the sample is considered to be a macroscopic fluid. According to this equation, it is obvious that the magnetization and diffusion of this nucleus can be correlated. Hence, if spatial and time changes of the magnetization are measured, the diffusion coefficient can be obtained.

In order to obtain T_2, Hahn developed the method called spin echo (SE) [97]. In the spin echo method, the pulse sequence, $\pi/2 - \tau - \pi$, is used. After 2τ from the first π pulse, the phase angles of the FID signals become equal again, and the echo signal of the magnetization is observed. The echo signal can be expressed by the following equation if it is considered as a function of 2τ:

$$A(2\tau) \propto \exp\left(\frac{-2\tau}{T_2}\right) \tag{2}$$

When a large magnetic field is applied, spin–spin relaxation and diffusion contribute to $A(2\tau)$ as is shown in Eq. (3) [98]:

$$A(2\tau) \propto \exp\left[\left(\frac{-2\tau}{T_2}\right) - \frac{2}{3\gamma^2 G^2 D \tau^3}\right] \tag{3}$$

Equation (3) indicates that, when G and T_2 are known, D can be determined in principle if $A(2\tau)$ is measured by varying τ.

In the PGSE method [99], D can be obtained by measuring spin-echo with the $\pi/2 - \tau - \pi$ pulse sequence, adding a time dependent magnetic gradient in a pulsed mode, and measuring the intensity changes of the obtained echo signal.

Figure 4 shows the pulse sequence of PGSE. The first pulse magnetic gradient G is given for time δ at t_1 time after each $\pi/2$ and π pulse. The echo signal intensity, $A(2\tau)$, which is obtained at t_2 time after the second pulse magnetic gradient is given, is provided by the following equation:

$$\ln\left[\frac{A(2\tau)}{A(0)}\right] = -\gamma^2 D \delta^2 \left(\Delta - \frac{1}{3\delta}\right) G^2 \tag{4}$$

where $A(0)$ is the echo signal intensity when magnetic gradient is not given, Δ the interval of the magnetic gradient application, and δ as the application time of the pulse magnetic gradient. The phase of magnetization M is randomized by the first pulse magnetic gradient, and then the π pulse reverses the direction of the phase randomization. The second pulse magnetic gradient unifies the phase. The effect of the first and second pulse magnetic gradient on the phase is the same. However, due to the

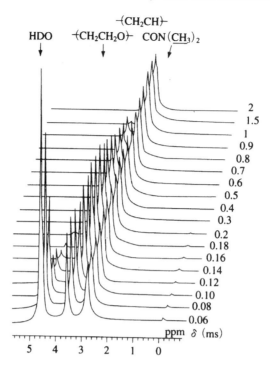

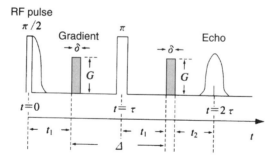

Fig. 4 The ¹H PGSE spectrum obtained from poly(N,N-dimethylacrylamide) gel (swelling ratio $q = 46.3$) to which poly(ethylene glycol) ($M_w = 20,000$) is added by varying the magnetic gradient application time (δ). The pulse sequence of the pulse magnetic gradient spin echo method is also shown.

reversal of the phase, the second pulse helps form the same phase. This indicates that the position of the nuclei in the sample is labeled with phase by the first magnetic gradient. When the nuclei change position through translational movement by diffusion during Δ, the phase will no longer unify even after application of the second magnetic gradient. The echo signal intensity $A(2\tau)$ decreases. Hence, the larger the D, the more prominent the way $A(2\tau)$ decreases, and the degree of reduction is related to D. As shown in Eq. (4) in the experiment, $A(2\tau)$ is measured while Δ, δ, or G is systematically changed and D is determined from the slope.

In a gel system, the movement of the materials being measured is often restricted and D and T_2 will then be small [100]. Thus, Δ and δ must be shortened during the measurement and this necessitates greater G.

2.5.1.3.2   Measurement examples by PGSE technique

Figure 4 depicts the echo signal of poly(N,N-dimethylacrylamide) (PDAAm) gel (swelling ratio, $q = 46.3$) that contains poly(ethylene glycol) (PEG) ($M_w = 20,000$) measured by the ^{1}H PGSE technique while changing δ [101, 102]. By using the ^{1}H PGSE technique, the individual D of the network polymer side chain ($-(C=O)N(CH_3)_2$), within the gel, solvent (HDO), polymer that is incorporated in the gel, and ($-(CH_2CH_2O)_n-$), can be simultaneously and quickly measured. From the plot of $\ln[A(2t)/A(0)]$ against $\gamma^2 g^2 \delta^2 (\Delta - \delta/3)$, the D of HDO and PEG is determined to be $2.02 \times 10^{-5}\,cm^2\,s^{-1}$ and $1.96 \times 10^{-7}\,cm^2\,s^{-1}$, respectively. The amount of D of PEG in the gel is approximately $1/100$ that of heavy water. Readers are referred to a detailed description of a comparative study of the translational movement and molecular weight dependence on the movement of PEG in a solution and gel, which are determined by the PGSE method, in Part 2, Chapter 2, Section 4; Transport and Permeability (Material Diffusion). Because the echo signal of network polymers does not decrease with varied δ, it is thought that their translational motion is small or the center of gravity is undergoing small fluctuation.

2.5.1.4   *High-resolution solid-state NMR technique*

Generally, sharp peaks can be observed and microstructure in solution can be discussed because the local magnetic field and chemical shift aniso-tropy diminishes as a time-averaged quantity due to the fast, isotropic molecular motion. The polymer chains in a gel are much more restricted in comparison to those in solution or a very broad peak, in an extreme case, nothing will be observed due to the dipole–dipole interaction or chemical shift anisotropy if the gel is measured using an ordinary solution NMR method. Generally the molecular motion of the polymer chains in a gel is heterogeneous. The solution NMR technique detects only the signal from the component with high molecular motion. Hence, the crosslink points or the polymer in their vicinity in the 3D networks of the gel will not be detected. Accordingly, due to these reasons, it is important to study the structure and dynamics of gels using the high-resolution, solid-state NMR technique, which is useful in structural analysis of a component having poor molecular mobility.

The methods used for a high-resolution, solid-state NMR technique will be discussed. First, in order to eliminate the strong dipole–dipole interaction, high-powered ^{1}H decoupling will be applied followed by magic angle spinning (MAS) to eliminate the chemical shift anisotropy. Combining this with cross polarization (CP) is the CP/MAS technique, which emphasizes the signal of the component with poor molecular motion where cross-polarization effectively takes place. On the other hand, the technique incorporating the enhancement of the sensitivity of the nuclei by the nuclear Overhauser effect is the pulse saturation transfer (PST)/MAS technique [103], which emphasizes the signal of the component with relatively high molecular mobility.

Generally, molecular motion in the gel state is much greater than in the solid state. Thus, obtaining the NMR spectrum with the PST/MAS technique is suitable. In this case, a spectrum with good S/N ratio can be obtained in a relatively small number of transients. However, the information on the crosslink points and their vicinity often cannot be obtained by the PST/MAS technique. In such a case, the CP/MAS technique is used. If the sample has good mobility as with gels, cross-polarization is inefficient and requires many transients in the CP/MAS technique to obtain a spectrum with good S/N ratio. Therefore, it is necessary to know in advance the optimum condition of cross-polarization. Also, the solvent in the gel leaks out during the fast magic rotation (3–4 kHz) if an ordinary rotor is used. Hence, a rotor with an O-ring is required.

Example of the Study on Structure and Dynamics of Poly(vinyl alcohol) (PVA) Gel by High-Resolution Solid-state NMR Technique
It is thought that hydrogen bonding plays an important role in the formation of crosslinking between polymer chains in the poly(vinyl alcohol) gel prepared by the repeated freeze/melting method. The NMR

provides information on the structure and dynamics on the order of the molecular level. It is very sensitive to the formation and morphology of hydrogen bonding. In this section, the example of crosslink structure analysis of PVA gel using solution ^{13}C NMR and high-resolution solid-state ^{13}C NMR techniques will be discussed. An example elucidating the dynamics of the PVA gel using the relaxation time measurement of the solid-state NMR technique will also be introduced.

As a sample, an atactic PVA with a degree of polymerization of 1700 and a degree of saponification of 99.9% is used. This polymer is dissolved in water, and a gel is formed by freezing it at $-20°$C and melting it at $25°$C 4 times. To avoid evaporation of water during the measurement, a rotor with an O-ring is used.

2.5.1.4.1 Analysis of crosslink structure of PVA gel

Figure 5 shows the ^{13}C NMR spectra of solution, gel and solid-state PVA [104]. In the ^{13}C NMR spectrum (Fig. 5(a)) of a PVA solution, the resonances of methene and methylene carbons show multiplexity due to stereoregularity. The triplet of the methine carbon is assigned to mm, mr, and rr triads from the lower magnetic field side. Here, m and r are meso and racemic dyads, respectively. The signal of the methylene carbon is split due to tetradic stereoregularity. In the ordinary spectrum (Fig. 5(b)) of a gel sample, the splitting of the methine carbon is observed due to stereoregularity. As expected, methine and methylene carbon peaks are both broadened. This is caused by the increased dipole–dipole interaction between ^{1}H and ^{13}C nuclei due to the restricted molecular mobility by the crosslink formation. Although the ^{13}C PST/MS NMR spectrum of PVA gel (Fig. 5(c)) is similar to the one obtained by the solution NMR technique, resolution is improved by the MAS that suppressed the broadening of the dipole–dipole interaction. The ^{13}C CP/MAS NMR spectrum of PVA gel (Fig. 5(d)) shows a broad peak that does not appear in the PST/MAS spectrum.

As this peak appears only with the SP/MAS technique, mobility is lower than the splitting peaks due to stereoregularity. Also, the chemical shift of this peak corresponds to peak II of the methine carbon in the ^{13}C CP/MAS NMR spectrum of solid PVA. In comparison to the difference in three chemical shifts of the methine carbon of 1.3 ppm, the three chemical shift value difference in solid PVA is as large as 6 ppm. Also, the relative intensity of the three peaks does not agree between solution and solid-state spectra. This is considered to be due to the low field shift by the

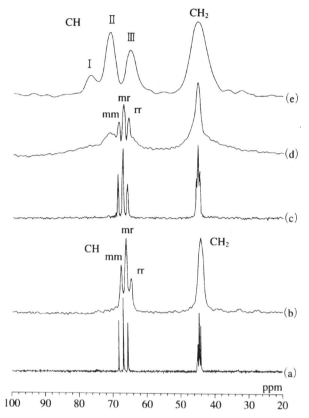

(a) A solution NMR spectrum of a PVA deuterium oxide solution
(b) A solution NMR spectrum of a PVA gel
(c) A PST/MAS spectrum of a PVA gel
(d) A CP/MAS spectrum of a PVA gel
(e) A CP/MAS spectrum of a solid PVA

Fig. 5   The ^{13}C NMR spectra of solution, gel and solid-state PVA.

intramolecular hydrogen bonding between the neighboring hydroxyl groups in the mesoconfiguration of the solid-state PVA. Peaks I, II, and III are assigned to the methine carbon that is forming two intramolecular hydrogen bonds with neighboring hydroxyl groups, the methene that is forming one intramolecular hydrogen bond, and the methene that has no intramolecular hydrogen bonds, respectively [105]. Based on this assignment of solid PVA, it is possible to analyze the crosslink structure of the gel.

Figure 6 depicts the ^{13}C CP/MAS NMR spectra of PVA gels with various water content [104]. As the water content decreases, the splitting of the methene carbon due to the stereoregularity decreases and the peak intensity of the I, II, and III peaks that appear in the solid state increases. In sample (d), the splitting peaks due to stereoregularity are completely diminished. In the gel, crosslinks are formed by intermolecular hydrogen bonding between different molecules. Therefore, peaks I, II, and III appearing in the ^{13}C CP/MAS NMR spectrum include not only the intramolecular but also intermolecular hydrogen bonds with two, one, and no connections. Hence, these peaks can also be assigned to the crosslink region of the PVA gel. The split peaks of the stereoregularity are assigned to the free polymer chains of the gel. From the spectral changes in Fig. 6, it has been found that the PVA gel with high water content mainly has the methene carbon with one hydrogen bond. On the other hand, the gel with low water content has the methene carbon with two hydrogen bonds in the crosslink region.

Accordingly, the crosslink structure of polymer gels has been identified using the high-resolution solid-state NMR.

2.5.1.4.2 Analysis of dynamics of PVA gel

From the measurement of ^{13}C spin-lattice relaxation time (T_1) measurement using the high-resolution solid-state NMR technique, analysis of PVA gel dynamics has been attempted [106]. Figure 11 shows that for the ^{13}C T_1, measurement by the pulse sequence of the PST/MAS technique combined with reverse recovery technique, recovered the three mm, mr and rr peaks with increased τ. From this, it is found that the free polymer chains possess almost the same T_1 regardless of the stereoregularity and its value is 0.3 s. The method that increases the magnetization intensity with the CP technique used is suitable for T_1 measurement of the portion with poor molecular mobility. As such a portion that has long T_1, the reverse recovery method that requires the intensity value of the magnetization at the equilibrium state requires an extremely long measurement time. Hence, the pulse sequence proposed by Torchia is used [107]. Figure 7 gives spectral changes for T_1 measurement using the Torchia method. With the increased τ, the mm, mr and rr peaks decreased at $\tau = 1$ s. Peaks II and III still remain even at $\tau = 5$ s. Upon analysis, the T_1 of peaks II and III are found to be approximately 7–8 s. As a result, it is found that the polymer chains that form the crosslink region of PVA gel are highly restricted in molecular motion compared with the free polymer chains.

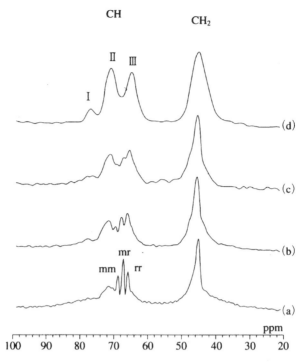

(a) A spectrum of gel a, which was obtained by the repeated freeze-thawing method.
(b) – (d) Spectra of gels, b, c, and d, which were prepared by evaporating moisture from gel a.

The polymer concentration of gels a, b, c, and d are 9.1, 11.8, 13.8, and 35.0 (wt/wt%) respectively.

Fig. 6   The ^{13}C CP/MAS NMR spectra of PVA gels.

Consequently, from the measurement of relaxation time with high-resolution solid-state NMR technique, both the relaxation time of each peak that reflects the difference in fine structure of the gel, and detailed knowledge of the dynamics of the gel are obtained.

2.5.1.5   NMR imaging

To investigate stimuli responsivity and evaluation of total properties as materials, it is essential to perform microscopic as well as macroscopic analyses. For a study of gel deformation that responds to external stimuli and the response of a gel that undergoes soft-hard changes a combined

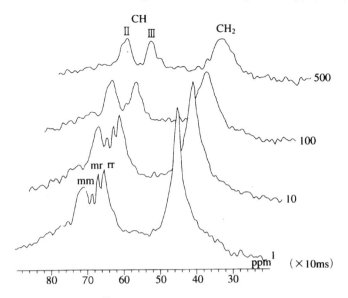

Fig. 7 Changes in the ^{13}C NMR spectra of a PVA gel using the Torchia method.

piece-by-piece and total approach is necessary. Imaging analysis is quite useful for this because NMR imaging is a technique that combines imaging analysis with the NMR technique, which is useful for the study of microstructure and molecular motion.

The characteristics of NMR imaging for polymer gels follow:

1. Information concerning the structure and motion of the sample (physical information) and information on the chemical environment (chemical information) can be obtained with respect to spatial position.
2. Stimuli response of gels and propagation of reaction can be macroscopically analyzed three-dimensionally. The sample can be analyzed spatially from any angle at any cross section.
3. As a special probe is not required to be inserted in the sample and damage is not inflicted on the sample, the sample can be analyzed as is.
4. A time-dependent process can be followed by a time-resolved analysis.

5. It is possible to selectively observe a portion of the gel (solvent, network polymer, counter ions, and other molecules in the gel) by the selection of the measuring nucleus and the development of a new pulse sequence.

On the other hand, there are shortcomings to NMR imaging: (1) measurements require several tens of seconds; and (2) spatial resolution is greater than $10\,\mu m$. For these, instrumentation and measurement techniques are actively being improved [108, 109].

There are many currently available NMR imaging techniques. We will discuss the combined technique of spin echo [110] and 2D Fourier transformation [111] techniques, which are used in both commercial instruments and gel analysis.

2.5.1.5.1   Principles of NMR imaging

In comparison to the homogeneous magnetic field of ordinary NMR, a sample is placed in the linearly graded magnetic field with respect to the position coordinate. In such a manner, a particular nucleus at a certain position experiences a specific static magnetic field unique to this position. Then the nuclear spin exhibits magnetic resonance at different frequencies depending on position, and thus the nuclear spin can be recorded as the resonance frequency. Furthermore, if a magnetic gradient is applied from a different coordinate direction, the position information in this axis can be recorded as the resonance frequency and phase difference. This process follows: (1) selective irradiation; (2) frequency encoding; and (3) phase encoding. Ordinarily, the selection of cross section is done by (1) selective irradiation, and the positional information on the X coordinate is recorded by (2) frequency encoding; positional information on the Y coordinate is recorded by (3) phase encoding.

If a linear magnetic gradient G_Y is applied for the time t_y in the Y axis direction, and linear magnetic gradient G_X is applied for the time t_X, followed by the observation of echo signal as a function of time, then at

$$\xi = \gamma G_X X t_X \tag{5}$$

$$\eta = \gamma G_Y Y t_Y \tag{6}$$

the value of 2D Fourier transform $F(\xi, \eta)$ on a linear line $\eta = \gamma G_Y Y t_Y$ is obtained as

$$s(t_X) = F(\gamma G_X X t_X, \gamma G_Y Y t_Y) \tag{7}$$

Furthermore, the magnetic field intensity in the Y direction and application time are systematically varied and repeated in the same manner. By determining the 2D Fourier transform on every point on the XY coordinates, and performing inverse Fourier transformations, the 2D image is obtained as a cross section on the Z axis. The spin warp technique [112], which is often seen in newer commercial instruments, maintains t_Y as the constant, systematically varies G_Y and applies phase encoding.

2.5.1.5.2 Measurement of polymer gels by NMR imaging

In gels in which 3D spread networks contain solvent, the mobility of the network polymers is greater than in the solid, and the network polymers and solvent individually become subjects of NMR imaging. However, the mobility of the network polymers in gels is restricted by the crosslinks compared with the corresponding linear polymer solution, and the line-width of the NMR signal broadens due to dipole–dipole interaction and chemical shift anisotropy. As a result, in the poly(methacrylic acid) (PMAA), which has a swelling ratio ($q =$ [the swollen mass of the gel]/[the dry mass of the gel]) of 5, the signal of the network polymers of the gel cannot be obtained with the solution NMR technique [113] and obtaining the image will be very difficult. Accordingly, in order to image the network polymers, it is necessary to modify the techniques used for solid-state NMR, such as increased decoupling power and magnetic field gradient, devising of the pulse sequence, and adoption of magic angle spinning [114]. While the motion of the solvent (water) in gels is also restricted, its ^{1}H T_2 is 20 ms at 27°C [115] and can be measured with a commercial NMR imaging instrument.

For example, the deformation generated by applying external stress to a polymer gel can be visualized by ^{1}H NMR imaging [116]. In this study, the signal from water is observed. For the measurement, the signal is recorded by employing Hahn's spin echo pulse sequence [110] and the image is two-dimensionalized by a 2D Fourier transformation [111]. The sample is placed in the static ^{1}H magnetic field of 270 MHz under the magnetic gradient of 2×10^{-2} T m^{-1}, and G_Y is applied by varying the intensity systematically based on the spin warp technique [112]. The application time for both π and $\pi/2$ RF is 1.5 µs, and the intensity of the π pulse is twice that of the $\pi/2$. The echo time is 36 ms, the time for phase encoding is 4 ms, and repetition time is 2–15 s.

When the PMAA gel with $q = 22$ is measured by NMR imaging without an applied stress, the image shown in Fig. 8 is obtained. The color

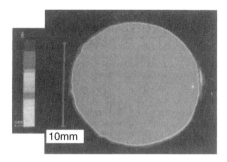

The cross section is 1.9 mm deep from the gel surface; the color is displayed using 256 steps of relative intensity ranging from purple to white.

Fig. 8 A ^{1}H spin density image of a PMAA gel.

scale indicates the relative proton density and the density increases as the color becomes brighter. From this figure, it can be seen that water is homogeneously included in this gel.

The image of the cross section normal to the stress, when the stress of 4.8 kPa is applied to a portion of this gel, is given in Fig. 9(a) and (b). By applying pressure, the spin density of the proton yields a gradient. In the data of Fig. 9(b), if the height of spin density (I) is taken as the Z axis and depicts the cross-sectional image as a 3D histogram, Fig. 10 is obtained. The histograms of XZ and YZ cross sections are also shown in the same figure. The signal intensity of the portion with applied stress is the weakest and as the position is moved from this stress center, intensity increases. Hence, from the fact that the spin density of water at the stressed portion decreases and a density gradient appears, it can be considered that water moves towards the rim and the inhomogeneity of the network polymer density appears. There is no such response with water and linear PMAA aqueous solution. The stress-inhomogeneity appearance is a unique phenomenon in gels where polymer chains are connected through crosslinking.

If inhomogeneity appears upon application of stress, the mobility of water is also expected to be influenced. In fact, according to the results of ^{1}H T_2 measurement using the Carr–Purcell–Meiboom–Gill technique (CPMG), when stress is applied to a PMAA gel, the ^{1}H T_2 is found to decrease as the stress increases [117, 118]. Thus, in order to investigate

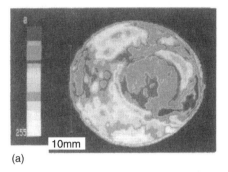

(a)

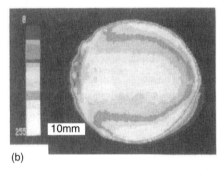

(b)

The cross section is perpendicular to the
applied stress; cross section depth is
(a) 1.0 mm and (b) 1.9 mm from the surface

Fig. 9 The ^{1}H spin density image of a PMMA gel on which 4.8 kPa stress is
applied.

the distribution of the ^{1}H T_2 of water in the gel under stress, the cross
section parallel to the application direction of the stress is imaged by NMR
and the obtained images are shown in Fig. 11(a)–(b). The density of water
having longer ^{1}H T_2 is higher when the color scale is brighter. Hence, the
mobility of water in the white portion is the highest. From the figure, it can
be seen that a gradient in the ^{1}H T_2 of water in the gel is generated and the
slope increases as the stress increases. It has been found that the motion of
water in gels is restricted and a distribution of the mobility of water
appears in the gel. Accordingly, the distribution and change of the density
and mobility of the solvent and network polymers under the external
stimuli can be observed by NMR imaging. Thus, new information that
could not be obtained in the past is becoming available.

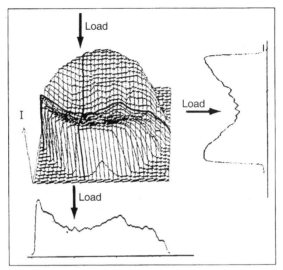

The arrow indicates the direction of applied stress. In the figure, the density distribution of the cross section that is cut along the thick solid line is also shown.

Fig. 10 A 2D histogram of ^1H spin density from the image analysis of Fig. 15(b).

2.5.1.5.3 Analysis of shrinking behavior of homogeneous polymer gels when direct current is applied

The crosslinked polymer gels that are swollen by water deform by applying a voltage.

This is because gels possess an ability to convert chemical, electrical and thermal energies into mechanical energy. Tanaka *et al.* reported the deformation of gels by voltage for the first time [119]. Recently, many phenomena using this deformation process have been studied and many works have been reported on this function. However, fundamental studies are reported only infrequently. Therefore, it is important to clarify macroscopic structure and mobility through microscopic information on the molecular level.

In order to elucidate the shrinking behavior by applying direct current, Shibuya *et al.* [120] obtained a spatial distribution image of ^1H spin density and ^1H T_2 of the water in a PMAA gel. The PMAA gel ($q = 40$) is analyzed using a rod with a diameter of $8.0\,mm$ and a length of $25.0\,mm$. The PMAA gel shrinks by expelling water as a voltage is applied. The shrinkage near the cathode is greater than it is near the anode.

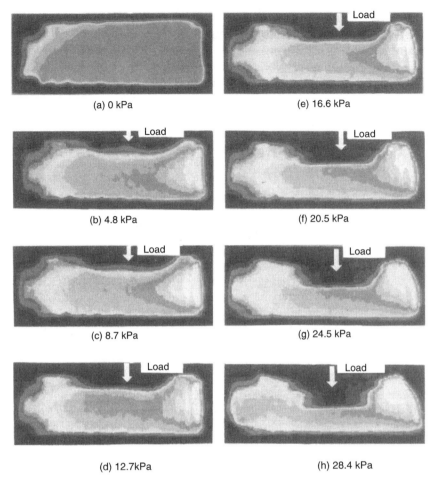

(a) 0 kPa

(e) 16.6 kPa

(b) 4.8 kPa

(f) 20.5 kPa

(c) 8.7 kPa

(g) 24.5 kPa

(d) 12.7kPa

(h) 28.4 kPa

The image is the cross section perpendicular to the applied stress that passes through the rod-like gel. As the image changes from (a) to (h), the stress is increased from 0 to 28.4 kPa.

Fig. 11 ^{1}H T_2 enhanced image of a PMMA gel with an applied stress.

As shown in Fig. 12, as the voltage is applied it shrinks as though it were a cork in a wine bottle.

(a) ^{1}H Spin Density Image In the ^{1}H NMR measurement, the signal of the PMAA gel cannot be observed due to the broadening effect of large dipole–dipole interaction if an ordinary imaging technique is used and only the water in the PMAA gel is observed due to the elimination of broadening by rapid molecular motion.

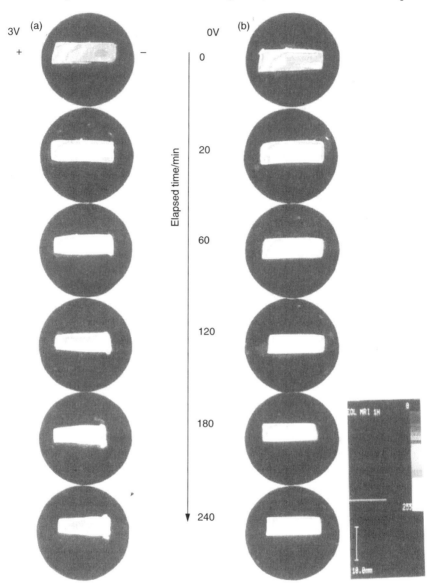

(a) Gel with applied voltage (b) Gel with no voltage

Fig. 12 ¹H spin density distribution images of the time-dependent changes of a PMMA gel.

The shrinking behavior can be highlighted by comparing the ^{1}H spin density image of the PMAA gel without voltage and with $3\,$V direct current. As shown in Fig. 12, for PMAA gels both with and without voltage, the dark region is the largest when elapsed time of $0\,$min ($T_e = 0\,$min), and the color is widely distributed from dark to white. Thus, ^{1}H spin density distribution can be seen as inhomogeneous.

In the case of the gel without voltage, the entire image becomes yellow, and the distribution narrows from light yellow to white. After this, there are no major changes until $T_e = 240\,$min. The difference between the maximum and minimum ^{1}H spin density narrows. It can be seen as homogeneous.

In the case of a PMAA gel with applied voltage, the image becomes yellow in the same way as in the PMAA gel, without voltage at $T_e = 20\,$min. At $T_e = 60\,$min, the shape of the gel changes. As the time changes from $T_e = 120$ to $T_e = 180\,$min, the red region gradually increases and the difference from that of the gel without voltage increases. As the color distribution narrows from dark yellow to white, color distribution becomes narrower than in the PMAA gel without voltage.

Accordingly, the following two point becomes apparent. First of all, the ^{1}H spin density distribution of the initially inhomogenous PMAA gel with voltage and the PMAA gel without voltage become homogeneous as a function of elapsed time. Second, the degree of homogeneity is larger for the PMAA gel with voltage than for the one without.

Figure 13 illustrates the 3D histogram of the ^{1}H spin density distribution. From the analysis, it is found that ^{1}H spin density distribution can be divided into four regions. The reduction of the ^{1}H spin density means the reduction of the number of water molecules in that region. If it is numbered 1, 2, 3, and 4 from the cathode, the first region is in the vicinity of the cathode and shows the highest spin density. The second region occupies the majority of the gel and possesses almost the same spin density. The third region shows the lowest spin density and finally the fourth region has the second highest spin density near the anode. This result indicates that the overall ^{1}H spin density homogenizes by applying voltage, although there is a characteristic spin density distribution upon detailed inspection. This distribution is the same even after a long application time.

(b) ^{1}H Spin–spin Relaxation Time (T_2) Enhanced Image   It is possible to obtain information on the mobility of water molecules in a PMAA gel.

3V (a)   0V (b)

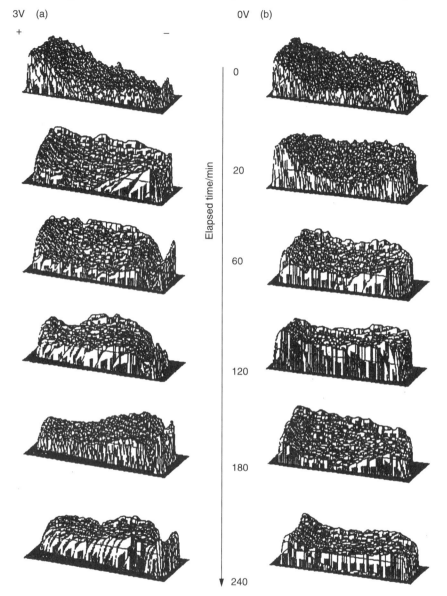

(a) Gel with applied voltage (b) Gel with no voltage

Fig. 13 A 3D ^{1}H spin density distribution histogram of the time-dependent changes of a PMMA gel.

Figure 14 shows the result of the ^{1}H T_2 measurement of a PMAA gel. In comparison to the almost constant ^{1}H T_2 distribution for a gel without voltage, a large distribution change is observed for the gel with voltage. Moreover, the ^{1}H T_2 near the cathode decreased markedly in comparison with the one near the anode. This indicates that the mobility of water near the cathode is restricted. Also, the distribution of ^{1}H T_2 can be divided into four regions in the same manner as the ^{1}H spin density. It is seen that the molecular motion is restricted in the region of high shrinkage. However, only the third region shows high molecular mobility despite relatively large shrinkage. In such a way, detailed information on the spatial distribution of the mobility of water in gels can be obtained by imaging of the ^{1}H T_2 distribution.

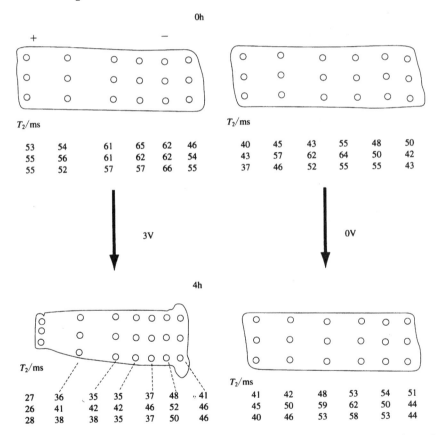

Fig. 14 The ^{1}H T_2 distribution of a PMMA gel with applied voltage and a gel with no voltage applied for 0 and 4 h.

2.5.1.5.4   Analysis of shrinkage of double-layered structured gel by voltage application

(a) Shrinkage Behavior of Double-Layered PMAA Gel with Internal Swelling Ratio of q = 26 and External Swelling Ratio of q = 28.   As shown in Fig. 15, a double-layered rod-like PMAA gel with internal swelling ratio of $q = 26$ and external swelling ratio of $q = 28$ is prepared [121]. By ^{1}H NMR imaging, the changes accompanying shrinkage can be studied in detail and comparison with a homogeneous gel is of interest.

For ^{1}H spin density distribution, Fig. 16 shows a voltage applied to a PMAA gel having an internal swelling ratio of 26 and an external swelling ratio of 28. From these figures, the gel can be seen to shrink as it repels water. As the rate of shrinkage of the internal gel with smaller swelling ratio is faster than that of the external gel, the internal gel shrank more at $T_e = 360$ min. The shrinkage is greater in the anode than in the cathode. From this, it can be found that the double-layered structure as a whole shows quite different shrinkage behavior, although the outside material exhibits similar behavior to the homogeneous gels.

Information on the mobility of water in the double-layered PMAA gel under applied voltage can be obtained by measuring a [1]H T_2 enhanced image. The variation of a [1]H T_2 enhanced image upon application of voltage is shown in Fig. 17. The intensity of the [1]H T_2 enhanced images of the inside and outside gels is approximately the same, meaning that the mobility of water in these regions is about the same.

However, upon application of a voltage, the intensity of the [1]H T_2 enhanced image increases more in the vicinity of the anode than of the cathode, indicating that the mobility of the water near the anode is greater than near the cathode. This trend is the same as for homogeneous gels.

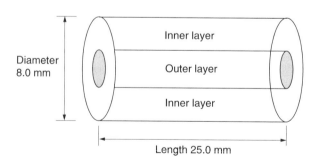

Fig. 15   Double-layered PMMA gel.

Outside   Inside   Outside

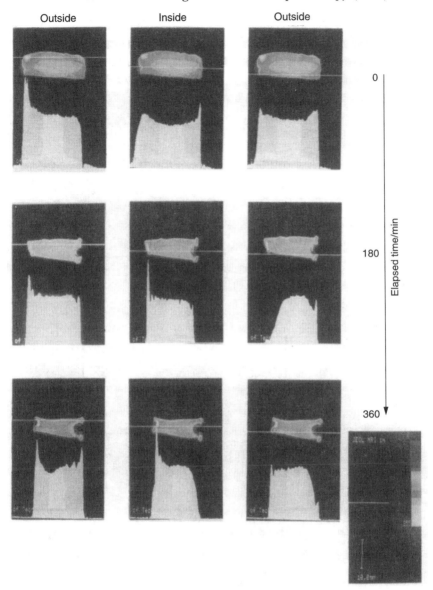

0

Elapsed time/min

180

360

Fig. 16   The ¹H spin density distribution images and intensity profile of double-layered PMMA gel (inner wall swelling ratio 26 and outer wall swelling ratio 28).

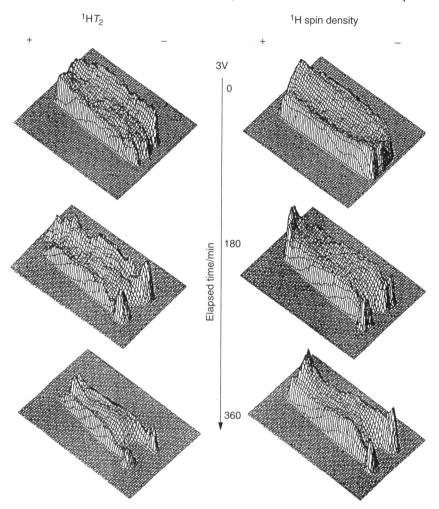

Fig. 17 The ^{1}H spin density distribution and 3D ^{1}H T_2 enhanced histogram of a double-layered PMMA gel (inner layer swelling ratio 26, outer layer swelling ratio 28).

(b) Shrinkage Behavior of Double-Layered PMAA Gel with Internal Swelling Ratio $q = 28$ and External Swelling Ratio $q = 26$ A double-layered PMAA gel that has the opposite degree of swelling (inside 28 and outside 26) from the previous section is prepared and the ^{1}H NMR image

^{1}H T_2 enhancement ^{1}H spin density

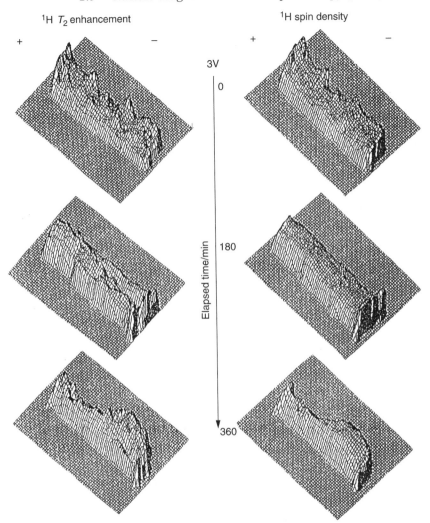

Fig. 18   The ^{1}H spin density distribution and 3D ^{1}H T_2 enhanced histogram of a double-layered PMMA gel (inner layer swelling ratio 28, outer layer swelling ratio 26).

is obtained upon application of dc voltage. Figure 18 depicts ^{1}H spin density distribution. It can be seen that the shrinking behavior upon application of voltage is completely different from the previous sample. The ^{1}H spin density distribution at the cathode and anode increases and

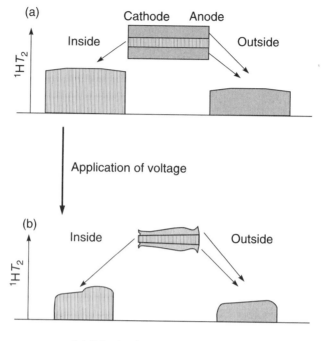

(a) Prior to the application of voltage
(b) After the application of voltage

Fig. 19 Schematic diagrams of ^{1}H T_2 distribution of a double-layered PMMA gel (inner layer swelling ratio 28, outer layer swelling ratio 26) upon application of voltage.

the shape becomes like the "cork of a wine bottle," which agrees with the behavior of homogeneous gels.

The ^{1}H T_2 enhanced image upon application of voltage is shown in Fig. 19. Prior to the application of voltage, almost the entire region shows similar T_2 and the mobility of the water molecules is approximately the same. After the voltage is applied for a long time, the ^{1}H T_2 of the inside and outside layers increases near the anode, reflecting the higher mobility of the water molecule in the vicinity of the anode. This higher mobility of water molecules nearer the anode than the cathode is the same behavior as that of homogeneous gels, indicating that the shrinking behavior of the double-layered gel and homogeneous gel are the same.

2.5.2 Natural Polymers

HAJIME SAITO

2.5.2.1 Introduction

Although gels have shapes and are solid-like in appearance, they are heterogeneous materials containing high mobility portions that are well swollen with a diluent. Natural polymers like polysaccharides, nucleic acids, and proteins often form gels under physiological or nonphysiological conditions via molecular association. In fact, many gelation processes, such as formation of muscle filament and coagulation of blood, play an important role in the body. There are also gels that relate to serious illnesses, involving denaturation of protein; in particular there are gels that form due to mutated protein formation. Moreover, attempts to improve the texture of food with gels go back many centuries. Thus, it is important to elucidate the mechanism of gel formation for the various natural polymer gels without using crosslinking agents from the viewpoints of acceleration or restriction of gel formation. Unfortunately, as can be seen in Clark and Ross-Murphy's work [122], the study of such molecular mechanisms is still insufficient.

2.5.2.2 Nuclear magnetic resonance (NMR) parameters

The NMR technique is suitable to study heterogeneous materials like gels (as is seen in its remarkable development in the medical field, including magnetic resonance spectroscopy (MRS) and magnetic resonance imaging (MRI)). There is no problem regarding scattering of light because NMR utilizes wavelengths much longer than the sizes of heterogeneous gels. Moreover, as listed in Table 1, information on molecular chain conformation and dynamics that differ in motion can be readily obtained by NMR parameters determined using signal intensity, conformation dependent chemical shifts, spin-lattice relaxation time and spin-lattice relaxation time in a rotating frame. The open circles in the table indicate that the technique can yield information on gel structure and dynamics.

2.5.2.2.1 Signal intensity

Depending on the degree of swelling by a diluent, gels are aggregates of regions with continuous variation of mobility that consist of high mobility regions like those of solution. There are also regions at or near the vicinity of crosslink points where they show solid-like behavior with restriction in

Table 1   Various NMR parameters to determine gel structures.

			Detectable region		
		Obtainable information	A Solution-like	B Solid-like, 10^{-8} s	C Solid-like, 10^{-5} s
Signal intensity	CP-MAS	Conformation/ dynamics		○	○
	DD-MAS	,,	○	○	
	Broadband decoupling	,,	○		
Conformation-dependent ^{13}C chemical shift		Conformation	○	○	○
Spin-lattice relaxation time (T_1)		Dynamics	○	○	
^{1}H spin-lattice relaxation time (rotating frame, $T_{1\rho}$)		Dynamics			○

molecular motion. However, it is not easy to obtain information on the structure of all individual regions. As is shown in Table 1, it is possible to distinguish three regions: A (solution-like); B (the solid-like region with correlation time of approximately 10^{-8} s); and C (the solid-like region with correlation time of approximately 10^{-5} s).

This mode of classification is equivalent to the classification based on a correlation time that is extremely sensitive to the change in spin relaxation time as shown in Fig. 1.

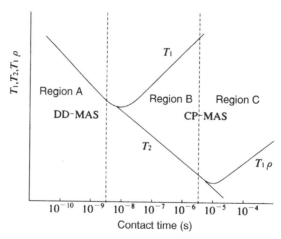

Fig. 1   The region which can be detected by various relaxation times.

The molecular chains in the solution-like region A have a modest degree of crosslinking that exists in gels having a high degree of swelling. The NMR signals can be detected by the ordinary spectrometers with broadband decoupling that are used in solution studies [123–128]. Synthetic polymer gels with chemical crosslinking agents belong to this category [125, 126]. On the other hand, it can be readily understood that the freedom of molecular motion is restricted and the correlation time of molecular motion will be long in polymer chains that possess a stereo-regular structure like that of a helix, as in natural polymers. However, as long as the molecular chain is isolated, it is possible to observe signals in the solution regime [123, 124]. However, if these molecular chains transform to multiple helices or associated chains via hydrophobic interactions, the correlation time of molecular motion increases to the regions B or C. NMR can no longer distinguish these samples from solids (the right-hand side of the minimum of the T_1 curve in Fig. 1). When the association proceeds further and the diluent is excluded (phase separated), this is practically a solid.

The high-resolution solid-state NMR analysis of solid-like materials often utilizes the standard, cross-polarization magic angle spinning (CP-MAS) technique [129–131]. In theory, it is desirable to use the high-power decoupling magic angle spinning (DD-MAS) technique for its reliability of signal intensity for both solution and solid regions. However, unlike CP-MAS, the signal from the solid region weakens due to the requirement of taking a much longer pulse interval than the ^{13}C T_1. Hence, measurement by the CP-MAS technique is more realistic. However, gels can be rubber and amorphous polymers, which are materials in an intermediate state between solid-like and liquid-like materials. Such materials exhibit high molecular motion and the NMR will not detect signals by cross-polarization. In this case, the DD-MAS technique will be effective. There are also occasions for which molecular motion is anisotropic and its broadening of the linewidth cannot be eliminated by broadband decoupling only. At any rate, it is nearly impossible to obtain a signal from the gel as a whole using either technique singly. Thus, both CP-MAS and DD-MAS measurements are essential. It is necessary to investigate the difference between the results with these two techniques.

Since the rate of magic angle spinning requires at least 4000 Hz, it is necessary to avoid leakage during high speed spinning, which will lead to drying of the gel and loss of spinning stability. To avoid these situations, the spinner can be sealed by an O-ring or an epoxy resin can be used on the cap of the spinner.

2.5.2.2.2   Conformation dependent chemical shift

In solution, a ^{13}C chemical shift does not necessarily reflect conformation. Even though conformation dependence exists, it is difficult to observe due to the fast molecular motion, which results in averaging. However, knowledge on molecular conformation that cannot be obtained in solution can be gained from solids due to freezing of the conformational fluctuation. If a chemical shift can be observed on the same molecular chain, and which can take different higher-order structures, that is, polymorphism, then it will be direct evidence of the existence of a conformation-dependent chemical shift. In fact, in ordinary molecules, ^{13}C chemical shifts of as large as 12 ppm, normally 8 ppm, have been observed [131, 132]. In polysaccharides or proteins, the dihedral angle around the glycoside or peptide bond determines the higher-order structure (see Fig. 2). Hence, chemical shifts of the carbons in its vicinity, namely, C-1 and C-X of polysaccharides and C_α, C_β, and C=O of proteins, can be used as probes to reflect sensitively the change of higher-order structures.

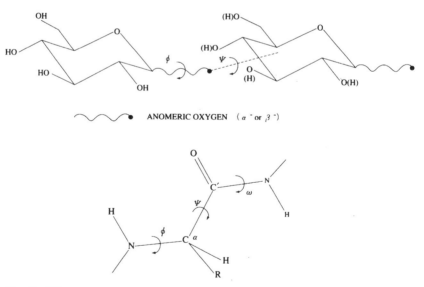

Fig. 2   Dihedral angle (ϕ, ψ) of the glycoside (upper structure) and peptide (lower structure) bonds that determine the higher-order structure of polysaccharides and proteins.

2.5.2.2.3   Nuclear magnetic relaxation time

From the relaxation parameters such as ^{13}C spin-lattice relaxation time (T_1) and ^1H spin-lattice relaxation time in the rotating frame, information on the molecular chain mobility in the various regions of network structures can be obtained (see Table 1 and Fig. 1).

As shown in Fig. 1, the T_1 value from the broadband decoupling or DD-MAS NMR originates from the solution-like region A at elevated temperature, which is on the left-hand side of the minimum. It is important to verify that the T_1 value increases as temperature increases. It becomes necessary to consider a distribution of correlation function that reflects molecular motion of the chains in the solution-like region of gels because of the increased entanglement of the networks. This is due to the linewidth of the signal, which increases drastically in the solution-like region of gels in comparison to that in homogeneous solutions [124, 125, 127]. Except for special cases, it is difficult to treat the motion in a solid in a quantitative manner based on the relaxation time measurement. Currently, determining qualitatively by which region, A or B, the motion is represented is the only area that can be judged. The same applies to whether the ^{13}C T_1 or ^1H $T_{1\rho}$ takes the value that corresponds to the minimum of each measurement. However, it would be desirable to identify in which region, high or low temperature, these values correspond.

2.5.2.3   *Conformation of gel networks*

2.5.2.3.1   Polysaccharide gels

Curdlan, which is a polysaccharide from a microbe (high molecular weight linear polymer $(1 \to 3)$-β-D-glucan) does not dissolve in water. However, if its suspension is cooled after heating to 50°C, a translucent, elastic gel (low-temperature gel) can be obtained. If the gel is heated to a higher temperature, it expels water, shrinks, and becomes opaque (high-temperature gel). Although the former sample exhibits substantial line broadening, the NMR signal from the solution-like region can be obtained by an ordinary spectrometer using broadband decoupling (see Fig. 3). Furthermore, the positions of the C-1 and C-3 peaks shift to the low magnetic field in comparison to a water-soluble low molecular weight oligomer (the average degree of polymerization is 14), exhibiting a conformation dependent shift [132, 133]. This indicates that conformation of the gel that is exhibiting signals in the solution-like region possesses some regular structure unlike that of the random structure of the low

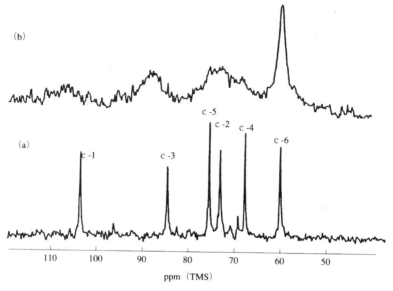

Fig. 3 Comparison of ^{13}C NMR spectra of (b) Curdlan gel and (a) its low molecular weight oligomer [123].

molecular weight oligomers. In fact, as can be seen later, this can be judged to be the same as the single helix structure seen in solid samples from chemical shift data.

Figure 4 shows spectral changes during the gelation process as a reference for the higher-order molecular chains of the gel, which are in the solid-like region [128, 129, 134]. Three samples are prepared and their spectra compared: (a) a powder Curdlan sample as a starting material; (b) one exposed to relative humidity at 96% overnight; and (c) a gel is heated at 150°C and later cooled gradually. From the powder x-ray diffraction data, the heat-treated sample was found to be a triple helix structure. Furthermore, the chemical shifts observed on the hydrated sample are assigned to the single helix structure based on their agreement with the solution-like spectrum of Curdlen gel and the difficulty of observing the signals of multiple helix structures in the solution-like region due to their rigidity. Therefore, the dehydrated sample from this single chain helix is termed a single chain. These two states mutually transform during hydration and dehydration processes. This single chain had been termed a single helix until 1987. However, in 1989 it was correctly called a single chain.

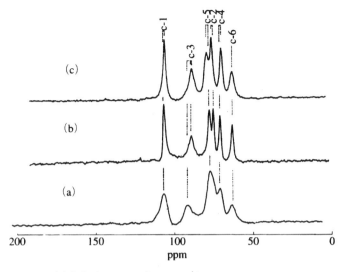

(a) Anhydrous powdery sample
(b) Its hydrated sample
(c) Sample heated to 150°C and gradually cooled

Fig. 4 The ^{13}C NMR spectra of polymorphic structure of solid Curdlan [134].

Shown in Fig. 5 are the comparison of spectra of: (a) hydrated solid state material; (b) and (c) the solution-like regions by broadband decoupling and DD-MAS, respectively; and (d) a solid-like region of caldron gel by CP-MAS [128, 135]. Interestingly, as more than 90% of the network structure is a single helix, the same spectrum as the hydrated sample shown at the bottom of the figure is obtained, although there seems to be a shoulder due to the triple helix. This indicates that the formation of a Curdlan gel proceeds with the formation of a low-temperature gel with opposing processes of the association of a hydrated single helix and swelling as the first step. Later, the gel becomes a stronger, high-temperature gel by transforming into a triple helix with high-temperature treatment [127, 128].

On the other hand, the fundamental backbone structure is $(1 \rightarrow 3)$-β-D-glucan, which is the same as Curdlan. Many branched glucans having side chains with a single glucose from the C-6 carbon at every 2 or 3 glucose residues have been isolated from fungi (e.g., mushrooms). In particular, schizophyllan and lentinan are being used clinically in Japan as

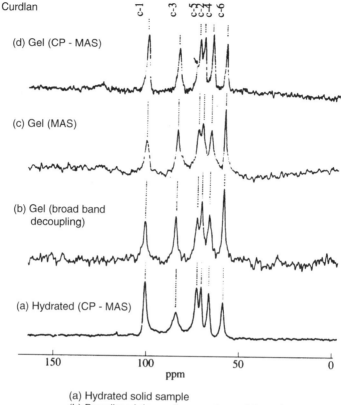

Curdlan

(d) Gel (CP - MAS)

(c) Gel (MAS)

(b) Gel (broad band decoupling)

(a) Hydrated (CP - MAS)

(a) Hydrated solid sample
(b) Broadband decoupling spectrum of the gel
(c) DP-MAS spectrum of the gel
(d) CP-MAS spectrum of the gel

Fig. 5 Comparison of ^{13}C NMR spectra of Curdlan obtained by various NMR methods [135].

tumor inhibitors. Unlike the linear chain Curdlan, these glucans form brittle gels. In these gels under neutral conditions, no signals can be detected from the solution-like region. However as indicated in schizophyllan, under the lower alkaline concentration of 0.06 M NaOH, which is at much lower concentration than the 0.3 M required for transformation from helical structure to random coil, the signals seen in the single helix are detected in the solution-like region as seen in the Curdlan gel [136]. In fact, the signals from the solid-like region of these branched glucans indicate triple helices [127, 128, 135–138].

Accordingly, the gel networks of linear Curdlan consist mainly of molecular chains with a single helix structure. The triple helix content is at most 10%. The crosslink structure is formed by association of these triple and single helices. On the other hand, the main structure of the branched glucan is a triple helix and the partial association of these chains functions as a crosslink structure. Thus, the model based on x-ray diffraction data of the thermally treated Curdlan, stating that the crosslink structure consists of a triple helix, is incorrect. This is why NMR data capable of providing information on structure *in situ* is desired. To be described, the dynamics data from nuclear magnetic relaxation time measurement can also provide information on the differences in these network structures.

Agarose gel, which exhibits elasticity similar to that of Curdlan, clearly provides signals from the solution-like region as shown in the upper ^{13}C DD-MAS spectrum of Fig. 6 [130]. However, a spectrum similar to this can also be obtained by the CP-MAS technique as shown in the bottom spectrum of Fig. 6 (the solid-like region). However, there appears to be a difference in the 77–78.5 ppm resonance, which is the characteristic region of polysaccharide chains. Unlike in the solid-like region, this difference is thought to arise from the random coil structure in

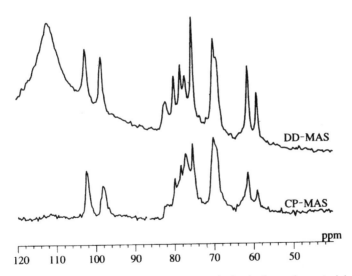

The signal at 102 ppm in the upper spectrum is due to the probe material

Fig. 6 Comparison of CP-MAS (upper spectrum) and DD-MAS (lower spectrum) NMR spectra of agarose gel [130].

the solution-like region. This is different from the association of the double helix model that is widely accepted as the network structure of agalose. In fact, it is not necessarily possible to distinguish a single chain from a triple chain by an x-ray fiber pattern [139]. Rather, it is more useful to use spectral changes that occur during the thermal treatment process as reference. This is true not only for agarose but also for other polysaccharides [128–131, 140–142]. Similarly, based on the chemical shift and T_1 data, the NMR signals from the solution-like region of a starch gel are found to originate from the random coil chains [130].

2.5.2.3.2   Protein gels

Natural polymers that form networks are not limited to the aforementioned polysaccharides. Many structures, such as globular proteins, also form gels under certain conditions. A gelatinous gel, which is the hydrolysis product of collagen, forms a gel by local triple helix formation, and many globular proteins, including silk fibroin, a structural protein, and insulin, are formed by β-sheet formation. The gelation of the latter materials is also termed amyloid formation [146]. Similar to the polysaccharide case, crosslink points are often caused by various stereoregular structures and represented by α-helix and β-sheet. It is convenient to characterize the structures in question through the conformation dependent chemical shifts that accompany structural changes. For this purpose, the chemical shift values for the triple helix structure of collagen at various amino acid residues [143–145], and the α-helix and β-sheet of polypeptides [132, 133], have been accumulated.

Here, the structural changes during hydration of C-hordenine, the barley protein, will be discussed as an example. This is an interesting problem since this subject relates to the changes in viscoelastic properties of grains like wheat during processing. The primary structure of this protein consists of the repeated structure, Pro-Gln-Gln-Pro-Phe-Pro-Gln-Gln, except for the N- and C-terminal groups. As shown in Fig. 7, the CP-MAS spectra of C-hordenine in dry, (a), and hydrated, (b) and (c) (samples (a) and (b) contain 25 mM urea), conditions differ significantly whether or not the samples are hydrated or urea exists. In particular, signal intensity is markedly reduced by hydration. In sample (b), the significant intensity reduction is seen on the main chain carbonyl and α-carbon signals [147]. Such reduction of signal intensity of the CP-MAS spectrum is due to molecular motion. This can be confirmed from the fact that the DD-MAS spectrum provides signals that are missing in the CP-MAS spectrum.

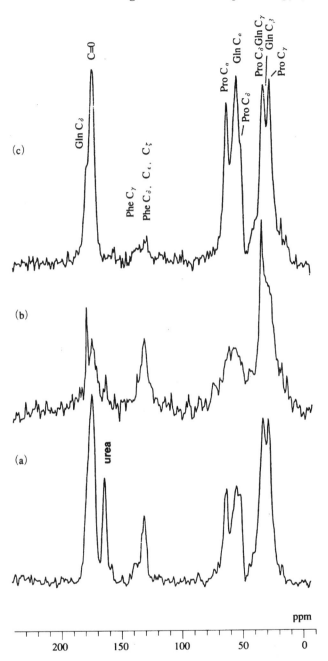

Fig. 7 The ^{13}C CP-MAS NMR spectra of (a) solid C-hordenine and its hydrated materials [147].

From the chemical shift and relaxation time measurements in these two spectra, information on local conformation can be obtained. Dynamics also can be determined based on the discussion in the next section. The formation of β-sheet structure in an amyloid protein is confirmed from not only the preceding empirical chemical shift data accumulation but also from the nonempirical interatomic distance determination [148, 149].

2.5.2.4 *Dynamics of network structure*

As already discussed in subsection 2.5.2.2.(3) here, whether or not the signals from the networks being studied currently correspond to A, B and C in Fig. 1 can be determined from various relaxation parameters.

For example, if ^{13}C spin–lattice relaxation time is measured by DD-MAS and CP-MAS, as shown in Table 2 the measured values are significantly different. This reflects clearly the molecular motion of the networks. In fact, it is obvious that the former corresponds to a solution-like region (A) and the latter to a solid-like region (B, C). In the case of the solution-like region, the correlation time and its distribution can be obtained by taking into account the values of the nuclear Overhauser effect (NOE) and spin–spin relaxation time (linewidth) [124, 125, 127]. On the other hand, as already described, it is impossible to analyze in detail the solid-like region. Nonetheless, in the $(1 \rightarrow 3)$-β-D-glucan gels described in subsection 2.5.2.3.(1) here, the proton $T_{1\rho}$ values differ markedly in Curdlan, which has a single helix, and schizophyllan and HA-glucan, both of which have a triple helix. As long as molecular motion with the time scale of approximately 10^{-5} s is concerned, the difference between the two can be distinguished.

2.5.3 Conclusions

The author has presented a discussion on how NMR can provide knowledge on the structure and dynamics of natural polymer networks and provided information on other research in the field. It is a bonus if readers understand how the NMR technique can provide unique information on these problems and, furthermore, develop an interest in such subjects.

Table 2 Spin–lattice relaxation time (s) of starch gel (33%) obtained by ^{13}C DD-MAS (upper rows) and CP-MAS (lower rows).

	C-1	C-4	C-3	C-2	C-5	C-6
Solution-like region	0.36	0.29	0.30	0.32	0.29	0.16
Solid-like region	9.2	11.8	11.9	11.9	11.9	2.1

Table 3 Comparison of proton spin–lattice relaxation time in the rotating frame of a $(1 \rightarrow 3)$-β-D-glucan gel.

	C-1		C-3		C-5		C-2		C-4		C-6	
	T_{CH} μs	T_{1_ρ} ms	T_{CH} μs	T_{1_ρ} ms	T_{CH} μs	T_{1_ρ} ms	T_{CH} μs	T_{1_ρ} ms	T_{CH} μs	T_{1_ρ} ms	T_{CH} μs	T_{1_ρ} ms
Curdlan	128	17.9	138	16.6	137	17.0	145	19.2	110	14.0	82.2	22.9
Schizophyllan	67.3	3.32	62.8	4.18	53.1	4.30	56.4	4.09	32.0	5.69	22.1	10.0
HA-β-glucan	56.6	5.04	35.4	5.38	64.4	6.27	47.8	5.34	50.7	5.80	53.0	7.28

2.6 SWELLING

HISAO ICHIJO

2.6.1 Introduction

Recent research on new materials has been conducted broadly on both the fundamental and the applied level. In many areas development is moving towards commercialization. In such a climate proper evaluation and improvements in reliability of those new materials are both indispensable. For these new materials to be accepted for actual use, establishment of a standard evaluation method is essential. Within companies internal standards are being established and at the industry or commercial level there are rules of association or industry standards. It is now difficult to develop new materials and products for an international market in a unified manner when the research has been done by only a single organization. Based on the determination of a new JIS, and through a new development towards ISO, evaluations that aid in standardization (at the international level) of new materials and products are being done.

Superabsorbent polymers are one of the new materials. Since 1987, when it was used for disposable baby diapers, production has increased each year. Production in 1995 reached 600,000 t [150]. Due to commercial applications like paper diapers and sanitary products, they are so familiar to us because they appear on TV almost every day. Other applications, such as a soil additive to retain moisture, materials for maintaining freshness, antifrost construction materials, gel perfumes, and controlled release drugs are under development. It seems important to measure and evaluate the superabsorbent polymers that have been used in many industries, accelerate their proper use, and establish guidelines for future material development and design.

In this section, measurement and evaluation of the swelling of superabsorbent polymers will be described with special emphasis on the investigation of standardization of reliability evaluation methods for new organic and composite material products. The work was conducted by the Polymeric Material Center as part of the requested project by the Japanese Ministry of Industrial Technology [151–153].

2.6.2   Present Testing Methods for Superabsorbent Polymers

Based on an inquiry made on behalf of the superabsorbent polymer manufacturers, performance testing items include appearance, smell, water uptake, water retention, rate of water absorption, particle size distribution, bulk density, pH, gel strength, durability, dry weights, residual monomer, and water soluble components. The testing items that should be standardized are water uptake and rate of water absorption [151]. Testing method preferences include:

1. Easy and objective testing method is desired.
2. No seasonal difference should be observed and any anomalies must be able to be fed back to producers immediately.
3. The method should not be application-specific.

2.6.3   Water Uptake Testing Method

Water uptake is defined as the water that is absorbed into the gel when a highly absorbent polymer has been swollen in solution (ion exchanged water or salt solution) and has reached equilibrium. General measurement methods involve immersion of the sample in a solution for a certain specified time to measure the weight or volume of the swollen gel (see Table 1) [154]. For example, there are methods to measure weight increase of a nonwoven cloth bag in which superabsorbent polymer is placed (e.g., like a tea bag) or to measure the volume of the solution by putting the polymer in contact with the solution.

There is a method to measure the change in UV absorption of a solution by immersing the superabsorbent polymer in a solution in which a UV absorber that cannot be absorbed by the gel is added. Upon UV measurement, swelling equilibrium is calculated. The advantage of this

Table 1

Measurement methods	Brief description of measurement
Tea bag method	The superabsorbent polymer is placed in a nonwoven cloth bag, then it is immersed in aqueous solution, and finally weighed after removing the nonabsorbed water
Centrifugal dehydration method	After the superabsorbent polymer is sufficiently swollen in an excess-aqueous solution, the swollen polymer is placed in a screen bag and water is expelled by centrifugal force. The recovered water is measured.
Filtration method	After the superabsorbent polymer is sufficiently swollen, excess solution is eliminated by a screen and water uptake is measured.
UV absorption method	The superabsorbent polymer is swollen in blue dextrin aqueous solution. Water uptake is calculated by the difference from the reference solution.
Microscope observation method	The size of the polymer before and after swelling is observed using a microscope and water uptake is calculated.
Capillary method, DW method	The superabsorbent polymer is placed on a porous plate or plate with small holes, put in contact with an aqueous solution for a specified time, and the change of the amount of solution is calculated.

method is that swelling can be measured without removing the gel from the solution.

Depending on the purpose, several measurement methods have been proposed, such as determining swelling force by how much solution is absorbed from the contact surface, or by the use of retention force to determine how to retain as much water as possible.

Each measurement method has different characteristics: (1) the tea bag method is easy and can be widely used; (2) the error of the filtration method is relatively large; and (3) the centrifugal method can easily plug up the screen. Using these characteristics and the results of inquiry, the tea bag method was chosen.

Four common superabsorbent polymers were examined for their water uptake at seven organizations under many different conditions. These included using two types of testing solution, two materials for tea bags, immersion time, temperature, measurement temperature, type of solution, conditioning of the sample, the amount of the sample, type, and particle size. The obtained results have been analyzed and the reproducibility from each organization and each measurement have been evaluated.

The water uptake measurement method (proposal) (in the proposal, the words superabsorbent resin are used) for superabsorbent polymers is thus organized as follows [152]:

Testing Method for Water Uptake of Superabsorbent Polymers (proposal, only the main text is extracted)

[1] Application range

This standard specifies the test method of water uptake of super-absorbent polymers using the tea bag method.

Remark 1. This test method specifies the test method for the superabsorbent polymers with water uptake 10 times the dry weight.

Remark 2. The particle size of the superabsorbent polymers to be used for the test is more than 60 μm. As particle size influences the experimental error, the particle size should be as narrow in distribution as possible. For particle sizes less than 60 μm, it is determined by agreement between the parties involved.

Remark 3. The reference specification will be listed as follows:

JIS K 0069 the test method for chemical remnants from the sieve

JIS K 6900 plastics terminology

JIS K 7100 the conditioning of plastics and standard conditions of testing laboratory

JIS K 8150 sodium chloride

JIS R 3503 glassware for chemical analysis

JIS Z 8401 method for rounding numbers

[2] Definition of terminology

The definition of terminology used in this specification is according to **JIS K 6900**. Additional terms are listed as follows.

(1) Superabsorbent resin: Resin that absorbs extremely large amounts and swells. Superabsorbent resin is hydrophilic and possesses a crosslink structure. It absorbs water upon contact and once water is absorbed, even if the resin is pressurized, dehydration is difficult.

(2) Water uptake: The absorbed water mass of a unit weight of superabsorbent resin.

(3) Deionized water: Water that has eliminated the majority of ions. Its electrical conductivity is $10\,\mu s/cm$.

[3] **Conditioning samples and test solutions, and test temperature and humidity**

[3.1] **Conditioning samples and test solutions**
Samples and test solutions are in principle conditioned for more than 24 h according to **JIS K 7100** under a standard temperature condition of the second degree at a temperature of $23°C \pm 2°C$ prior to the test.

Remarks: The conditioning of superabsorbent polymers is done by placing the polymer in a closed container and adjusting the temperature. As necessary, changes in dry weight are examined.

[3.2] **Test temperature and humidity**
The test is conducted in a room that is adjusted to a standard temperature condition of the second degree (temperature $23°C \pm 2°C$) according to **JIS K 7100**.

[4] **Testing apparatus, test solutions, and tea bag**

Testing apparatus, test solutions, and tea bag are specified as follows.

[4.1] **Testing apparatus**

(1) Beaker: The beaker is used to fill test solutions in which a tea bag that contains a polymer sample is immersed. It is identified as a 1 L beaker according to **JIS R 3503**.

(2) Balance: A balance is used to measure the mass of the sample; its sensitivity should be 1 mg.

[4.2] **Test solutions**

The test solution is to let the polymer sample absorb water. According to **JIS K 8150**, a sodium chloride solution is prepared using sodium chloride and deionized water at 0.9 ± 0.001 w/v%. Another solution, containing only deionized water, is also used.

[4.3] **Tea bag**

The tea bag is to contain a polymer sample and be immersed in the test solution. In principle, it is made of nylon cloth with hole size of 57 μm (255 mesh), and is 10 cm wide and 40 cm in length. As shown in Fig. 1, the longer side is folded and both sides are heat sealed to make a bag.

[5] **Samples**

[5.1] **Particle size and mass of samples**

The particle size of the sample is more than 60 μm. The mass of the sample is approximately 0.20 g for a deionized H_2O test solution and 1.00 for a 0.9 w/v% sodium chloride test solution.

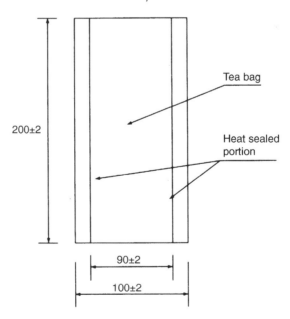

Fig. 1 The shape and size of the tea bag (unit mm).

[5.2] **Sampling**
Sampling is done according to the following procedures.

(1) The sample should be collected at random as much as possible so that the particle size distribution represents the mother group.

(2) When particles of less than 60 mm exist, they will be removed by the dry sieve as specified by **JIS K 0069, 3.1**.

Remarks: When the sample is sieved, it should be conducted in a room adjusted to the standard temperature condition (temperature $23°C \pm 2°C$, relative humidity $50\% \pm 5\%$) according **to JIS K 7100**, or sealed. Measurement of water uptake by a sample with special particle size distribution is made after agreement between the two parties involved.

[5.3] **Number of tests**
The tests were done five times.

[6] **Procedures**

The procedure described here should be followed.

(1) The test solution is deionized water and approximately 0.20 g of the sample is weighed to within a 0.001 g accuracy. This weight is specified as "a." For the 0.9 w/v% salt solution, approximately 1.0 g is sampled within a 0.001 g accuracy and also specified as "a."

(2) The test solution, which has been measured for its pH and electrical conductivity, is placed in a 1 L beaker.

(3) The sample is placed uniformly at the bottom of the tea bag. The tea bag is lowered quietly so as not to produce aggregates of the samples. The tea bag is immersed 150 mm from the bottom of the solution for the specified time. The immersion time is 3 h or 24 h for deionized water and 1 h or 3 h for the 0.9 w/v% salt solution.

(4) After immersion for the specified time, the tea bag is taken out of the solution, hung diagonally for 10 min to eliminate excess water, and immediately measured for its weight. This weight is specified as "b."

(5) An empty tea bag (a control) is immersed in the test solution for the specified time. After procedure (4), the weight of the empty tea bag is measured and specified as "c."

(6) Procedure (4) is repeated five times. Loss of sample during measurement should be carefully checked. If leakage is found, repeat the experiment.

[7]   Calculation

[7.1]   Water uptake
The water uptake is calculated according to Equation (1).

$$W = \frac{b - c - a}{a} \tag{1}$$

where W is the water uptake (g/g); a is the mass of the sample (g); b is the mass of the tea bag that has been immersed in the test solution for the specified time and excess water eliminated; and c is the mass of the empty tea bag that has been immersed in the test solution for the specified time and excess water eliminated.

[7.2]   Rounding of test results
Each test result is calculated individually and its average is rounded off to two significant digits according to **JIS Z 8401**.

[7.3]   Standard deviation and coefficient of variance
When standard deviation and coefficient of variance are needed, they are rounded off to two significant digits according to **JIS Z 8401**.

2.6.4 Test Method for Rate of Water Absorption

Depending on the type of superabsorbent polymers, measurement of the rate of water absorption is possible using the tea bag method and the DW method.

Table 2 Method for determining the rate of water absorption.

Measurement method	Brief description of the measurement
Vortex method	A specified amount of solution in a beaker is stirred while a specified amount of superabsorbent polymer is added. The time for the liquid surface to become flat or for the stirrer chip to hide beneath the liquid is measured.

However, when the rate of water absorption is very fast because of a certain type and particle size, such a measurement is very difficult. Therefore, the use of a method to determine the force to eliminate fluidity (gelation force) is regarded as a reasonable approach. Due to its ease of use, the vortex method (see Table 2) is selected.

In a manner similar to the method described in the water uptake information four types of superabsorbent polymers were used as common samples and eight organizations determined the rate of water absorption by varying conditions that included: particle size, the amount of sample, the method of sample addition to the solution, stirring condition, method to identify the end point, two types of solution, the amount of solution, measurement temperature, and the number of measurements. Based on the results and their analysis, the method for measuring the rate of water absorption for superabsorbent polymers is proposed as follows.

Testing Method for the Rate of Water Absorption of Superabsorbent Polymers (proposal, only the main text is extracted)

[1] Application range

This standard specifies the test method for measuring the rate of water absorption of superabsorbent polymers.

Remarks: A similar expression as for the water uptake measurement method is used for the water uptake, particle size, and reference specifications of superabsorbent polymers.

[2] Definition of terminology

In addition to the main definition of the terminology used in this standard according to **JIS K 6900**, the following are also used.

(1) superabsorbent polymers: the same definition as the one used in the water uptake test.

(2) rate of water absorption: the mass of the solution per unit weight of the sample that is absorbed from the time of starting to the end point of the test specified in this standard.

(3) deionized water: the same definition as the one used in the water uptake test.

[3]   Conditioning of sample and test solution, and test temperature and humidity

Conditioning of the sample and test solution, and test temperature and humidity are the same as in the description used in the water uptake test.

[4]   Testing apparatus and test solutions

[4.1]   Testing apparatus

(1) Magnetic Stirrer: The rotation speed of the magnetic stirrer should be 600 ± 60 rpm. The stirrer chip has a central diameter dimension of ϕ 8 mm, terminal diameters of ϕ 7 mm, and length 30 mm with fluorinated resin coating.

Remarks: The rotation speed of the magnetic stirrer should be confirmed by measuring the rotation speed of the axis using a tachometer. Also, the magnetic strength and torque need to be periodically checked to ensure proper operation.

(2) Beaker: The beaker is a container to hold the test solution with the sample, which is placed on the magnetic stirrer with a rotating stirrer chip. The beaker specified in **JIS R 3503**, a flat-bottomed beaker with 100 mL capacity, should be used.

(3) Balance: The same as the one used for the water uptake test is used.

(4) Timer

[4.2] **Test solution**

The same as the one used for the water uptake test is used.

[5] **Samples**

[5.1] **Particle size and mass of samples**

The particle size of the sample is more than 60 mm. The mass of the sample is approximately 0.50 g for deionized water and 2.00 g for the 0.9 w/v% sodium chloride solution.

Sampling [5.2] and the number of tests [5.3] are the same here as the ones used in the water uptake test information that begins on page 297.

[6] **Procedures**

The following procedures should be followed.

(1) For the deionized test solution, a sample of approximately 0.50 g is very quickly sampled to an accuracy of 0.001 g. For the 0.9 w/v % sodium chloride solution, a sample of approximately 2.00 g is very quickly sampled to an accuracy of 0.001 g.

Remarks: If the end point of this test is difficult to observe, it becomes easier by dyeing the test solution with a small amount of a dye.

(2) Fifty grams of the test solution, that has been tested for its pH and electrical conductivity, and a stirrer chip are contained in the beaker and the beaker is placed on the magnetic stirrer.

(3) After adjusting the rotation speed of the magnetic stirrer to 600 ± 60 rpm, generation of a stable vortex should be confirmed.

(4) The sample is carefully and homogeneously added near the beaker wall and the timer is started.

Remarks: When the amount of absorption is low or the rate of absorption is small, it is difficult to measure using this vortex method. In this case, the amount of the sample can be increased. When the rate of water absorption is reported, the amount of sample must be specified.

(5)   The end point of the test is identified as the point at which the stirrer chip is covered by the test solution. The time required to observe the end point is recorded to an accuracy of seconds.

[7]   Calculation

[7.1]   Rate of water absorption

The rate of water absorption is calculated by Equation (1):

$$V = \frac{M/a}{t} \tag{1}$$

where V is the rate of water absorption (g/g/s); M is the mass of the test solution (g); a is the mass of the sample (g); and t is the time to the end point (s).

2.6.5   Conclusions

For standardization of the water uptake test, multiple organizations conducted the round robin test using common samples and procedures. Based on their results, methods (proposal) to determine water uptake and the rate of water absorption are being developed. A written inquiry was also distributed to the member companies of the Polymer Materials Center regarding water uptake. Currently, the standard for water uptake has not been established anywhere in the world. It seems important first to establish the standardization of the water uptake testing method for superabsorbent polymers and then to make it the international standard. The testing methods (proposal) shown here are still under discussion and when it is standardized by JIS, there may be partial modification.

2.7   ELECTROCHEMICAL TECHNIQUES

MASAYOSHI WATANABE

2.7.1   Introduction

Hydrogel undergoes drastic volumetric changes including volumetric transition due to changes in the environment that include solution composition, temperature, and pH [155–157]. As a consequence, the amount of solvent that is enclosed in the gel phase, the diffusion coefficient of the solute in the gel, and further, the polarity of the gel network itself changes significantly [155–157]. Such sudden changes draw attention not only from a theory-based perspective but also from the application point of view. Potential uses, such as in chemomechanical transducers, actuators, for material separation, condensation, and controlled drug delivery, have been proposed [155–157]. Today, an understanding of the volumetric changes of gels at the equilibrium conditions are fairly advanced; however, a quantitative understanding of other factors that accompany these volumetric changes is poor. In this section, electrochemical techniques that provide important information concerning the diffusion coefficient of the solute in the gel and changes in the distribution of the solute in the gel during volumetric changes will be discussed with emphasis on experimental methodology.

Electrochemical techniques are quite accurate and relatively easy methods to determine the diffusion coefficient of the solute in solution [158]. However, it has been difficult to apply these techniques to the solute in polymer gels for the following reasons. First of all, when a polymer gel is immersed in a solution with a solute concentration C_0, and its diffusion coefficient D_0, the solute and solvent in the solution are distributed in the gel as shown in Fig. 1. At equilibrium, they will take a certain solute concentration C and the diffusion coefficient D. Unfortunately, using an ordinary electrochemical technique, D and C cannot be simultaneously determined. Second, the electrical resistivity of the gel increases by swelling. In particular, this influence is significant in the unswollen state. Thus, it is difficult to obtain electrochemical responses that can be used for quantitative analysis. Third, unless a proper method to fix the gel onto the electrode is used, it is difficult to obtain information within the gel due to the very large changes in volume.

The authors [5–7] reported on the possibility of measuring the material distribution into a gel accompanying the phase transition, and the change of the material diffusion in the gel using an ultramicroelectrode on which hydrogel is chemically fixed [159–161]. As shown in Fig. 1, if the gel-treated microelectrode is immersed in the solution with redox molecules and electrochemcial measurement is performed with various time scales, the diffusion coefficient D and the concentration C of the redox molecules in the gel can be independently determined. Moreover, by employing the gel-treated microelectrode, not only the information at equilibrium but also dynamic changes during the change in degree of

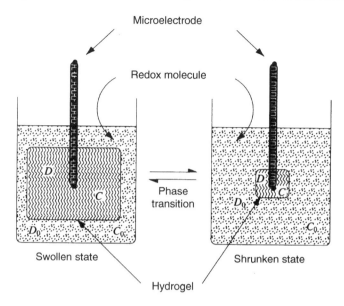

This diagram is schematic and, thus, omits reference electrode and counter electrode, which are necessary for actual measurements

Fig. 1 The measurement principle of the diffusion coefficient and distribution coefficient (concentration) of the solute in a gel using electrochemical methods.

swelling can be detected. Here, procedures will be described using a typical hydrogel, polyacrylamide (PAAm).

2.7.2 Characteristics of Microelectrode

In traditional electrochemistry, electrodes with typical size on the order of mm have been used. In comparison, electrodes of micrometer size will be called microelectrodes. Their characteristics include [162–167]:

1. small size;
2. independent detection current for long term electrolysis;
3. small diffusion layers;
4. reduction of iR drop; and
5. feasibility of rapid electrochemical measurements.

When an electrode functions as a probe for electrochemical analysis, local analysis is possible by reducing the size of the electrode. In other words, it increases spatial resolution. Using these characteristics, a microelectrode can be used as a detector for reactions within the body.

Electrode reaction is controlled by mass transport, that is, the diffusion to and from the electrode. The thickness of the diffusion layer is roughly given by $(Dt)^{1/2}$, where D is the diffusion coefficient and t is the electrolysis time. As the diffusion coefficient of an ordinary solution is around 10^{-5} cm^2/s, the thickness of the diffusion layer becomes 0.1 mm after electrolysis for 10 s. If the size of the electrode is sufficiently larger than the diffusion layer, the area of the diffusion layer is the same as the area of the electrode. Namely, only the diffusion normal to the surface of the electrode (1D diffusion) is effective. When the size of the electrode becomes comparable to the thickness of the diffusion layer, 2- and 3D diffusion become dominant. Material supply due to diffusion in the diagonal direction also becomes possible and current increases. When the diffusion becomes completely dominated by the 3 D diffusion, a constant current is observed after sufficiently long constant voltage electrolysis. This also leads to a constant thickness in the diffusion layer. Hence, if a cyclic voltammogram is recorded with a slow sweeping rate, a flat limiting current without a maximum will be observed. When the sweeping direction is reversed, it traces back the same voltammogram in the reverse direction. The current–voltage curve independent of time clearly expresses the electrode reaction and, thus, is easy to interpret.

Figure 2 indicates an equivalent circuitry of an electrochemical cell system where WE, RE, and CE express acting electrode, reference electrode and counter electrode, respectively, C_{dl} indicates the capacity of the electric double layer of the acting electrode, and R_z expresses the resistance for electron charge transfer. The current that passes through R_i is a Faraday current i_f, and the current which is required to charge C_{dl} is capacitative current i_c. In addition, R_s is the solution resistance between WE and RE. Voltametry, which is widely used in electrochemical analysis, is the technique of adjusting the voltage between WE and RE so as to maintain a constant voltage between WE and CE, and measure the current between WE and CE for each adjusted voltage. In the usual slow potential sweep, the current becomes extremely small, that is, the current that passes through R_s is small, and the potential drop (iR drop) at that position becomes extremely small. As a result, when the microelectrode is used, the potential drop in the solution becomes small and the electrode voltage can be measured more accurately even if the solution resistance is large. Hence, measurement in an organic solvent solution with high resistance

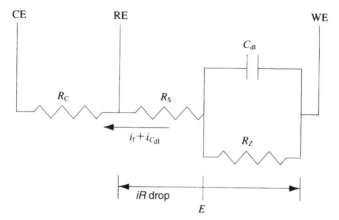

WE, acting electrode; RE, reference electrode; CE, counter electrode; R_s, R_c, resistance of solution (gel); R_z, charge transfer resistance; and C_{dl}, electrical double layer capacity

Fig. 2   Equivalent circuitry of an electrochemical cell.

[168], electrochemistry under ultralow temperature [169], and electrochemical measurement of polymers like hydrogen [159–161] or ionic conducting polymers [170–172] become possible with the use of microelectrodes.

The influence of C_{dl} is significant in rapid voltametry measurement. The equivalent circuit in Fig. 2 can be regarded as a simple RC circuit with a time constant, $\tau = C_{dl}R_s$, especially when no redox material exists ($R_z = \infty$). In other words, the current signal response flattens due to the limitation imposed by the time constant during a fast sweep. However, for the microelectrode, $\tau = \rho a$ where ρ is the specific resistance of the solution and a is the radius of the electrode. Hence, unlike with the use of an ordinary electrode, fast sweep is possible.

For microelectrode characteristics, readers are referred to several excellent review articles [162–165] and monographs [166, 167].

2.7.3   Preparation of Microelectrode and Fixation of Gel

Microelectrodes can be classified into point electrodes as in the case of microdisk electrodes and line electrodes, as in microtubular electrodes, based on structure, size, mass transport, and application. The point electrode means that the electrode appears as a point at a distance. On the other hand, when it looks like a line, it is a line electrode. Although it sounds like a superficial classification, this classification is fundamental from the viewpoint of the magnitude of current and mass transport. A microdisk electrode is easy to manufacture and its theoretical treatment has been established. Here, we will discuss the manufacturing method for microelectrodes and the fixation method for gels. The methods discussed in this section are applicable to other microelectrodes.

A microdisk electrode can be manufactured as follows. A fine platinum wire (it is easy to use a wire 10–50 μm in diameter) is enclosed in a pyrex capillary with a lead copper wire that is fixed by silver/epoxy paste. After sanding the glass surface with an emery paper and exposing the platinum wire, the electrode surface is mirror polished by a diamond paste followed by alumina (see Fig. 3).

Inserting the microdisk electrode into a gel or polymerizing a gel on this electrode surface will not allow electrochemical measurement and no information on the solute in the gel can be obtained. The authors chemically fixed a gel onto the platinum surface following the procedure indicated in Fig. 4 and succeeded in making an electrochemical measurement of the solute within a gel. After introducing hydroxyl groups on the platinum surface, the surface is converted into vinyl groups by a silane coupling agent. By free radical polymerization of the gel monomer on this electrode, the gel can be chemically fixed onto the electrode.

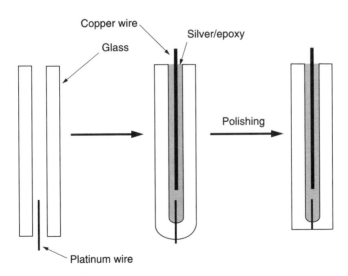

Fig. 3 The manufacturing procedure of a microdisk electrode.

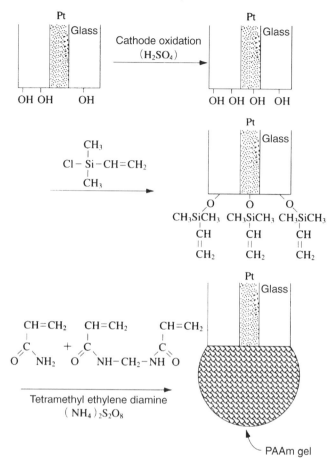

Fig. 4 Manufacturing method for a hydrogel treated with a microelectrode.

Specifically, the prepared microelectrode surface is washed in a concentrated sulfuric acid for 5 min, washed by ion-exchanged water, and then cathode oxidized at 1.9 V against SCE (saturated Calomel electrode) in 0.5 M sulfuric acid aqueous solution for 5 min. Next, cyclic voltametry is repeated until the voltammogram becomes steady (~ 2 h). After this, the electrode is kept at 1.1 V against SCE and left until the current becomes constant (~ 5 min). By this method, hydroxyl groups can be introduced onto the platinum electrode surface. The electrode is washed carefully with ion-exchanged water and dried. To treat the surface with a coupling agent, the anode is oxidized under an argon atmosphere and a dried electrode is immersed into the 0.1 M iso-octane solution of chloro-dimethyl vinylsilane. This will allow introduction of vinyl groups onto the electrode and glass surfaces. The electrode surface is washed by hexane and methanol. By preparing PAAm gel via free radical polymerization on this surface treated electrode, namely by placing the monomer solution onto the electrode dropwise in a glovebox under an inert gas and polymerizing it, the gel is chemically fixed onto the electrode surface. The gel is purified by immersing the gel-coated electrode into distilled water and extracting the unpolymerized materials. Using this electrode, debonding can be avoided even if the gel is swollen or shrinks in a solution with a solute.

2.7.4 Measurement Principles

When a solute is electrolyzed at a constant voltage that is sufficiently higher than its oxidation potential or sufficiently lower than its redox potential, the current is proportional to $t^{-1/2}$, over a very short time. In other words, it follows the so-called Cottrell equation. As time passes, the current approaches steady state. This change can be obtained by solving the time-dependent diffusion equation.

The current change is expressed by the following approximation [173]:

$$I = 4nFCDa\{0.7854 + 0.4431a(Dt)^{-1/2} + 0.215 \exp[-0.391a(Dt)^{-1/2}]\} \tag{1}$$

where I is the current, n is the number of reacting electrons, F is the Faraday constant, C is the solute concentration in the bulk, D is the diffusion coefficient of the solute, and a is the radius of the disk. For the very short electrolysis time where the thickness of the diffusion layer is sufficiently thinner than the radius of the electrode, the diffusion mode becomes 1D and the current can be expressed by the Cottrell equation:

$$I = nF\pi^{1/2}D^{1/2}a^2Ct^{-1/2} \tag{2}$$

On the other hand, when the thickness of the diffusion layer becomes sufficiently thicker than the radius of the electrode, the diffusion layer spreads hemispherically around the electrode and the mode of diffusion becomes radiative. As a result, as a steady-state diffusion layer of the solute concentration that is inversely proportional to the distance from the electrode surface is formed, the current also becomes steady state. This steady-state current value (limiting current value) is expressed as follows:

$$I = 4nFCDa \tag{3}$$

The thickness of this steady-state diffusion layer is at most 10 times that of the electrode radius [166].

If the thickness of the chemically fixed gel is more than 10 times that of the electrode radius, the observed electrolysis current value reflects the concentration and the diffusion coefficient of the solute within the gel. This allows us to ignore the concentration of the solute and its diffusion coefficient in the solution in which the gel is immersed. If the measurement is taken in such a way as to follow the time scale of Eqs. (2) and (3), that is, if the slope in the relationship between I and $t^{1/2}$ in the region where the time dependence of the electrolysis current follows Eq. (2), C and D can be independently determined. This is possible because the unknown parameters in Eqs. (2) and (3) are only C and D. When the diffusion coefficient is sufficiently fast, for example, the diffusion coefficient of the solute in a swollen gel is around 10^{-6}–10^{-5} cm^2/s, the time scale which follows Eq. (2) is on the order of milliseconds and, thus, measurement becomes difficult. In an ordinary electrochemical measurement, the mode of diffusion becomes a combination of 1D diffusion and radiative diffusion. The response of cyclic voltametry for this mixture of diffusion modes has already been analyzed.

When cyclic voltametry is performed using a microdisk electrode with very slow sweep of the potential, a steady-state wave (s-shaped curve without a maximum) without hysteresis is observed. This steady-state current agrees with the value in Eq. (3). As the sweeping speed increases, the peak starts appearing. Upon reversing the potential, hysteresis is also observed. Further increase in sweeping speed provides the same wave as the one obtained with a large electrode at a slow sweeping rate. Thus, the shape of the wave relates to the diffusion coefficient D of the solute, sweeping rate v, and electrode radius a (the degree of mixing of 1D diffusion and radiative diffusion), which can be expressed by the following parameter:

$$p = \left(\frac{nFa^2v}{RTD}\right)^{1/2} \tag{4}$$

where R is the gas constant and T is the temperature. The characteristic quantity of the wave, peak current I_p, can be approximated by the following equation [174]:

$$I_p = 4FCDa\left(0.34\, e^{-0.66p} + 0.66 - 0.13\, e^{-11/p} + 0.351p\right) \tag{5}$$

When p is small (slow sweep, small radius of the electrode), I_p is the same as in Eq. (3). Therefore, by performing cyclic voltametry at various sweeping rates and obtaining the steady-state current, which is expressed by Eq. (3) for a sufficiently slow sweeping rate, and the peak current, which is expressed by Eqs. (4) and (5) for a fast sweeping rate, it is relatively easy to determine the concentration and diffusion coefficient of the solute in a gel.

2.7.5 Electrochemical Measurement of Solute in Gels

As shown in Fig. 5, the volumetric degree of swelling (V/V_0) of poly-acrylamide (PAAm) gel in a water/acetone mixed solvent changes from a

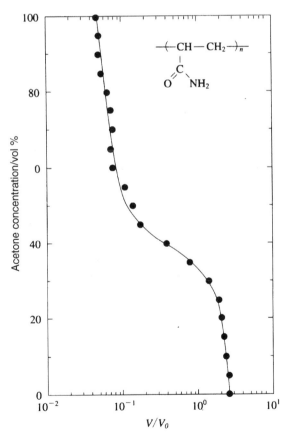

Fig. 5 Volumetric swelling changes of PAAm gel in water/acetone mixed solvents at 25°C.

swollen state to a shrunken state as the acetone concentration increases. The results of electrochemical measurements of the diffusion coefficient and distribution coefficient of the solute will be shown using this PAAm gel as an example.

Prior to the measurement, it is important to select the right active solute and coelectrolyte to add to the solution so as to increase ionic conductivity. Among the common characteristics required for a redox active solute and coelectrolyte is that they have sufficient solubility for the solvent used and at the temperature range to be measured. When change in the degree of swelling of the PAAm gel is to be studied, the coelectrolyte must be soluble regardless of the composition of the mixed water/acetone solvent. N-Ferrocenylmethyl-N,N,N-trimethyl ammonium hexafluorophosphate (Fc) and $LiClO_4$ are an appropriate redox active solute and coelectrolyte, respectively.

In addition to solubility, a redox active solute should possess an electrochemically reversible redox reaction, chemical reactions accompanying the addition of this material should not occur, and there should be no adhesion on the electrode surface regardless of the degree of swelling of the gel. The aforementioned Fc satisfies all these requirements when it is used with PAAm gel (see Fig. 6).

For the actual measurement, the mixed water/acetone solution has had added to it 1 mM of a solute Fc, which functions as a probe molecule to measure the diffusion coefficient and concentration of the solute in the gel and a coelectrolyte, $LiClO_4$, at 0.1 M (it is desirable to add 100 times the concentration of the probe molecule). Figure 7 shows the structure of a cell used for electrochemical measurement. Ordinarily, electrochemical measurement is performed by a 3-probe type. However, with the microelectrode, current becomes of the order of nA and thus measurement using a 2-probe type is also possible. The SCE is used as a reference electrode and it is connected to a potentiostat and function generator; the measured value is recorded with an X-Y recorder. All electrochemical measurements

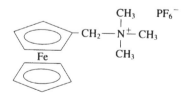

Fig. 6 Chemical structure of Fc.

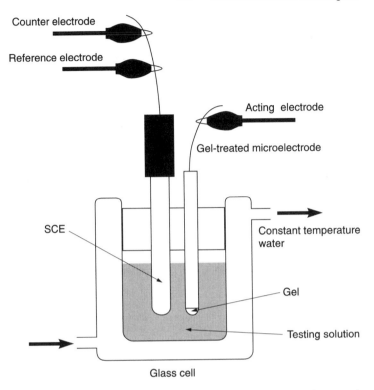

Counter electrode

Reference electrode

Acting electrode

Gel-treated microelectrode

SCE

Constant temperature water

Gel

Testing solution

Glass cell

Fig. 7 Electrochemical measurement methods using a gel-treated micro-electrode.

should be done in a Faraday cage in order to minimize the electrical noise. The gel-fixed electrode is first inserted into the test solution. After the gel reaches swelling equilibrium and the solute Fc achieves distribution equilibrium between the gel and solution, electrochemical measurement begins. Although the time necessary for achieving these equilibria depend on the size of the fixed gel on the electrode, when the hemispherical gel radius was 2.5 mm in the swollen state, it was left standing for 48 h.

Figure 8 shows the results on the diffusion coefficient of Fc in a water/acetone mixed solution obtained from the cyclic voltammogram using a microelectrode without the fixed gel and Eq. (3). This diffusion coefficient corresponds to the diffusion coefficient D_0 of Fc in the external solution in Fig. 1. As the viscosity of acetone is less than that of water, D_0 is larger in acetone. However, the D_0 value does not change linearly with respect to the solvent composition and it shows a minimum at an acetone

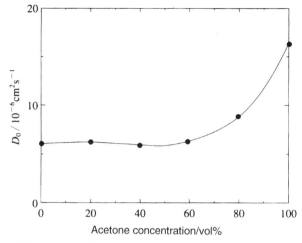

Fig. 8 The diffusion coefficient of 1 mM Fc in the water/acetone mixed solvent with 0.1 M $LiClO_4$ at 25°C.

concentration of 60 vol%. Figure 9 depicts the electrochemical responses of Fc in the gel that is immersed in such a solvent. In the order of A (A') ⇒ D(D'), the concentration of acetone increases; A–D are cyclic voltammograms when sweeping rate is slow. A steady-state current response without hysteresis is observed. The limiting current value for this condition is given by Eq. (3).

On the other hand, A'–D' are the results with faster sweeping rate. They show a clear current peak. The value of this peak current is given by Eqs. (4) and (5). When the acetone concentration became more than 60 vol%, cases C(C') and D(D'), the current value became extremely small. In addition, the limiting current and peak current values could not be obtained. This is possibly due to the reduction of the diffusion coefficient or the reduction of the solubility of Fc in the gel as it shrinks.

Table 1 lists the average values of D and C of Fc in the individually synthesized three PAAm gel samples under the same conditions. The values of D_0 and $C_0 = 1$ mM in the external solution; W/W_0 indicates the weight-based degree of swelling of PAAm. Accordingly, by employing the technique discussed here, it is possible to obtain information on the free volume and change in polarity of the gel networks. Furthermore, the diffusion coefficient and distribution coefficient of the solute that reflect directly the structure of the solvent in the gel networks also can be independently obtained.

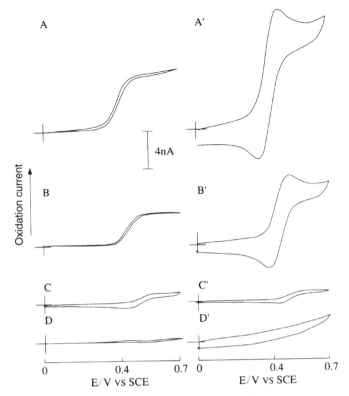

Sweep rate: A–D, 1mV/s; A′–D′, 300mV/s; solvent
composition (acetone concentration): A, A′,0 vol%;
B, B′, 40 vol%; c, C′, 60 vol%; D, D′, 80 vol%

Fig. 9 The cyclic voltammogram of Fc measured by a PAAm gel treated
microdisk electrode (diameter 50 mm). The gel was at equilibrium swelling
state in a water/acetone mixed solvent with 0.1 M Fc, 0.1 M LiClO$_4$ at 25°C.

Table 1 Electrochemical measurement results of PAAm gel at 25 °C.

	Water/acetone (volume ratio)					
	1/0	**0.8/0.2**	**0.6/0.4**	**0.4/0.6**	**0.2/0.8**	**0/1**
W/W_0^*	20.0	18.5	6.4	2.1	1.9	1.6
$D \times 10^6$ /cm^2 s^{-1}	3.3	4.5	1.3	—	—	—
C/mM ($C_0 = 1$ mM)	1.4	1.0	1.3	—	—	—
$D_0 \times 10^6$ /cm^{2}s^{-1}	6.3	6.2	5.7	5.9	8.8	16.3

* W/W_0 is the weight based degree of swelling.

Finally, a unique electrochemical analytical method of gel using microelectrodes will be briefly introduced. This method involves measurement during changes of the degree of swelling using microdisk array electrodes as shown in Fig. 10 [175]. A disk array electrode is manufactured by using a fine glass capillary to encase the platinum wire and fuse several of these glass capillaries. With this array electrode there is the possibility that an individual electrode can serve the purpose. As the size of the individual microelectrode is on the order of micrometers, the phenomenon of spatial inhomogeneity within the gel can be detected. Furthermore, by conducting the measurements as a function of time, time-resolved spatial changes can be studied.

Figure 10 shows the PAAm gel that is swollen in a solution with 1 mM Fc and 0.1 M LiClO$_4$, which is later immersed in an acetone/water mixed solvent system (70/30) that contains LiClO$_4$. This figure depicts the time change of the electrolysis current when 0.7 V, based on the SCE potential, which is much higher than the oxidation potential of Fc, is applied. As the radius of the hemispherical gel that covers the array electrodes is approximately 2.5 mm, the time scale in Fig. 10 is at the beginning stage of gel swelling. The current value in Fig. 10 displays the time-dependent diffusion limited current of the central and external electrodes in the array electrodes. This current value is the time-dependent limiting current value expressed by Eq. (3), which is the change of the value proportional to the product of D and C of Fc within the gel.

Neither the current detected at the central electrode (I) nor the electrode near the rim (O) reduces monotonically as the gel shrinks.

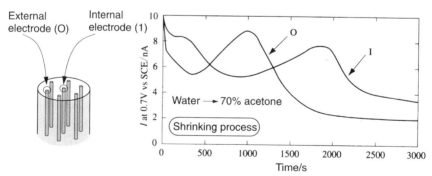

Fig. 10 Time dependence of the limited current when the PAAm gel that was at equilibrium swelling state in a water/acetone mixed solvent with 1 mM Fc, 0.1 M LiClO$_4$ at 25°C was transferred to the water/acetone mixed solvent (3:7).

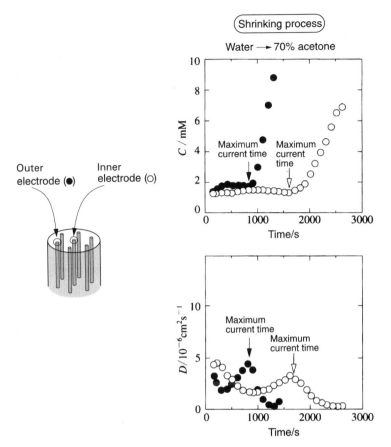

Fig. 11 Time-dependent change of the concentration and the diffusion coefficient of Fc in the PAAm gel in the same condition as in Fig. 10.

Rather, they show a specific peak in the middle. The time to exhibit this peak is shorter in the O electrode, and longer in the I electrode. Figure 11 shows the results of the individually determined D and C from the peak current value by fast sweep rate cyclic voltametry as a function of time, changing the solvent from water to 70% acetone solution and using the same gel-treated electrode. The current peak corresponds to the maximum of the diffusion coefficient of Fc and the time needed for this maximum to appear is less in the O electrode and more in the I electrode. Although it is not yet clear as to why the maximum appears in the diffusion coefficient, it is hypothesized that it relates to the exchange process of the solvent from water to acetone in the gel. As shown in Fig. 8, the diffusion coefficient of Fc increases rapidly in the region of the acetone concentration, by more than 60%. In the gel, the solvent exchanges from water to acetone starting from the outer layers. As a result, the gel shrinks and the gel networks suddenly shrink. However, with acetone entering the gel, the diffusion coefficient increases temporarily. Due to these opposing effects, the maximum in D is thought to appear. In fact, after the peak of D, a rapid increase in Fc's local concentration, which is thought to correspond to the sudden shrinkage of the gel networks, has been observed.

2.7.6   Conclusions

To date, the diffusion coefficient of a solute in a gel has been determined by the membrane technique, magnetic gradient nuclear magnetic resonance (NMR) spectroscopy, and fluorescence polarization relaxation. Compared to these techniques, the characteristics of the method described in this section include: (1) the information obtained by this method is the value near the electrode and the electrode itself acts as a microprobe; (2) the diffusion coefficient and distribution coefficient of the solute can be simultaneously determined; (3) not only the gel at equilibrium state but also a non-equilibrium state gel can be studied; and (4) relatively simple apparatus can be used to make measurements. Future developments that will build on these advantages is desired.

On the other hand, this technique has a shortcoming: it is necessary to chemically fix the gel onto the electrode surface. Therefore, the swelling and shrinking of the gel is not isotropic and one side is restricted. A question also remains as to whether the structure of the gel at the electrode interphase is the same as in the bulk.

2.8   VISCOELASTICITY, MECHANICAL PROPERTIES

2.8.1   Viscoelasticity and Superposition Principle

KUNIO NAKAMURA

2.8.1.1   Introduction

Studies on the gel state and the sol state have attracted many colloidal chemists. It has been the subject of important research and many studies have been reported on the internal structure and origin of elasticity [176].

An interesting point on the mechanical properties of gels is that, although gels appear to flow on earth due to gravity, elasticity attempts always to maintain the shape of gels. Figure 1 compares the shapes of a poly-acrylamide-water gel under ordinary gravity and microgravity. Here, the structure of various gels and their relationships with the viscoelastic and mechanical properties will be described.

2.8.1.2   The simplest mechanical models of sols and gels

There are a number of simple mechanical models for sols and gels [177]. One mechanical model, the Zehn model, is well known [178]. This model consists of a parallel connection of the Maxwell model and a spring with modulus G_e, whereas the Maxwell model has series connection of a spring

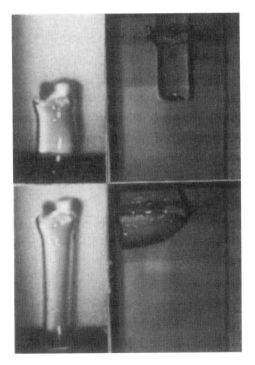

Upper photograph: under normal gravity
Lower photograph: under microgravity
The left side is when the gel is placed on a flat surface
The right side is when the gel is hanging
Sample: acrylamide gel
Experiment: JAMIC drop tower (Kami-sagawa shi)

Fig. 1   "Recovery" of a soft gel under microgravity.

with modulus G and dashpot with viscosity η. Complex modulus $G^*(\omega)$ is expressed by the following equation:

$$G^*(\omega) = G_e + \frac{G\left(\dfrac{i\omega}{\omega_0}\right)}{\left[1 + \left(\dfrac{i\omega}{\omega_0}\right)\right]} \tag{1}$$

where ω is the angular frequency, ω_0 is the characteristic relaxation rate, and its inverse τ_0 is the characteristic relaxation time. In this mechanical model the relationship $G\tau_0 = \eta$ holds. The ratio of the viscosity η and modulus G is τ_0. When τ_0 and angular frequency ω have the relationship $\omega\tau_0 = 1$, then the loss modulus G'' exhibits the maximum. Thus, ω_0 is the median frequency of relaxation. Figure 2 shows the frequency dependence of a complex modulus based on the data of an aqueous solution of agarose. The complex compliance and complex viscosity are also calculated assuming linear viscoelasticity. The storage modulus shows a characteristic maximum as complex compliance in the low frequency region where frequency dependence is small. Thus, the viscoelastic properties of gels are often expressed by complex compliance. The Maxwell model is used as a model for sol, corresponding to $G_e = 0$ in Eq. (1) (Fig. 3). Such a simple model often precludes the ability to describe the relaxation in the sol-gel transition of polysaccharide aqueous solutions. This makes a multielement model a necessity. Aside from the

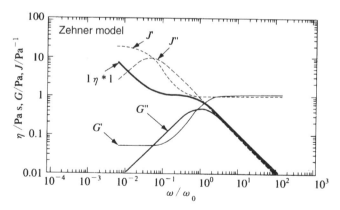

Fig. 2   The viscoelastic behavior of a gel with the Zehner model.

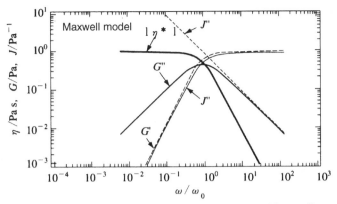

Fig. 3   The viscoelastic behavior of a gel with the Maxwell model.

use of these models, various empirical equations on the mechanics and dielectric relaxation phenomena have been reported on for years.

There is an ongoing attempt to find a proper function that fits the viscoelastic property data and to calculate the strength of mechanical relaxation and parameters related to relaxation and its distribution. The method used to describe linear viscoelastic behavior by these mathematical models is described in detail in the monograph by Tschoegel [179]. There is an example of the analysis of viscoelastic behavior data obtained using this method where the sol-gel transition of a gellan aqueous solution was the subject [180].

2.8.1.3   Superposition principle

For the temperature changes of viscoelastic relaxation, there have been many studies and the analysis of viscoelasticity using the temperature-time superposition principle has been done many times. Sol-gel transition is, on appearance, a simple transition from the liquid state to the solid state. It cannot be expected for the time-temperature superposition to be applicable in a wide temperature range when there are complex structural changes within the gel.

2.8.1.3.1   Time-temperature and time-concentration superposition principle: crosslinked rubbers and swollen crosslinked polymer gels

If chemically crosslinked rubber is considered a xerogel of a swollen network polymer, its viscoelastic properties exhibit fundamental mechanical properties [181]. Assuming that G_e in Eq. (1) is the modulus of the rubbery state and G is the modulus of the glassy state, the majority of viscoelastic behavior can be expressed. When angular frequency approaches zero, the compliance J_e is proportional to the degree of polymerization between two crosslink points Z_0. The angular frequency shows that the maximum of J'' is inversely proportional to the product of Z_0^2 and the monomer frictional coefficient ζ. The superposition principle that takes into account the changes of Z_0 is called f-shift. The viscoelastic properties of rubbers change significantly by temperature and their mode of change has certain rules. When Z_0 is independent of the temperature, the complex compliance shows the time (frequency)-temperature super-position principle:

$$J_T^*(\omega) = \left(\frac{T_0 \rho_0}{T \rho}\right) J^*(\omega a_T) \tag{2}$$

where T_0 is the arbitrarily decided reference temperature, and ρ_0 and ρ are the densities at T_0 and T, respectively. Although the change in density at temperature T and the correction of viscoelasticity, taking into account the temperature dependence of the rubber elasticity, are needed, the majority of viscoelasticity is determined by the shift factor $a_T = \zeta_T / \zeta_{T0}$, which is a function of only the temperature.

Poly(2-hydroxyethyl methacrylate) (PHEMA), which is a typical swollen crosslinked polymer gel, is an interesting material from the viewpoint of its application as a medical polymer as well as from its interaction with water because it is both a hydrophilic/hydrophobic polymer. Its gel structure depends strongly on the concentration of the diluent and crosslinking agent used for crosslinking polymerization. When water is used, the effect is especially noteworthy. When the water content in the polymerizing solution is less than 40%, the gel becomes homogeneous. On the other hand, if the water content exceeds this critical dilution, a heterogeneous, porous gel is obtained. For the swelling dependence of viscoelasticity of the homogeneous gel, the following time-temperature superposition principle holds:

$$J_c^*(\omega) = \left(\frac{v_2}{v_{2,0}}\right) J^*(\omega a_c) \tag{3}$$

The master curves of J' ($\bigcirc$) and J'' ($\bullet$) for the polymers with various dryness are shown in Fig. 4 [182]. As the volume fraction of the polymer v_2 increases, the master curve shifts towards lower frequency without significantly changing shape. The temperature dependent shift factor a_T can be explained by free volume. Figure 4 suggests that the concentration dependent shift factor can be treated similarly. As with time/temperature, time and concentration are equivalent. The PHEMA changes from a glassy to a gel state gradually as a function of water content. The magnitude of this effect depends on the free volume, namely, the viscosity, of the pure solvent. As the water content at equilibrium swelling of PHEMA is 40% and it is a relatively hard gel, this polymer is used for contact lens application. The examples stated thus far are strong gels with covalent bond crosslinking for which the concentration superposition principle applies.

On the contrary, for physical gels formed by so-called secondary bonds, such as entanglement, hydrogen bonding, hydrophobic interaction and static interaction, this superposition principle does not apply over a wide temperature and concentration range. In fact, by examining whether this working hypothesis applies or not, gels can be classified as either chemical gels or physical gels.

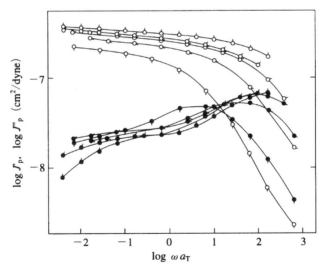

The volume fraction of water from upper curve to lower was
$v_1 = 0.475, 0.434, 0.418,$ and 0.378

Fig. 4 Master curves of storage and loss compliances for PHEMA gel.

2.8.1.3.2   Network structure changes and superposition principle: network gels and biogels by microcrystallites

Various superposition principles for the physical gels whose network structure alters as a result of external environmental factors such as temperature and concentration have been proposed. Poly(vinyl chloride) (PVC) is a commodity polymer that is in a glassy state at room temperature. Without special chemical crosslinking, a soft, strong gel is formed upon the addition of a large amount of plasticizer and this polymer then can be used in such applications as hoses. Plasticization effects on PVC cause a dilution effect that lowers the glass transition temperature and filler effect due to the increased number of microcrystallites. For weak gels crosslinked by microcrystallites, the following time-temperature superposition principle holds:

$$J_{bl}^*(\omega) = q J_{bl,0}^*(\omega a_f) \tag{4}$$

where a_f is the shift factor taking into consideration the aforementioned shift factor by concentration and free volume changes due to the existence of microcrystallites and q relates mainly to the concentration of micro-crystallites. From Eq. (4), if loss tangent $\log\big(J_{bl}''(\omega)/J_{bl}'(\omega)\big) = \log \tan\delta_{bl}(\omega)$ is plotted against $\log \omega a_f$, a master curve can be obtained. The J'' master curve drawn by using such a shift factor a_f is shown in Fig. 5 [183]. Due to the melting of the microcrystallites as temperature increases, plateau compliance increases. From the magnitude of the vertical shift, q can be evaluated. Furthermore, depending on the degree of plasticization, the shape is different and the relaxation mechanism changes, this resulting in nonapplicability of a simple time-temperature superposition.

The gels described to this point have been soft but with sufficient rigidity to maintain their shape under normal gravity. We will next consider gels of much lower modulus. Egg albumen exhibits a modulus of about 10 Pa shortly after the egg has been laid. This turns into a sol if it is kept at room temperature. This is called the waterization phenomenon. If the number of crosslink chains per unit volume n_c becomes f times by going from the standard storage time S_0 to S, $J_s(\omega)$ can be generally written as,

$$J_s^*(\omega) = \left(\frac{1}{f^{2.4}}\right) J_{s,0}^*(\omega a_s) \tag{5}$$

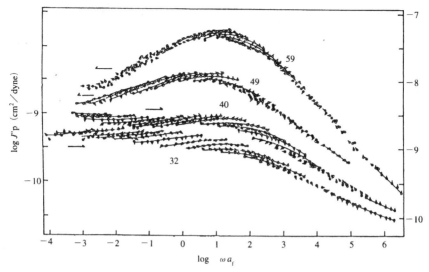

The number in the graph is the concentration of the plasticizer in wt%.
The reference temperature for the 32% sample is 35.5°C; for 40%,
19.6°C; for 49%, −0.4°C; and for 59%, 25°C.

Fig. 5   The master curves of loss compliance of plasticized PVC gel (PVC-DOP system) using Eq. (4).

where a_s is the shift factor, which relates to free volume change accompanying storage time S. This is similar to Eq. (4), wherein superposition by loss tangent holds. Figure 6 shows the change of viscoelastic properties during the waterization process of egg albumen gel obtained from the egg immediately after it was laid by keeping the gel at 25°C [184]. The loss tangent shows superposition at a high-frequency region. However, at a low-frequency region, the mismatch becomes obvious.

In the same figure, the master curves of J_s' and J_s'', which are obtained by vertical shifting, are also displayed. There is a region around $\log \omega a_s = 2$ where the maximum of J_s'' is observed (main relaxation). Below this frequency, there is another region where J_s'' again increases (subrelaxation). The viscoelasticity of the main relaxation of the sample obtained immediately after the egg was laid can be mostly expressed by a simple 3-element model (see Fig. 2). The superposition of Eq. (5) holds well for the main relaxation region. However, for the subrelaxation region, the degree of superposition gradually worsens as a function of storage time due to the increased relative intensity of the

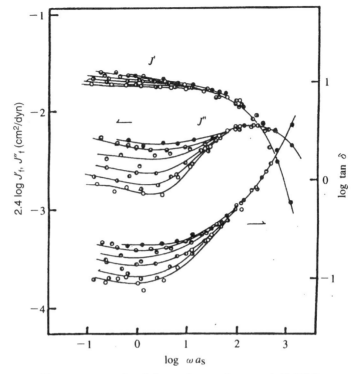

The measurement and storage temperatures are both 25°C;
the reference state is set at $s = 0$ days

(○) 0 days, (◑) 1 day, (◐) 3 days, (◓) 6 days, (◒) 9 days, (●) 17 days

Fig. 6 Master curves of loss tangent and complex compliant of a concentrated egg albumen (Shaver species) at various storage times and s.

subtransition compared to the main relaxation. The loss of networks by waterization accompanies not only the change of f but also the intensity increase of the subtransition caused by the flow of the nonfunctioning chains.

2.8.1.4 Viscoelasticity during gelation

2.8.1.4.1 Sol-gel transition
Because changing a viscous fluid into an elastic solid is chemical cross-linking, the sample exhibits a viscoelastic response during the reaction. As a wide variety of structures appear during the intermediate stage, the relationship between the molecular structure and viscoelasticity has been

extensively studied. By simplifying these complex structural changes, recent work has attempted to analyze viscoelasticity by correlating the polymerization process of multifunctional monomers and percolation. It is difficult to analyze gel molecules using an ordinary chemical analysis technique. This originates from the fact that, whether the gel is strong or weak, the chemical structures of the starting materials (sol molecules) and reaction products (gel molecules) are essentially the same. The main difference is that the sol molecule has finite size whereas the gel molecule extends to the entire volume of the container. Thus, the measurement of macroscopic physical quantities is relatively easy. Furthermore, it has the advantage of clearly reflecting the sol-gel transition. The viscosity of a solution increases as it approaches the gel point and becomes infinite at the gel point, where 3D infinite networks are formed (loss of fluidity). As soon as the gelation is passed, elasticity due to the 3D networks is observed (appearance of solid). Assuming that the extent of reaction at the gel point is p_c, this indicates that viscosity and elasticity have the following relationships with the normalized extent of reaction, $\varepsilon = |p - p_c|/p_c$:

$$\eta_0 - \left(\frac{|p - p_c|}{P_c}\right)^{-s} = \varepsilon^{-s} \text{ for } p > p_c \tag{6}$$

$$G_0 - \left(\frac{|p - p_c|}{P_c}\right)^{t} = \varepsilon^{t} \text{ for } p < p_c \tag{7}$$

Winter and Chambon [185] investigated in detail the viscoelastic behavior of poly(dimethyl siloxane) (PDMS) near the gel point by stopping the crosslinking reaction at various stages. At $p > p_c$, a frequency independent modulus is observed when the frequency approaches zero (Fig. 2). At $p < p_c$, the relationship is characterized by $G' - \omega^2$, and $G'' - \omega$ when the frequency similarly approaches zero [186]. They reported for the first time that the following relationship holds in a wide frequency range at $P = P_c$:

$$G^*(\omega) = G\left(\frac{i\omega}{\omega_0}\right)^{n}, \quad \tan\left(\frac{n\pi}{2}\right) = \frac{G''}{G} (0 \leq n \leq 1) \tag{8}$$

where n is $\frac{1}{2}$.

Figure 7 illustrates the viscoelastic behavior near the critical gel concentration of gellan whose crosslink points are thought to form by

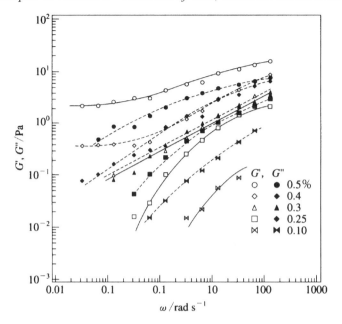

Fig. 7   Frequency dependence of complex moduli of gellan aqueous solution at various concentrations at 25°C.

association of helices. Viscoelastic behavior is characterized by the power law with $n = \frac{2}{3}$ at the critical gel concentration, $\phi_c = 0.3\%$. When it is higher than this concentration, then a constant modulus is observed at the low-frequency region; when it is lower, a constant viscosity ($\eta' = G''/\omega$) is observed.

The value of the powers, superscripts s and t, in Eqs. (6) and (7) are 1.3 and 1.8, respectively, by replacing p with the polymer concentration. There are many careful measurements reported on the modulus near the gel point. Tokita and Hikichi [187] measured the modulus of dilute agarose aqueous solution as a function of temperature and determined the value of t to be 1.9 and the critical temperature, $T_c = 80°C$. The concentration dependence of the agarose aqueous solution at 20°C was also measured and obtained as $\phi_c = 0.0137$ g/100 mL, $t = 1.93$. Gauthier–Manuel *et al.* [188] measured the dynamic viscoelasticity ($f = 10^{-3}$ Hz) of silica particle suspension as a function of reaction time and obtained $t = 2$. The experimental value is closer to the index 1.9 of the 3D percolation theory than is $t = 3$ of the Flory–Stockmayer theory. Accordingly, the modulus of the gel in the vicinity of the transition can be well expressed with the percolation theory. The fact that the temperature, concentration and reaction time-dependence indices are all approximately 2 is the common characteristic of the structural aspects of gels. This does not depend on the mechanism of crosslink formation. Rather, it is considered to be due to the similarity of the gel structure, leading to the universality of the critical phenomena.

Figure 8 is the plot of the viscosity and modulus, which are obtained from the index for the aforementioned aqueous solution of gellan, against the polymer concentration. The characteristic viscoelastic properties of the gel point follow the exponential law. In other words, if the frequency approaches zero, the viscosity increases limitlessly following ω^{n-1} and the modulus decreases in proportion to ω^n. Under ordinary frequencies, this solution has finite viscosity and modulus. If $n = \frac{1}{2}$, from Eq. (8), $G''/G' = 1$ provides a critical gel condition. This agrees with the conclusion from the gelation study of the gelatin aqueous solution by Djabourov *et al* [178]. If $n = \frac{2}{3}$, $G''/G' = 1.73$ is the critical condition. In the case of condensation polymerization products of tetraethoxysilane (TEOS), Hodgson and Amis have reported $n = 0.72$ [189]. They proposed the superposition principle of complex modulus G_p^*, which takes into

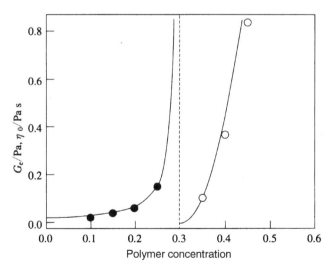

Fig. 8 Change in mechanical properties of gellan aqueous solutions near the gel point.

account the extent of reaction:

$$G_p^* = \left(\frac{1}{a_v}\right) G_{p,0}^*(\omega a_h) \tag{9}$$

where the vertical axis shift factor can be expressed as $a_v = \varepsilon'$, and the horizontal axis shift factor is $a_h = e^{-s-t}$. It is interesting to be able to determine indices t and s from the combination of the viscoelasticity data at various reaction stages.

2.8.1.4.2   Description of sol-gel transition by exponential law

Aside from the empirical expression by Eq. (1), which describes the changes of viscoelasticity accompanying the sol-gel transition, application of Eq. (8) is also possible. That is, when $n = 0$ the loss tangent becomes zero and pure viscoelastic properties like $G'' = 0$, and $G'(\omega) = G$ can be obtained. On the other hand, if $n = 1$, the loss tangent becomes infinite and the behavior of a pure viscous material with $G' = 0$, and $\eta'(\omega) = G/\omega_0$ can also be obtained. Xanthan aqueous solution remains to be expressed by exponential law relaxation behavior rather than the mechanical relaxation indicated by the 3-element model despite the changes in the external environment [190]. With the addition of a salt, the value of n approaches 0.2 [191]. Exponential law type relaxation behavior is often seen among biogels. To date, no gels with n from 0 to 1 have been found [192]. However, the 3D infinite networks, which possess a characteristic length defined by the molecular weight between crosslinks, will clearly be elastic gels.

Exponential type gels are possibly observed when the networks assume a hierarchical structure, which does not possess characteristic length. Clarke and Ross-Murphy [193] termed this gel-like behavior.

We have discussed the experimental analysis of viscoelastic behavior accompanying the sol-gel transition. It has been widely known that the temperature dependence of relaxation time can be expressed by the WLF equation when the glass transition phenomenon is treated as a relaxation phenomenon. Similarly, in the sol-gel transition, it will be possible to predict the wide range of viscoelastic behavior if the temperature dependence of relaxation is known. In addition, viscoelastic behavior in the vicinity of the gel point will continue to be an interesting problem.

2.8.2   Viscoelasticity and Its Evaluation Technique of Vibration Damping Gels

HIDEO YAMAZAKI AND SHIGERU KOSHIBE

2.8.2.1   Introduction

Recent industry trends include development of multifunction and high-density devices, which then lead to new problems regarding vibration and impact. For example, for electronic devices such as the OA device and CD-ROM drive, approaches for reduced vibration are needed for accuracy of movement, speed-up, and prevention of wrong operation. However, as the instruments continue to be reduced in size, approaches to handle vibration and impact are becoming increasingly difficult. In the semiconductor industry, development of next generation technology such as miniaturization, high accuracy, automation, and large wafer size are constantly demanded, and problems caused by vibration and impact are attracting attention. In the health and medical fields, prevention of injury from sports-related impact and reduction of discomfort by vibration and noise are also desired.

Traditionally, selection of molecular structure, additives, and complex materials suitable for vibration damping using polymers, such as rubber, elastomer, thermoplastics and thermosetting resins, have been studied and commercialized. However, traditionally used materials have shortcomings, including high modulus, low vibration damping ability, high-temperature dependence, and poor long-term stability. Therefore, these materials cannot cope with many newer problems.

In this section, silicone gels, which are attracting attention as new vibration-absorption materials, will be explained with respect to the viscoelasticity appearance mechanism and its evaluation methods.

2.8.2.2   Mechanism of vibration absorption

Typical approaches for vibration are: (1) vibration prevention; (2) controlled vibration; (3) dynamic vibration absorption; and (4) impact reduction.

The goal of vibration prevention is to lessen the vibration to and from the external environment. For example, when vibration prevention is achieved by a simple vibration system with one degree of freedom, the vibration transferability (the ratio between the vibration that is transferred to the instrument and the external vibration) can be expressed by the following equation (see Fig. 1) [194]:

$$T = \frac{\sqrt{1 + (2\xi\lambda)^2}}{\sqrt{\left(1 + \lambda^2\right)^2 + (2\xi\lambda)^2}} \tag{1}$$

$$T\,(\text{dB}) = 20\ \log_{10} T \tag{2}$$

$$\lambda = \frac{\omega}{\omega_n}$$

$$\xi = \frac{c}{2\sqrt{mk}} = \frac{1}{2}\ \tan\delta \tag{3}$$

$$\omega_n = \sqrt{\frac{k}{m}} \tag{4}$$

where T is the vibration transferability; ξ is the damping ratio; $\tan\delta$ is the loss factor; c is the viscosity reduction coefficient; ω_n is the resonance frequency; and k is the spring constant. In the vibration insulation region with vibration transferability, (dB) ≤ 0, the vibration transferred from the environment to the instrument decreases. In order to achieve vibration protection over a wide frequency range, the spring constant of the vibrating system must be small and resonance frequency must also be low (see Fig. 2).

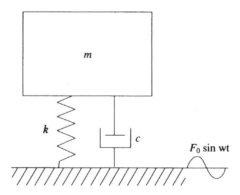

Fig. 1   A vibrating system with one degree of freedom.

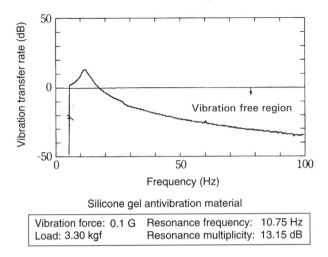

Fig. 2 Frequency dependence of the rate of vibration transfer.

Resonance transferability (resonance harmonics) at resonance frequency is also important. When there is no damping component such as a metallic spring, resonance harmonics become very large, leading to malfunctions and faulty operation. It also prevents effective vibration isolation due to the generation of first-, second-, third- ... nth-order resonances caused by the surging in the high-frequency region. As traditional antivibration rubber has high modulus and low damping coefficient, sufficient antivibration performance cannot be expected for newer integrated small parts. For antivibration materials, an optimum balance of spring constant and damping coefficient is desirable for application purposes.

Controlled vibration reduces vibration by adhering a material with high internal friction to the vibration transfer portion. The effect of controlled vibration can be expressed by the loss coefficient (η) of the vibrating system as a whole. For example, in the case of nonrestrictive controlled vibration materials, the loss coefficient of the vibrating system as a whole by adhering the controlled vibration material to the substrate can be expressed as follows [195]:

$$\eta \simeq 14 \times \tan \delta^2 \times \frac{E_2}{E_1} \left(\frac{H_2}{H_1}\right)^2 \qquad (5)$$

The controlled vibration effect depends on the loss tangent (tan δ) of the controlled vibration material, the modulus ratio of the substrate and controlled vibration material (E_2/E_1), and the thickness ratio of the substrate and controlled vibration material (H_2/H_1). To improve the controlled vibration effect, it is effective to use material with a large loss tangent, increase the modulus compared with the substrate, and increase the thickness of the vibration damping material. A soft silicone gel is suitable for controlled vibration of small, lightweight parts with low substrate modulus. As a peak of tan δ appears near the glass transition temperature, an ordinary controlled vibration material adjusts the glass transition temperature to the use temperature by means of a polymer blend or additives. However, poor stability of a controlled vibration effect due to the strong temperature dependence of the viscoelastic properties is still a problem.

Dynamic vibration absorption handles vibrational energy and reduces the vibration of the main vibrating system by a dynamic vibration absorption device with a mass (m), spring constant (k), and viscosity damping coefficient (c). As dynamic vibration absorption is effective only under optimum conditions, it is necessary to optimize the dynamic vibration system prior to use by analyzing the mode of the vibration of the source. The optimization can be achieved by controlling the characteristic frequency, damping factor, mass, and installation location (see Fig. 3).

Optimum conditions:

$$\frac{\omega_d}{\omega_n} = \frac{1}{1 + \mu} \tag{6}$$

$$\xi = \sqrt{\frac{3\mu}{8(1 + \mu)^3}} \tag{7}$$

$$\mu = \frac{m}{M} \tag{8}$$

where ω_d is the characteristic frequency of the dynamic vibration absorption device, ω_n is the characteristic frequency of the main vibrating system, μ is the mass ratio of the dynamic vibration absorption device and main vibrating system, and ξ is the damping ratio.

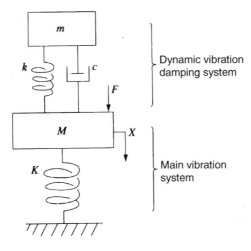

Fig. 3   Mechanical model for the dynamic vibration damping device [196].

Impact reduction reduces impact by changing the impact energy into heat (W_a) using the internal friction of the damping material and reducing the recovery energy (W_b) (see Fig. 4):

$$E = \frac{W_a}{W_a + W_b} \tag{9}$$

$$A = \frac{W}{\left(F_{max}/X_{max}\right)} \tag{10}$$

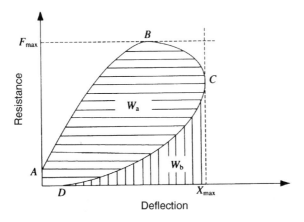

Fig. 4   Resistance-deflection diagram of vibration damping material.

where E is the energy absorptivity; A is the reduction efficiency; F_{max} is the maximum impact strength; and X_{max} is the maximum bending.

In coping with vibration problems, there are various approaches. Depending on the application and object, selection of the modulus of the material and loss tangent will be important. A viscoelastic material such as soft silicone gel is suitable for recent lightweight, precision devices.

2.8.2.3 Characteristics of vibration damping gels

Gel is a polymer and its swollen material has 3D networks insoluble to any solvents. Depending on whether the gel is swollen by water or an organic solvent, gels can be classified as hydrogel or organogel. Hydrogel becomes a gel when the polar groups form hydrogen bonds and retain water. Hydrogels are not suitable as vibration absorption gels due to freezing of water molecules and changes in their properties that result from drying. Silicone gel is an organogel that consists of networks with low crosslink density (solid portion) and unreacted silicone oil that remains inside the gel (liquid portion) (see Fig. 5). As silicone gel uses thermally and chemically stable silicone oil, the effect on its properties by its environment is small and maintains gel conditions over a wide temperature range. Gels can be formed by crosslinking via chemical bonding through covalent bonds or physical bonding such as in aggregation or entanglement. In order to support a device with a gel, chemical bonding is required to maintain shape. Without chemical bonding, the gel will undergo plastic deformation from external forces. Silicone gels are synthesized by reacting an organosiloxane polymer with vinyl end groups and organosiloxane oligomers with SiH groups in the main chain using a noble metal catalyst like platinum, leading to the formation of 3D

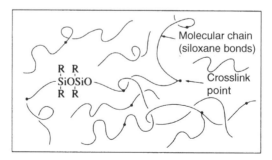

Fig. 5 Structure of silicone gel.

networks. The structure of the raw material, molecular weight, and composition will allow control of the crosslink structure and the amount of unreacted silicone oil. This indicates that the mechanical and viscoelastic properties can be adjusted and the hardness of the gels can be changed from a rubbery to a liquid state. Because the reaction is an addition reaction, the structure after hardening is stable with no byproducts formation. Thus, the various advantages of silicone can be utilized. Silicone gels possess inorganic siloxane bonds and organic groups like the methyl group and, thus, have a very different main chain from ordinary organic polymers. When a siloxane bond and a carbon–carbon bond are compared, the siloxane bond has longer bond length than the carbon–carbon bond, and the bond angle is wider at 140°. This indicates that the dimethylsiloxane chain can spread wider and the freedom of bond rotation is higher than in the other polymers (see Fig. 6) [197].

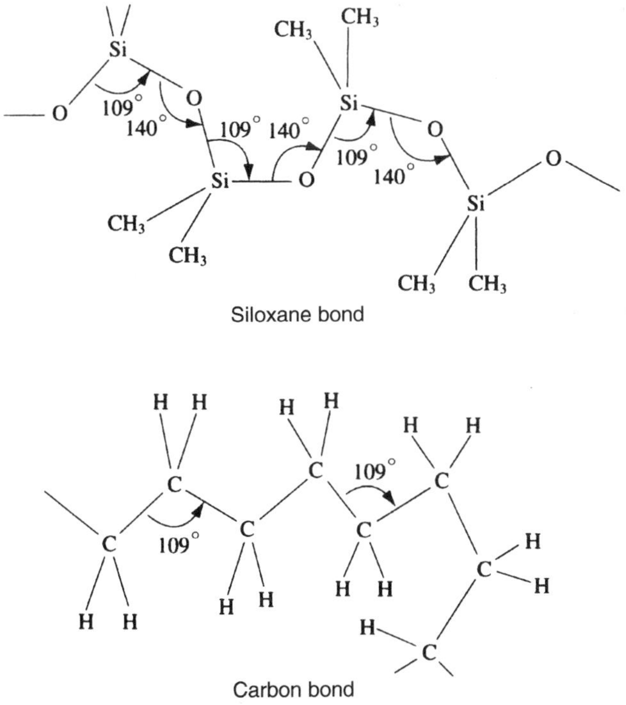

Fig. 6 Comparison between the siloxane and carbon bonds.

Therefore, the siloxane bonds that form the silicone gel backbone is highly flexible and functions like a joint. Moreover, as silicone gels possess a lesser number of crosslink points than rubbers, the structure restriction factor is small and properties can be changed by changing the organic groups on the side chain. Silicone gels also exhibit different thermal properties from synthetic organic rubbers. In the bond energies of the siloxane bonds that compose the main chain of the silicone polymers and the carbon–carbon bond of organic rubbers, the former is 106 kcal/mol, and the latter is 85 kcal/mol. Due to the difference in bond energies, silicone polymers have superior thermal and degradation resistance as compared with synthetic organic rubbers. Silicone gel has no significant property loss even at 200°C in air, and can be used for a long time. Regarding lower temperature, silicone gel has a melting temperature of −50°C and a glass transition temperature of −123°C. Therefore, it maintains a stable gel from approximately −50°C to 200°C, a temperature range of 250 degrees Centigrade.

The low-temperature properties of silicone gel can be further improved by forming a copolymer with a repeat unit having a group that prevents crystallization as in the phenyl group. Furthermore, with the addition of various additives, the modulus, mechanical strength, viscoelastic property, impact damping property, density, electrical property, and magnetic properties are also improved.

2.8.2.4   *Viscoelastic properties of vibration absorbing gel*

When an external vibrational force is applied to a matter it deforms and the intermolecular distance changes, thereby resulting in an increased intermolecular potential (storage of elastic energy). In an elastic body, the mechanical energy that was exerted by the external force will transfer in the form of elastic deformation. Thus, when deformation recovers, the energy given by the external energy is released to the outside of the body and returns to the beginning state. On the other hand, in the case of viscoelastic materials, the intermolecular interaction is not fixed in the same way as for fluids. Part of the elastic energy stored will be consumed as heat due to the friction caused by the slippage of molecules. Hence, after one vibration cycle, part of the mechanical energy will not be returned to the external source. Accordingly, due to a simultaneous occurrence of energy storage and energy dissipation, a relaxation of viscoelastic material is observed. When a strain (e) is applied to a viscoelastic material, the corresponding stress (σ) will not appear spontaneously, but appear after with phase (δ):

$$e = e_0 \exp(i\omega t) \tag{11}$$

$$\sigma = \sigma_0 \exp\{i(\omega t + \delta)\} \tag{12}$$

The stress corresponding to the sinusoidal strain can be divided into the stress component that is the same phase as the strain (storage modulus: G') and the 90° off phase from the strain (loss modulus: G''). The complex modulus G^* is expressed as follows (see Fig. 7):

$$G^* = \frac{\sigma}{e} = (G' + iG'') \tag{13}$$

The angle to express the time lag between the strain and stress is given by the loss factor ($\tan \delta$):

$$\tan \delta = \frac{G''}{G'} \tag{14}$$

The loss factor is the measure for the energy dissipated as heat and the maximum storage energy during one cycle. The loss modulus (G'') is proportional to the energy (ΔE) that is dissipated during one cycle [198]:

$$\Delta E = \int \sigma \, de = \int_0^{2\pi/\omega} \sigma \frac{de}{dt} \, dt = \pi G'' e_0^2 \tag{15}$$

The vibrational characteristics of viscoelastic polymeric materials are influenced strongly by the molecular structure, average molecular weight, molecular weight distribution, and polar structure, and by the shape of the particle, particle diameter, and surface structure of added fillers. When stress and strain are added to a polymeric material, vibration damping

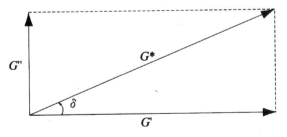

Fig. 7 Complex modulus.

takes place by changing the vibrational energy into thermal energy due to the intermolecular friction and interfiller friction caused by macro-Brownian motion of the main chain and micro-Brownian motion of the side chain. In the case of silicone gel, due to the high degree of freedom of siloxane bonds, the friction caused by micro-Brownian motion and the energy absorption is higher than for other polymers. In other words, the viscous term is larger than the elastic term, which in turn gives a larger loss factor (tan δ). As the temperature is reduced while the frequency and amplitude of vibration are kept constant, the storage modulus (G') suddenly increases at a certain temperature. This temperature is called the glass transition temperature (T_g) and a peak of loss factor (tan δ) appears in its vicinity. This phenomenon is caused during the transition from the rubbery state to glassy state by the freezing of the macro-Brownian motion (free rotation) of the polymer chain (see Fig. 8) [199]. The transition shifts to higher temperature at higher frequency. If the frequency increases while the temperature and amplitude are kept constant, viscoelastic curves similar to the case of temperature reduction, as already stated here, can be obtained. At higher frequencies, the polymer exhibits a glassy behavior. Glass transition temperature (T_g) is higher for those polymers with rigid main chain, voluminous side chain with high steric hindrance, and increased polarity of the substituents.

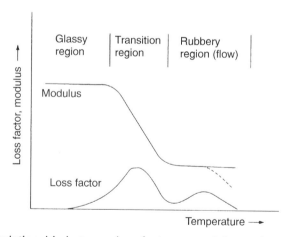

Fig. 8   The relationship between loss factor or modulus and temperature of a polymer.

Just prior to the glass transition temperature, the modulus is high. In the transition region, the tan δ peak appears. However, as both modulus and loss factor drastically change as a function of temperature, careful temperature control is necessary and the application environment is thus limited. Therefore, when a silicone gel is used as a vibration absorption gel, it is used in the soft gel region above the melting temperature, which is higher than the glass transition temperature. In the gel-like region, the modulus and loss factor (tan δ) are stable as a function of temperature. The viscoelastic properties of a silicone gel can be controlled by the crosslink structure and the amount of unreacted silicone oil. Type I silicone gel is prepared by increased crosslink density to make the 3D networks tight and the networks are swollen by a large amount of unreacted silicone oil. For the frequency dependence of viscoelastic properties of type I silicone gel, the storage modulus (G') is almost frequency independent, and loss modulus (G'') and loss factor (tan δ) increase with increased frequency (see Fig. 9) [200]. In the case of type I silicone gel, the 3D network is the origin of elasticity and the unreacted silicone oil makes it soft. As the networks are stable, the friction due to the deformation of the networks is small at the low-frequency region. As the frequency increases the loss factor (tan δ) increases due to the friction caused by the silicone oil and the main chain with micro-Brownian motion. In the type II gel, on the other hand, a smaller number of crosslink points and quasi-crosslink points form soft networks. Thus, the networks easily deform and friction due to the micro-Brownian motion by side chains and unreacted silicone oil can take place effectively from the low-frequency region. Type II silicone gel is slightly inferior to type I silicone gel in ability to maintain shapes. Therefore, when it is used for vibration damping, a combined use with metallic spring will be useful.

We will now consider the viscoelastic properties of silicone gels with a mechanical model. It is possible to express complex mechanical behaviors by properly connecting a spring and dashpot, which are the mechanical models for modulus and viscosity, respectively. The Maxwell model, which is the model to connect a spring and dashpot in series, continues to deform upon application of external force. It is therefore liquidlike and is convenient to express the mechanical properties of a sol, which has elastic properties. In contrast, the Voigt model, which is the model to connect a spring and dashpot in parallel, reaches a finite deformation and exhibits equilibrium. Hence, it is solidlike and is convenient to express a gel that shows loss in mechanical energy [201].

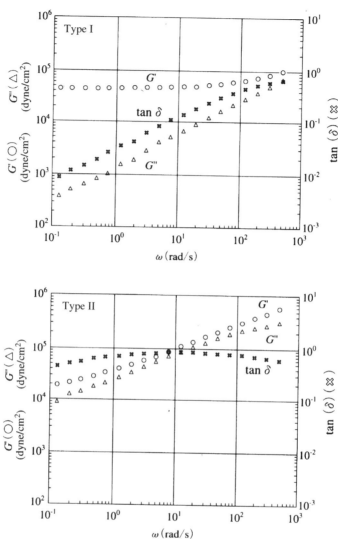

Fig. 9   Frequency dependence of dynamic viscoelasticity of silicone gel.

The measured viscoelastic properties of type I silicone gel do not accurately agree with the G' and G'' calculated by the Voigt model as shown by the broken line. However, the generalized Voigt model does agree with the experimental values as shown in Fig. 10. In this model, Eq.

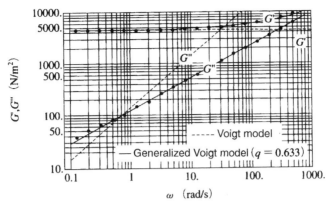

Fig. 10   Modeling of dynamic viscoelasticity of silicone gel (generalized Voigt model).

(16), which is the relationship between the stress (σ) and strain (e), is replaced by Eq. (17) [202–204]:

$$\sigma(t) = G_0 e(t) + \eta \frac{de(t)}{dt} \tag{16}$$

$$\sigma(t) = G_0 e(t) + \eta \frac{d^q e(t)}{dt^q} \tag{17}$$

The Fourier transform of Eq. (17) is

$$\bar{\sigma}(\omega) = G_0 \bar{e}(\omega) + \eta \left[(i\omega)^q \, \bar{e}(\omega) \right] \tag{18}$$

From this, the complex modulus $G^*(\omega) = G'(\omega) + G''(\omega)$ is expressed as follows:

$$G^*(\omega) = \frac{\bar{\sigma}(\omega)}{e(\omega)} = G_0 + \eta \omega^q \left[\cos\left(\frac{q\pi}{2}\right) + i \, \sin\left(\frac{q\pi}{2}\right) \right] \tag{19}$$

Thus, the real part G' and imaginary part G'' of $G^*(\omega)$ are

$$G'(\omega) = \mu \omega q \, \cos\left(\frac{q\pi}{2}\right) + G_0 \tag{20}$$

$$G''(\omega) = \mu \omega q \, \sin\left(\frac{q\pi}{2}\right) \tag{21}$$

Consequently, the viscoelastic properties of silicone gels can be expressed by the generalized Voigt model [202].

2.8.2.5   Evaluation methods of viscoelasticity for vibration damping gels

2.8.2.5.1   Vibration damping gels

In order to determine the viscoelastic properties of silicone gel materials, vibrational measurement techniques are used. The principle of vibrational measurement techniques involves determining the viscoelastic quantities by observing the responding mechanical behavior when cyclic deformation is applied repeatedly. Dynamic viscoelastic property measurements are performed with a dynamic mechanical spectrometer such as the RDA-II from Rheometrics. This instrument measures the complex modulus G^* and loss factor $\tan \delta$ from the torque measured by the transducer equipped at the upper portion of the sample chamber when shear deformation is applied to the sample at a specified shear strain and frequency from the actuator at the bottom of the sample chamber.

The dynamic properties of a silicone gel as a vibration damping material can be obtained from the vibration experiments (see Fig. 11). In this equipment, a silicone gel vibration damping material is placed on top of the vibration table, a disk-shaped weight is placed on top of the sample, and acceleration detectors are placed on the disk weight and the table. These detectors are connected to an FFT analyzer through an amplifier and the acceleration signal is sent. The acceleration frequency response function (vibration transferability) is measured from the acceleration wave function of the vibration table and disk weight by sweeping the frequency of the vibration table. The acceleration frequency response function $G(\mathrm{i}\omega)$ can be divided into a real part G_R, which is in-phase with the added vibration, and an imaginary part G_I:

$$G(\mathrm{i}\omega) = G_R + \mathrm{i}G_I \tag{22}$$

The frequency dependency of spring constant $v(\omega)$ and frequency dependency of loss factor $\eta(\omega)$ of the silicone gel material can be expressed by the G_R and G_I:

$$v(\omega) = \frac{G_R(G_{R-1}) + G_I^2}{(G_{R-1})^2 + G_I^2}\, \lambda^2 \tag{23}$$

$$\eta(\omega) = \sqrt{-1 + \frac{(2G_{R-1})\lambda^2}{(G_{R-1})v(\omega)} - \frac{G_R \lambda^4}{(G_{R-1})v(\omega)^2}} \tag{24}$$

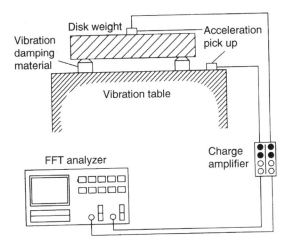

Fig. 11 Flow chart of vibration experiment.

Furthermore, an attempt is made to predict theoretically the dynamic properties of silicone gel vibration damping material by finite element analysis (FEM) using the viscoelastic parameters obtained from dynamic mechanical analysis [205]. It is possible to obtain the frequency response function of silicone gel vibration damping material by incorporating the frequency dependence of the complex modulus and Poisson's ratio as determined by dynamic viscoelasticity experiments. The experimental values and theoretical prediction agree relatively well. However, for soft silicone gels, the experimental values deviate from the theoretical values because the disk weight causes initial strain within the vibration damping material. Hence, analyses were conducted using a composite vibration damping material made of a spring and silicone gel by supporting the disk weight and reducing the initial strain (see Fig. 12) [206]. As a result, in the analysis of the composite vibration damping material, the experimental and FEM values agree well even if a soft and high damping gel is used (see Fig. 13). A similar analysis evaluates vibration control.

2.8.2.5.2 Gels for dynamic vibration absorption devices

It is necessary to install a dynamic vibration absorption device in the optimum position in such a way as to obtain the best conditions for the mass, spring constant and loss factor, because the device absorbs a specific vibration energy at a specified frequency. For this, the vibration mode of the vibration damping material is first obtained using the transfer matrix

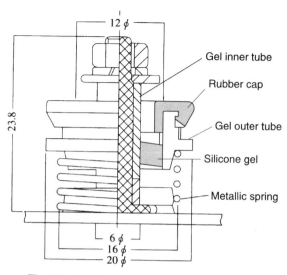

Fig. 12   Coil/spring composite insulator.

method and experimental mode analysis. Then, the viscoelastic properties of the dynamic vibration damping device are determined by specifying the vibration mode of interest. Application examples include metallic baseball bats and a dynamic vibration absorption device that is installed at the head of the golf club [207, 208]. For example, during the use of a metallic bat, the vibration caused when a ball is missed is troublesome. By the analysis of vibration mode, the vibration can be absorbed by installing a dynamic vibration damping device at the grip end (see Fig. 14). The viscoelastic

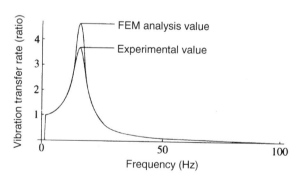

Fig. 13   Frequency response function of acceleration for the composite type insulator [206].

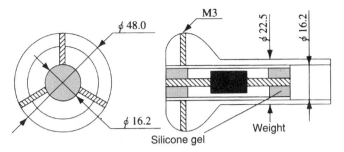

Fig. 14 Attachment condition of a dynamic vibration damping device [207].

properties of the dynamic vibration damping device are obtained from the acceleration frequency response function in the vibration experiment. By installing this silicone gel dynamic vibration damping device into a metallic bat and a golf club, the resonance peak of the target frequency was reduced and vibration was controlled at an early stage (see Fig. 15).

2.8.2.5.3   Gels for impact reduction

The evaluation of impact reduction gels is done by the falling weight testing method (see Fig. 16). The abilities of these impact reduction materials are expressed by the ratio of the maximum impact energy and the energy that is converted to the thermal energy in the course of the cycle (energy absorptivity). To reduce maximum impact, it is necessary to

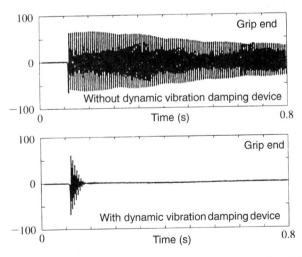

Fig. 15 Effect of dynamic vibration damping device [207].

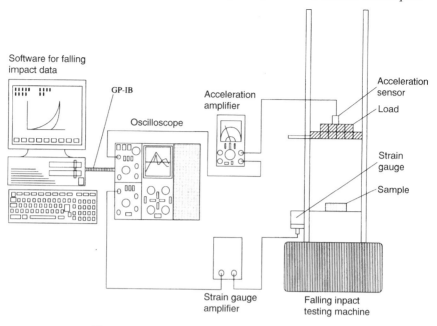

Software for falling impact data

GP-IB

Oscilloscope

Acceleration amplifier

Acceleration sensor

Load

Strain gauge

Sample

Strain gauge amplifier

Falling inpact testing machine

Fig. 16   Falling impact testing system [209].

adopt springlike properties to improve impact reduction efficiency. For silicone gels, an attempt to improve impact reduction efficiency by adding organic balloons has been made. To increase energy absorptivity, use of high internal friction gel and optimum frequency response are necessary. Regarding the viscoelastic behavior of impact reduction gels, the frequency dependence of the spring constant and loss factor is obtained by Fourier transforming the deformation-acceleration response waveform taken from the fallen weight test. As a result, it was found that the frequency dependence of the spring constant and loss factor differ depending on the gel [210]. An effort is now underway to evaluate the optimum viscoelasticity from the viewpoint of the high impact reduction effect. An evaluation method for impact reduction materials involving the quasi-dynamic analysis of shoes is being tested. In order to dampen the impact upon landing, various impact reduction materials have been used. However, there has been little analysis on the continuously changing load caused by walking. Only recently has a quasi-dynamic analytical program for nonlinear material by 3D finite element analysis been developed. By combining this method with a pressure distribution measurement instrument, the system to perform the quasi-dynamic analysis of shoe soles is being evaluated [211].

2.8.2.6 Future challenges

The viscoelastic properties of vibration absorption gels are being revealed but information is at this point inadequate. For the development of new viscoelastic properties, research on subjects that include new fillers and the addition of electrical responsivity are being tried. The future direction of vibration damping gels involves targeting the materials for improvements in the specialized performance and improved precision areas. For this purpose, further detailed analysis on optimization of vibration damping gels is desired.

2.9 THERMAL PROPERTIES

HIROHISA YOSHIDA

2.9.1 Introduction

The viscosity of a material suddenly changes and loses fluidity at the gel point. Techniques to follow this phenomenon as a function of temperature are called thermal analysis techniques. According to the definition of the International Confederation of Thermal Analysis and Calorimetry, thermal analysis is a series of collective techniques to measure the physical properties of a material (or a reaction product) by changing the temperature according to a certain program [212, 213]. There are various thermal analyses depending on the physical properties to be measured. In this section, differential scanning calorimetry (DSC), which is the technique to measure heat capacity of the sample, and thermomechanical analysis (TMA), which measures the viscosity or modulus, will be discussed.

2.9.2 Measurement Techniques

As shown in Fig. 1, DSC measures the temperature and heat flux of the sample when it is heated or cooled at a constant rate (dT_f/dt) when a sample is placed in the sample cell and an inert material is placed in the reference cell [214]. Assume that the temperature of the furnace, reference cell, and sample cell are expressed as T_F, T_R and T_S, respectively. Similarly to T_F, the T_R and T_S also increase the temperature at the same

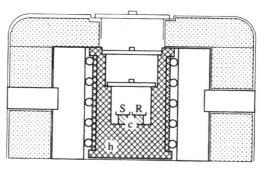

S: Sample cell, R: Reference cell,
c: Thermocouple, h: Heat sink

This DSC is superior to ordinary DSC in thermal
insulation and sensitivity to heat flux is higher

Fig. 1   Structure of high sensitivity DSC.

rate shortly after the temperature begins to rise and becomes $T_F > T_R > T_S$. Under this steady state, the following holds:

$$\frac{dT_F}{dt} = \frac{dT_R}{dt} = \frac{dT_S}{dt} \tag{1}$$

At a certain time t under the steady state, $T_F - T_S$ is constant, the amount of heat that transfers from the furnace to sample cell per unit time, that is, heat flux (dQ_S/dt) can be expressed by Eq. (2) using the thermal resistance between the furnace and the sample cell:

$$\frac{dQ_S}{dt} = \frac{1}{R}\left(T_F - T_S\right) \tag{2}$$

Because the added heat heats the sample and the sample cell, if the heat capacity of the sample and the sample cell are expressed as C_S and C_{C_S}, then the heat flux to the sample under the steady state can be given by the following:

$$\frac{dQ_S}{dt} = \left(C_{C_S} + C_S\right)\left(\frac{dT_F}{dt}\right) \tag{3}$$

The same equation also holds on the reference cell side. When the equations for heat flux of the sample and reference cell are subtracted, the heat capacity and weight of the sample cell and reference cell are equal, and the reference cell is empty, then the heat flux to the sample per unit time can be expressed by a simple equation (4) that depends only on C_S:

$$\frac{dQ_S}{dt} = -\frac{1}{R}\,\Delta T = \frac{C_S}{dt}\frac{dT_F}{dt} \tag{4}$$

$$\Delta T = T_S - T_R \tag{5}$$

In DSC, the ΔT is measured by a thermocouple or thermopile and its output is a function of the heat capacity of the sample. Detecting the amount of heat per unit time makes it difficult to study a slowly occurring phenomenon because the ΔT will be small. The sensitivity of DSC equipment is determined by the sensitivity of the thermocouple or thermopile used. Generally, the ΔT of a gel sample is small and difficult to study with ordinary DSC equipment. In such a case, a high-sensitivity DSC must be used.

As shown in Fig. 2, TMA is a technique to measure the sample deformation as a function of temperature or time by applying a static or dynamic load to the sample through the probe in the sample cell [212]. Various measurements are possible by changing the type of the probe. In recent instruments, cyclic load can be applied at multiple frequencies simultaneously. For gel samples, by modifying the sample cell and measuring with a vibrating mode, dynamic shear modulus can be obtained in a manner similar to a dynamic viscoelastic instrument. For example, the probe system that can be expressed by the equation of motion, Eq. (6), will be considered as follows:

$$F = F_S + kx + \mu\frac{dx}{dt} + m\frac{d^2x}{dt^2} \tag{6}$$

where F, F_S, k, μ, m are the stress signal, the load that is applied to the sample, recovery force of the probe system, viscosity coefficient, and the mass of the probe, respectively. Also, x and t are displacement and time, respectively. When a sinusoidal shear deformation is added to the sample that shows Newtonian viscosity expressed by Eq. (7) using the probe as described here, the stress signals can be expressed by Eq. (9):

$$F_S = \alpha\eta^*\frac{dx}{dt} = \alpha G^*x \tag{7}$$

$$x = x_0\,e^{i\omega t} \tag{8}$$

$$F = x_0 e^{i\omega t}\{k + i\omega(\alpha\eta^* + \mu) - \omega^2 m\} \tag{9}$$

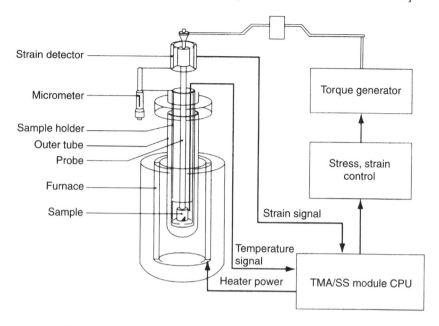

For the study of gels, a sample container is fixed inside the sample holder, probe is inserted 10 mm into the gel, the stress response is measured by 10 μm microdynamic mechanical deformation, and the stress response is detected

Fig. 2   Construction of TMA.

where α, η^* and G^* are the shape factor, complex viscosity, and complex modulus of the sample, respectively, and X_0 is the amplitude of the displacement and ω is the angular frequency. The viscosity of the sample can be expressed as the real part of η^* and is expressed by the following:

$$\eta = \frac{1}{\alpha}\left(\frac{1}{\omega}\frac{F_0}{x_0}\sin\delta - \mu\right) \qquad (10)$$

where F_0 is the amplitude of the stress signal and δ is the delay of the stress signal for the corresponding strain signal in radians. Hence, if the instrument constant is predetermined using a standard sample of known viscosity, the dynamic rigidity of a gel sample can be evaluated as a function of temperature.

2.9.3   Measurements of Gelation Process

Polymer gels can be classified into chemical gels and physical gels depending on the formation mode of crosslink points. Table 1 lists the phenomena that are the subjects of thermal analyses. As gelation accompanies a large change in viscosity, it is easily observed by TMA. However, use of DSC is limited to the phenomenon that accompanies thermal exchange, such as phase transition. Also, as described here, DSC will have difficulty in measuring the phenomenon with small dQ_S/dt. In either method, only the gelation process is observed for a chemical gel as it is an irreversible phenomenon. By contrast, in the case of a physical gel, sol-gel transition can often be observed as a reversible phenomenon. It is possible to observe the exotherm of covalent bond formation for the chemical gel. For the physical gel, if the crosslink structure consists of the association of polymer chains or crystallites, the melting phenomenon of those structures can be observed by DSC.

2.9.3.1   *Crosslinking by association of molecular chains*

In general, gel formation of polysaccharides takes place via formation of molecular association. There are various mechanisms as listed in Table 1. In reality, various modes are combined to become the mechanism of gel formation.

Figure 3 shows the DSC thermograms of agarose aqueous solution at various concentrations during the heating and cooling processes. The endothermic peak observed during the heating process is assigned to the gel-sol transition and the exothermic peak observed during the cooling process is attributed to the sol-gel transition [215]. Agarose is thought to

Table 1 Gelation phenomena which can be measured by thermal analysis.

Type of gels	Gelation mechanisms	Measurement methods
Chemical gels	**Gelation by covalent bonds**	**DSC, TMA**
Physical gels	Association molecular chains: polysaccharides, proteins	DSC, TMA
	Association of helices	DSC, TMA
	Conformation changes of molecular chains	DSC
	Ion pair formation	TMA
	Phase separation	DSC, TMA
	Microcrystals: PVA, PE	DSC, TMA
	Solvent complex: PS/carbon disulfide	TMA
	Nodule crosslinking: block copolymers	TMA

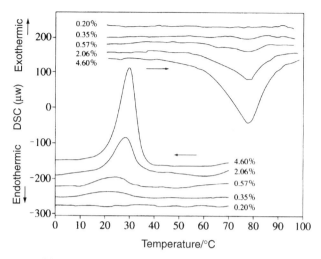

Measurement was made by Seiko Electric DSC 120;
amount of sample 40 mg and scanning speed 2°C/min.

Fig. 3 Heating and cooling DSC thermograms of various concentration agarose aqueous solutions.

form gel by associating several helical molecules. The endothermic peaks observed by DSC likely indicate melting of crosslink structure and the exothermic peak probably indicates the crystallization process. There are a number of helical molecules among polysaccharides other than agarose. Carrageenan, gellan gum, and xanthan gum form double helices, and thyzophyllan is thought to form a triple helix. These aqueous solutions of helical molecules exhibit thermoreversible gel-sol transition above the critical concentration (in many polysaccharides, it is approximately above 1–2 wt%). From the small angle x-ray scattering data of the gel state, it is a crosslink model with hexagonal packing consisting of 6–7 helical molecules [216]. In other words, in polysaccharide gels, the association of helical molecules is observed by DSC as a thermoreversible gel-sol transition [217].

Eldridge and Ferry [218] used the van't Hoff equation for the study of gelation assuming that the gelation is a chemical equilibrium:

$$\frac{\partial \ln K}{\partial T} = \frac{\Delta H_{\mathrm{m}}}{RT^2} \tag{11}$$

They expressed the relationship between the melting temperature and concentration of gel as follows [218]:

$$\frac{\partial \ln C}{\partial T} = \frac{\Delta H_\mathrm{m}}{RT^2} \tag{12}$$

where C, T, and ΔH_m are the concentration of gel, sol-gel transition temperature, and enthalpy necessary to form 1 mole of crosslink points, respectively.

The plot of the reciprocal melting temperature against the log concentration of agarose with various molecular weights, the so-called Eldridge–Ferry plot, is shown in Fig. 4 [215]. From the plot in Fig. 4, it is found that the enthalpy for the crosslink point elimination is independent of the molecular weight and is approximately 1.3×10^3 kJ/mol. The ratio of the enthalpy between the elimination and formation of the crosslink

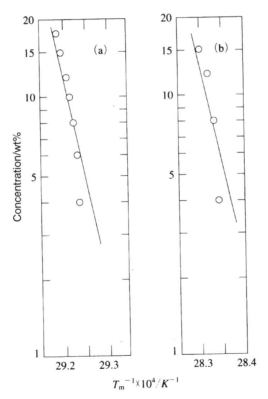

Fig. 4   Eldridge–Ferry plot of the melting temperature of agarose solution.

point calculated from the Eldridge–Ferry plot from the DSC results in Fig. 3 on the low concentration agarose gel is approximately 1.4. This ratio is about the same as the enthalpic ratio between the endothermic and exothermic peaks, suggesting the correctness of the Eldridge–Ferry plot. Also suggestive is the fact that the crosslink point of agarose gel consists of molecular association with a specific structure.

The association of helical structure in polysaccharides is due to interactions stronger than van der Waals forces, such as intermolecular hydrogen bonding and the static interaction of counter ions.

To understand gelation via the association of helical molecules, Nishinari [219] and others have analyzed the number of chemical repeat units associated through intermolecular interaction (such as hydrogen bonding) and the number of associating molecules by analyzing DSC thermograms using the zipper model, which is used for the melting behavior of deoxyribonucleic acid (DNA).

2.9.3.2 Crosslinking accompanying conformational changes

In an aqueous solution of helical molecules, the molecules that are associated with each other separate by heating and a gel-sol transition is observed. However, among polysaccharides, multiple helices to a single helix transition or conformation transition are sometimes observed. Xanthan gum is known to gel upon mixing with other polysaccharides such as galactomannan [220]; however, it has long been thought that xanthan gum alone will not gel. A 2 wt% xanthan aqueous solution kept at 40°C for 24 h exhibits a viscous appearance but also fluidity. It is closer to a plastic fluid than to a gel. When the xanthan solution is heat treated at 90°C, which is higher than the aforementioned temperature above which conformational changes occur, and subsequently cooled, no fluidity is observed and the solution gels [221]. In the temperature change of the dynamic storage modulus (G') obtained by TMA as shown in Fig. 5, a thermoreversible sol-gel transition is observed. When the thermal treatment time at 90°C is short, the transition temperature is low and the G' value of the gel is also small. As the thermal treatment time is extended, the transition time increases and the G' value as gel phase also increases. Furthermore, hysteresis during the transition will be small. From the small angle x-ray scattering data of xanthan gel, it is believed that about 6 helical chains associate with each other. The conformational changes mentioned here are thought to be due to the double helix to single helix transition. Xanthan is a polysaccharide that originates from a microbe and is in the

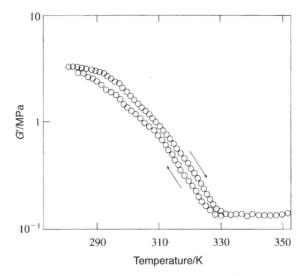

Data obtained with Seiko Electric TMA/SS2200;
measurement frequency was 0.05 Hz, scanning
speed was 1°C/min, and the sample was
treated at 90°C for 5 min and cooled prior to
the measurement

Fig. 5 The temperature-dependent dynamic modulus of 2 wt% xanthan aqueous solution.

various associated states of double helix molecules. From this state, a homogeneous aqueous solution cannot be obtained. Rather, due to the suspension of molecular association, the solution behaves as a plastic fluid. Above the transition temperature, the molecular association changes to single helical molecules and molecular reorientation becomes possible. When cooled from this state, association of a single helix occurs and crosslink points are formed. However, G' of xanthan gel depends on the frequency. It is not a stable gel but rather it is in a quasi-gel state.

2.9.3.3 Crosslinking by ion pair formation

Arginic acid, a seaweed polysaccharide, is a block copolymer of gluronic acid and manuronic acid. If a multivalent ion such as calcium is added to this solution, it gels. For the crosslink point structure, an egg-box model where manuronic acid surrounds the calcium ion has been proposed. In general, similar to what happens with arginic acid, polysaccharide electrolytes gel upon ion pair formation. It is difficult to observe this process by a temperature sweep-type DSC, thus an adiabatic calorimeter must be used.

2.9.3.4   *Crosslinking accompanying phase separation*

Methylcellulose dissolves in water at low temperature. However, when its aqueous solution is heated, it becomes turbid and gels. This gelation takes place thermoreversibly and when cooled it returns to an aqueous solution once again. Figure 6 depicts the DSC thermograms of 2 wt% methylcel-lulose aqueous solutions with various molecular weights of the polymer. Gelation during heating is observed as an endothermic peak around 60°C. The gel-sol transition during cooling is seen as an exothermic peak with almost no supercooling. This indicates that the crosslink points are not made of a specific molecular association structure like that of crystallites. The fact that the gelation temperature reduces as the molecular weight

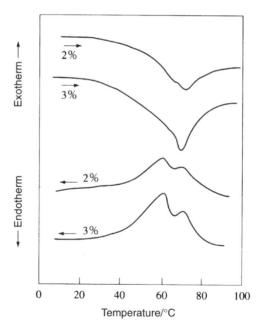

Data obtained by Setaram MicroDSC with sample
weight of 9.50 mg and scanning speed of 1°C/min

Fig. 6 Heating and cooling DSC thermograms of methyl cellulose aqueous solution.

increases is in good agreement with the molecular weight dependence of the binodal curve of a phase diagram.

The gelation of the methyl cellulose aqueous solution is regarded as the result of phase separation by heating. Specifically, the polymer chains that phase separated at an elevated temperature aggregate via a hydrophobic interaction and form a gel through entanglement. Hence, in the aqueous solutions of other cellulose derivatives with lower critical solution temperature (LCST)-type phase diagrams, similar gelation can be observed. Such gelation behavior by phase separation has been explained from the statistical mechanics point of view [222].

2.9.3.5 Crosslinking by microcrystallite formation

When the semicrystalline polymer solution is quenched, newly appearing crystallites become crosslink points and the solution gels. In such a solution, crystallization and phase separation take place concurrently and, thus, the gelation behavior will be complex. A phase diagram with special emphasis on a sol-gel transition of a semicrystalline polymer solution is illustrated in Fig. 7 [223]. The sol-gel transition curve approaches an asymptotic concentration, ϕ_{∞}^{*} and above this concentration, the solution gels; however, above the melting point of the microcrystallites, the solution will be a sol. Gel is not formed below the concentration where the spreading polymer chains overlap, $\phi_{R_g}^{*}$. In a real solution, this curve overlaps with both the binodal and the spinodal curve due to the phase separation. The gelation rate, that is, the crystallization rate, increases with greater supercooling. As gelation takes place homogeneously or faster than the crystallization above the spinodal decomposition temperature region, a transparent gel is formed if the crystallite size is small. However, below the spinodal curve, the spinodal decomposition occurs faster than gelation. In this temperature region, phase separation takes place earlier than gelation and a turbid gel is formed. Such examples include poly(vinyl alcohol)/water, polyethylene/xylene, and isotactic polystyrene/decalin systems. Figure 8 illustrates the DSC thermograms of syndiotactic polystyrene/dichlorobenzene solutions. During heating, an endothermic peak is observed at 120°C, which corresponds to the gel-sol transition. Upon investigation by temperature-dependent infrared spectra, it is found that the TTGG tetrads have been eliminated at this transition temperature and the β crystals form crosslink points [224].

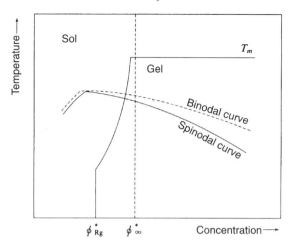

Fig. 7 Phase diagram of crystalline polymers (gelation takes place as a competing phenomenon of phase separation and crystallization) [223].

2.9.4   Glass Transition Temperature

When a gel is cooled, the solvent crystallizes and the gel networks freeze into a glassy state. Due to the fast crystallization rate of water in hydrogels, the majority of water crystallizes. However, a portion of water vitrifies along with the gel networks [225]. The vitrified water is thought to be directly or indirectly restricted by the polymer.

When vitrified water is examined by heating using DSC, the water in the gel shows a glass transition temperature as a cooperative phenomenon

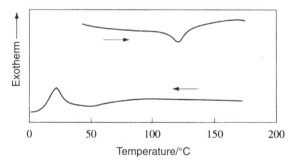

Fig. 8 The DSC curves of syndiotactic polystyrene/*o*-dichlorobenzene solutions [224].

between the gel networks and water. In other words, despite the low glass transition temperature of water at $-138°C$, the vitrified water is kept frozen until several tens of degrees higher than this temperature, at which Brownian motion of the gel networks begins. Among polysaccharides, hyaluronic acid can restrict a large amount of water and vitrification of water can be observed at concentrations greater than 30 wt%. A similar phenomenon can be observed in xerogel using organic solvents with slow crystallization.

2.9.5   Condition of Water

As the condition of water in hydrogels relates to the structure of the gel and function, it has been studied from various points of view. The DSC allows quantitative observation of water in a gel in various states [226] Nakamura and Hatakeyama [227] reported DSC results of various polymer/water systems and attempted quantitative analysis of water using the crystallization exotherm of water calculated from the crystallization enthalpy of pure water. In the hydrophilic polymers that absorbed a small amount of water, the crystallization temperature of restricted water by the polymer is observed at a lower temperature than occurs with ordinary water. Hence, they classified water in the polymer into three categories: free water; restricted water; and unfrozen water. From other measurements, polymer/water systems are also shown to have at least three different states of water. For the quantitative analysis of water by DSC, there are occasions where pure water cannot be used for the standard enthalpy. Figure 9 shows the melting enthalpy with a wide range of water content for the hyaluronic acid/water systems [228]. From the slope of each straight line, the standard melting enthalpy can be obtained. In xanthan/water systems, the enthalpy of water in the gel is the same as pure water (333.1 J/g) in all water contents [229]. However, it requires caution when the restricted water is quantitatively analyzed in hyaluronic acid because the value of the melting enthalpy depends on the water content.

When a large amount of water is contained like a hydrogel, it is difficult to separately determine the free water and the restricted water. Assuming that the water that melts below 0°C is considered to be restricted water, an example of a curve resolving in a DSC melting thermogram is shown in Fig. 10 [230]. Using this method, the aforementioned three different water categories can be quantitatively determined. In hard gels, which do not deform because of water crystallization, it is also

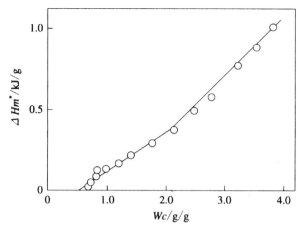

The measured melting enthalpy is normalized (ΔH_m^*) by
the polymer weight and plotted against the water content
(Wc = water weight/polymer weight); the slope is the
standard melting enthalpy

Fig. 9   The method to determine standard melting enthalpy of water in a gel.

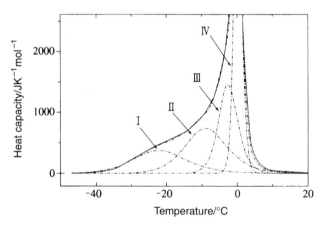

The peak at 0°C is due to the free water; peaks below
0°C are all classified as due to restricted water; from
the total enthalpy of melting, the water that crystallizes
can be obtained and subtracted from the total water
content to obtain the quantity of nonfreezing water

Fig. 10   Quantitative curve resolving of free water in lipids [230].

possible to estimate the micropore size distribution of the gel from the crystallization curve of water [231].

2.10   DIELECTRIC PROPERTIES

YOSHIRO TAJITSU

2.10.1   Introduction

Dielectric measurement has many advantages such as a wide range of properties and wide applicability regardless of the type of materials [232–235]. In particular, it has long been used in the area of polymer characterization, and especially as a convenient technique to obtain information about the mode and rate of movement of solid-state materials. Characteristic of gelation is the formation of giant networks that extend to the entire system. Liquid as sample prior to gelation has no particular shape; however, upon gelation, a particular shape can be formed [236–243]. Nevertheless, from the results of x-ray diffraction studies, the structure is closer to liquid or amorphous materials even after gelation. A main property of a gel is microscopic in scale, it is close to a liquid, but in macroscopic scale, the system exhibits elasticity [236–243]. However, in reality, the structure of gels differs markedly depending on the manufacturing method, and there are homogeneous and heterogeneous gels [236–243]. As a result, there is difficulty in understanding the dielectric properties of gels. The sol-gel transition is an ideal subject in polymer physics that uses difficult mathematical procedures such as percolation and fractal geometry [236, 242, 243]. In this section, the measurement principle and operation that require caution will be described using typical dielectric measurement techniques that are applied for polymer gel studies. Also, several interesting examples of dielectric measurements will be introduced.

2.10.1.1   Complex dielectric constant
In general, the polarization response of a material to an electric field occurs with time lag. This is caused by the mass effect and viscous resistance with dipole rotation and change in electric charge, which are the causes of polarization. Hence, a distribution of responses will be observed

depending on the measurement frequency, and various spectra can be observed by different polarization mechanisms. When the frequency (time) dependence of the dielectric constant is taken into consideration, it is called the dynamic dielectric constant. If the electric field E of the amplitude E_0 and angular frequency ω is expressed as a complex quantity, then the following equation can be used:

$$E = E_0 \exp(i\omega t) \tag{1}$$

If this electric field is applied to a sample, change of electricity D with phase lag δ is observed:

$$D = D_0 \exp[i(\omega t - \delta)] \tag{2}$$

From the ratio of D to E, complex dielectric constant ε^* can be obtained:

$$\varepsilon^* = \frac{D}{E} = \varepsilon' - i\varepsilon'' \tag{3}$$

$$\varepsilon' = \frac{D_0}{E_0} \cos \delta \tag{4}$$

$$\varepsilon'' = \frac{D_0}{E_0} \sin \delta \tag{5}$$

where ε^* is the frequency response function. The electric energy accumulated by a dielectric material with a dielectric constant e under the electric field is simply expressed as $(1/2)\varepsilon E^2$. As the real part ε' of ε^* basically corresponds to this ε, it is also called the storage dielectric constant. Also, the imaginary part ε'', which dissipates as thermal energy, is called the loss dielectric constant. The ratio between the real and imaginary parts is called the loss tangent and δ is termed the loss angle. (See References [232–235] for more information on complex dielectric constants.)

$$\tan \delta = \frac{\varepsilon''}{\varepsilon'} \tag{6}$$

2.10.1.2 Relaxation-type dispersion and orientation polarization

Both ionic polarization (where the positive and negative charges move in opposite directions) and electronic polarization (where the electrons move relative to the nuclei) exhibit a resonance-type dispersion spectrum, which exceeds the frequency range of 1 GHz. Hence, this region has little to do with polymer gels and thus will not be referred to in this section. However, a relaxation-type dispersion spectrum may be expressed using angular frequency ω

$$\varepsilon^* = \varepsilon' - i\varepsilon'' = \varepsilon_{in} + \frac{\Delta\varepsilon}{1 + i\omega\tau} \tag{7}$$

$$\Delta\varepsilon = \varepsilon_{eq} - \varepsilon_{in} \tag{8}$$

where ε_{eq} is the dielectric constant at equilibrium, and ε_{in} is the spontaneous dielectric constant, which is the contribution from electronic polarization or ionic polarization that appear in a shorter time scale (high-frequency side) than the distribution polarization. Also, t is the relaxation time and $\Delta\varepsilon$ is the relaxation strength. According to this equation, the real part ε' of ε^* decreases as the frequency increases, whereas the imaginary part ε'' of ε^* exhibits frequency dependence with a maximum at $\omega = 1/\tau$. This equation is generally called the Debye function. A Debye-type relaxation spectrum is generally seen when orientation polarization by dipoles appears. See Fig. 1 for a dielectric spectrum.

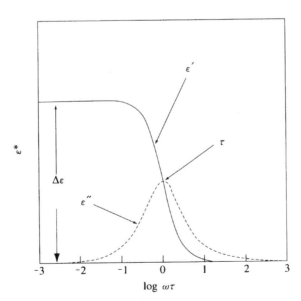

Fig. 1 Dielectric spectrum.

If it is taken into consideration that the potential energy of a dipole in an electric field is much smaller than the thermal energy, then the time variation of the orientation polarization P of N dipoles having dipole moment μ is found to follow a linear differential equation:

$$\tau \frac{dP}{dt} + P = \frac{N\mu^2}{kT} E \tag{9}$$

This equation can be readily solved to give:

$$\varepsilon = \varepsilon_{in} + \Delta\varepsilon \left[1 - \exp\left(-\frac{t}{\tau} \right) \right] \tag{10}$$

$$\Delta\varepsilon = \varepsilon_{eq} - \varepsilon_{in} = \frac{N\mu^2}{kT} \tag{11}$$

Hence, the relaxation strength $\Delta\varepsilon$ reflects the increase of the dielectric constant by the orientation polarization. If this equation is Fourier transformed, the following frequency response function is obtained:

$$\varepsilon^* = \varepsilon_{in} + \frac{\Delta\varepsilon}{1 + (i\omega\tau)^\beta} \tag{12}$$

This equation is in the same form as the relaxation-type spectrum. From a slightly different point of view, the rotation of a rigid body is influenced by the inertia, recovery force, and viscous resistance, whereas in the case of dipole moment, the effect of the inertia is small, which corresponds to the large viscous resistance. The τ in Eq. (12) corresponds to the viscous resistance against the recovery force. Consider that a dipole is shuttling between the different states separated by the wall of the potential. In order to overcome this potential wall, the thermal energy of the system and the electrical energy given by the external field are needed. In this case, τ relates to the ratio of the potential wall to thermal energy. Also, the relaxation-type dispersion does not only take place by the orientation of the dipole, but also in a manner apparent in heterogeneous materials with respect to the dielectric constant and electrical conductivity [232–235]. Therefore, it is necessary to pay careful attention to the cause of the dispersion.

2.10.1.3 Empirical equation for relaxation-type dispersion
The spectrum of an actual material often shows a smaller slope than the aforementioned equation. This originates from the difference in environments of dipoles, which individually orient microscopically. This corre-

sponds to the distribution of relaxation rather than a single relaxation time. The following is the Havriliak–Negami equation that incorporates the distribution of relaxation:

$$\varepsilon^* = \varepsilon_{\text{in}} + \frac{\Delta\varepsilon}{\left[1 + (i\omega\tau)^\beta\right]^\alpha} \tag{13}$$

where α and β are positive constants. When $\alpha = 1$ and $\beta < 1$, the equation is called the Cole–Cole equation, and when $\alpha < 1$ and $\beta = 1$, it is called the Davison–Cole equation. If ε^* is plotted on a complex coordinate, it will give a semicircle when $\alpha = 1$ and $\beta < 1$ (see Fig. 2). The center of the semicircle where the imaginary part becomes the maximum can be readily calculated to be $\omega\tau = 1$, and it obviously provides relaxation time. If $\alpha < 1$ and $\beta = 1$, then the shape becomes asymmetric, called a lemniscate shape. In this ease, the maximum of the imaginary part is different from the relaxation time. In any case, the α and β in the Havriliak–Negami equation only express the distribution of relaxation time indirectly and there is little physical meaning.

Amorphous polymers are known as examples of dielectric dispersion of dipole orientation type [235]. This originates from micro-Brownian motion of polymer chains. When a molecule can rotate above the glass transition temperature, the dielectric constant will be large due to the contribution of the orientation polarization if the voltage with frequency slower than the molecular motion is applied. In contrast, at higher frequencies, the orientation polarization will not contribute and the dielectric constant will be small. As a result, in general, the dielectric dispersion expressed by the Havriliak–Negami equation is followed. Information on the mode and speed of motion can be obtained by evaluating τ and $\Delta\varepsilon$ of Eq. (13) from experimental results. Also, if the

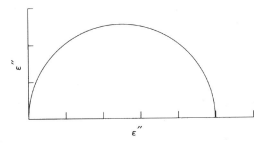

Fig. 2 Cole–Cole plot.

temperature increases, the spectrum shifts to a higher-frequency side due to the increased motion. The result at higher temperature is equivalent to the low-temperature spectrum that is shifted to higher frequency. This is called the frequency-temperature superposition principle [232–235].

2.10.2   Dielectric Measurement Techniques

For the dielectric measurement of gels, an alternating current is usually applied. Depending on the frequency ranges, the measurement methods can be divided into the concentrated constant circuit and the distributed constant circuit. First, we will discuss low-frequency range methods (concentrated constant circuit) for which many commercial instruments are available, and then high-frequency range methods (distributed constant circuit) [232–235].

2.10.2.1   Measurements in the low-frequency range

As a measurement method in the low-frequency range, the balance method that utilizes a bridge circuit is used from 10 Hz to 1 MHz, and the resonance method, for its accuracy from 100 kHz to several tens of megahertz, has been used. The balance method is used to connect a capacitor and resistance with the sample in parallel and to obtain directly the capacitor and resistance components of the sample. On the other hand, the resonance method connects a variable capacitor and inductance component in parallel with the sample, activates resonance by changing the variable capacitance or frequency, determines the resonance frequency or Q value, and finally obtains the capacitance component of the sample. Because these two methods are quite popular, readers who are interested in a detailed description of the techniques are referred to the many available monographs [232–235]. In contrast to these methods, commercial instruments adopting the so-called direct method are available. This technique applies directly sinusoidal electric field to the sample, detects the current that passes through the sample as the voltage signal using an I/V converter, and detects the in-phase and $90°$ out-of-phase components with the applied voltage by a lock-in amplifier. In addition, this technique will allow the measurement of complex dielectric constants. With the development of electronics elements, the accuracy of this equipment has improved and, unlike the traditional balance method or resonance method, the error originating from inexperience is much less.

2.10.2.2 Measurements in the high-frequency range (emphasis on dielectric measurements from 1 MHz to 1 GHz and impedance)

When the measurement frequency range increases and approaches the megahertz range, the problems that were not serious in the low-frequency measurements become important [232–235]. When the measurement frequency increases and the length of the signal wire becomes comparable to or longer than the wavelength, the voltage and current will not be merely a simple function of time but a function of time and position. In this case, it is necessary to treat the problem as the distributed constant circuit where the wire constant distributes along the wire [244, 245]. This can be understood by considering the next scenario. For instance, when the voltage propagates the measurement circuit with a frequency, $f = 100\,\text{Hz}$. In this case, because the phase constant is $\beta = (2\pi/3) \times 10^{-6}(1/\text{m})$ even for the 3-m circuit, the phase lag is only $(3.6 \times 10^{-4})°$. In contrast, for $f = 200\,\text{MHz}$, because $\beta = 4\pi/3(1/\text{m})$, even for a circuit of 1 m, a phase lag as large as $270°$ will result and there will be a significant difference in current distribution. In this case, the wire constant cannot be considered as being concentrated on a single point and it is necessary to handle this situation as a distributed constant circuit. The distributed constant circuit is one of the most widely studied in electric circuitry and many introductory monographs describing its fundamentals as well as applications have been published [244, 245]. The readers are referred to these monographs regarding the concept of distributed constant circuits and others.

For the dielectric measurement of gels, the required properties of the cell include the necessity of changing the temperature and also maintaining the gel and sol conditions. There are many cases where an instrument is used beyond the recommended limit set for commercial instruments to protect the cell, in particular, at the high-frequency range. To prevent such an occurrence, we will briefly introduce in this section the principle of dielectric measurement at high-frequency ranges using the distributed constant circuit concept [245]. When the incident wave enters from the wire with characteristic impedance Z_{01} and propagation constant γ_1 into the wire with characteristic impedance Z_{02} and propagation constant γ_2, the entered wave reflects and returns to the original terminal from which the wave entered. Assume that the voltages of the incident and reflected waves at junction $x = 0$ are written as V_{io} and V_{ro}, respectively. The voltage V_i and current I_i of the incident wave and the voltage V_r, and current I_r, of the reflected wave can be expressed as follows:

$$V_i = V_{io} \exp(-\gamma_1 x) \tag{14}$$

$$I_i = -\frac{V_i}{Z_{01}} \tag{15}$$

$$V_r = V_{ro} \exp(-\gamma_1 x) \tag{16}$$

$$I_r = -\frac{V_r}{Z_{01}} \tag{17}$$

Also, if the voltage at the junction is written as V_{to}, the voltage V_t, and current I_t, of the transmitting wave through the junction are expressed as follows:

$$V_t = V_{to} \exp(-\gamma_2 x) \tag{18}$$

$$I_t = -\frac{V_t}{Z_{02}} \tag{19}$$

From the boundary conditions, the voltage and current are continuums. Thus,

$$V_{io} + V_{ro} = V_{to} \tag{20}$$

$$\frac{V_{io}}{Z_{01}} - \frac{V_{ro}}{Z_{01}} = \frac{V_{to}}{Z_{02}} \tag{21}$$

Organizing these equations, and letting the reflection coefficient of the voltage and current at the junction be Z_{01}^* and Z_{02}^*, respectively, we obtain

$$\Gamma_v = \frac{V_{ro}}{V_{io}} = \frac{Z_{02} - Z_{01}}{Z_{01} + Z_{02}} \tag{22}$$

$$\Gamma_v = -\Gamma_i \tag{23}$$

Assuming that the input and output complex impedance are expressed as Z_{01}^* and Z_{02}^*, respectively, the reflection coefficient at the discontinuous point in the complex form is

$$\Gamma_v^* = -\frac{Z_{01}^* - Z_{02}^*}{Z_{01}^* + Z_{02}^*} \tag{24}$$

If the sample is placed at the receiving terminal of the wire with impedance Z_0, then we obtain the complex impedance of the sample as

$$Z_s^* = Z_0 \frac{1 + \Gamma^*}{1 - \Gamma^*} \qquad (25)$$

Accordingly, if the voltage of the input and output waves can be measured accurately, the reflection coefficient in the left-hand term can be determined, leading to the determination of the complex impedance of the sample. Using this principle, dielectric instruments at the high-frequency range from 1 MHz to 1 GHz are commercially available as impedance analyzers. As is apparent from the foregoing discussion, the observed value by the instrument will be quite different from the sample by the influence of parameters such as connection at the junction, impedance other than the sample and circuit wire, and transmission coefficients. Hence, it is important to design the cell so as not to violate these two points.

2.10.3 Examples of Dielectric Relaxation Spectra

There are many dielectric measurements of gels and numerous papers have been published [246–265]. The subjects vary widely from the sol-gel transition to the coulombic field of polyelectrolytes, some of which will be introduced in this section.

2.10.3.1 Coulombic field of polyelectrolytes

There have been many recent additions to knowledge regarding the coulombic field of polyelectrolytes gained from analysis of dielectric relaxation spectra [258–261]. The high electron charge of a polymeric ion forms a strong coulombic field in its vicinity. Therefore, the counter ions that have opposite charge from the polymer ion gather around the polymeric ion and form a restricted layer. As a result, these counter ions lose the function of a free charge. Then, as a restricted charge, the contribution is known to shift from conductance to dielectricity. At this point, it is expected that an anomalous coulombic field will be formed. The information on the micromorphology of the coulombic field around this central polymer ion and the gel can be obtained from dielectric spectra. The spectra shown in Fig. 3 were obtained from the dielectric measurement of sodium acrylate gel [259–261]. Especially noteworthy is the observation of a dielectric relaxation peak around 1 MHz. Determining the relationship between the relaxation time τ and crosslink density of the

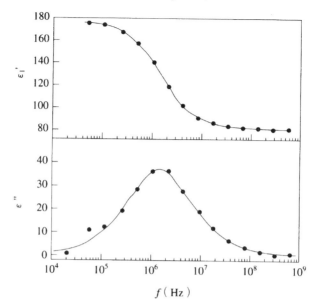

Fig. 3   Dielectric spectrum of sodium acrylate gel.

gel C_c from these relaxation spectra, we obtain the following equation as indicated in Fig. 4:

$$\tau \propto C_c^{-2/3} \qquad (26)$$

Rewriting this equation with the distance between the polymer ions, or the distance between the crosslink points L, the following equation also holds:

$$\tau \propto L^2 \qquad (27)$$

This equation is the same as the one for the entanglement of polyelectrolytes [259–261], which suggests that the relaxation mechanism is the same as the sodium acrylate gel.

Several researches [259–261] also arrived at the conclusion regarding the morphology of gels that the magnitude of thermal fluctuation is almost the same size as the network. This high-frequency relaxation is regarded as a useful technique to study the motion of local counter ions, and future development of investigation in this field is expected.

2.10.3.2   Sol-gel transition of polyelectrolyte gels

In this section, the information on the sol-gel transition of polysaccharide electrolyte gels obtained by dielectric relaxation spectra will be introduced

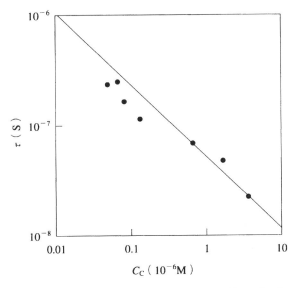

Fig. 4 The relationship between crosslink density C_c and relaxation time τ.

[237–241]. Unlike chemical gels that are formed by a chemical reaction using a crosslinking agent, the characteristic of polysaccharide electrolytes is to form gels by physical bonds through intermolecular forces among polymer chains. A polysaccharide electrolyte gels during the cooling process of the sol. One of the polysaccharide electrolytes is κ-carrageenan, a polymer with a molecular weight from several hundred thousands to several million. It is a linear polymer of galactan that consists of D-galactose, 3,6-anhydro-D-galactose, and sulfite groups. Carrageenan can be classified into κ-, λ-, and μ-type depending on the concentration of the sulfite groups, and it has a high reactivity with protein. Upon addition of a small amount of a cation such as K^+ or Na^+, its ability to form gel drastically increases and it is often used as an anti-drying agent. For the gelation mechanism of carrageenan, the mechanism seen in Fig. 5 was proposed by Rees and Morris *et al* [257]. The first intermolecular association of carrageenan is the coil-domain transition. Upon cooling, the concentration of the domain (double helix) increases. When it is heated, it returns to the coil structure. In the presence of a cation, such as an alkali metal ion, a further aggregated structure is formed. For some time, κ-carrageenan has been thought to gel without passing through the domain structure. On the other hand, it has been reported that a single

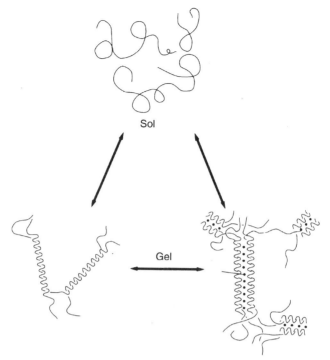

Sol

Gel

Fig. 5   Schematic diagram of the phase transition of carrageenan [237–241].

helix is formed without first forming a double helix [258]. Several researchers [262, 263] used dielectric relaxation to study the sol-gel transition. Figure 6 shows the complex dielectric constant, $\varepsilon^* = \varepsilon' - i\varepsilon''$, of carrageenan.

Usually, the dielectric spectrum of carrageenan shows a monotonic increase as the frequency decreases due to the ionic conductivity caused by the cations. However, by carefully analyzing the spectrum, it was found that there is a large relaxation phenomenon below the gel point as shown in Fig. 6. This relaxation phenomenon cannot be explained by the dielectric relaxation phenomena such as dipole orientation because the relaxation intensity is as high as 1000. This large relaxation phenomenon suggests the development of needle-like domains below the gel point, which is worthy of attention in relation to the sol-gel transition mechanism shown in Fig. 5. It seems certain that carrageenan gels by association of helical structures that form a crosslink domain. However, many unsolved problems remain.

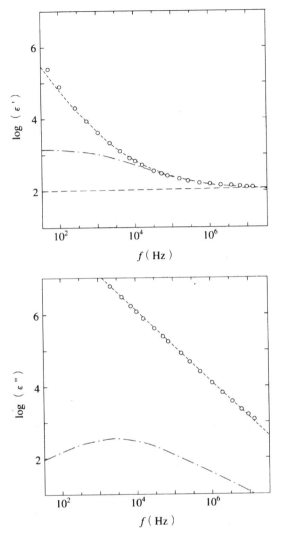

Fig. 6   Dielectric spectrum of carrageenan [262, 263].

2.10.3.3   *Movement of water molecules in polymer gels: lecithin gels*

In order to obtain knowledge on the behavior of water molecules that are enclosed in a lecithin gel, low-frequency dielectric spectra have been studied [237–243]. For example, Fig. 7 depicts the dielectric spectra from

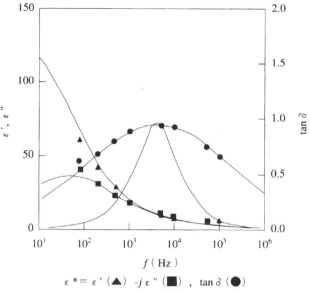

$$\varepsilon * = \varepsilon ' \ (\blacktriangle) \ -j \varepsilon '' \ (\blacksquare) \ , \ \tan \delta \ (\bullet)$$

The solid line is obtained by curve fitting
Debye function to the measured values

Fig. 7   Dielectric spectrum of a lecithin gel.

100 Hz to 100 kHz. Measurements were made as a function of temperature and the spectra were analyzed using curve fitting based on the Debye-type simple relaxation function shown in the following and individual relaxation was evaluated:

$$\varepsilon^* = \varepsilon_{in} + \frac{\Delta \varepsilon}{1 + i\omega\tau} \tag{28}$$

Based on the results and using the gel model (see Chapter 2), it was concluded that the diffusion of water molecule in the gel took place following the orientation of dipoles of lecithin in the micelle [262].

2.10.3.4 Phase transition and state of networks of polymer gels

Both ionized and nonionized polyacrylamide gels have been studied for the dielectric behavior of their networks and as a solution in mixed solvents of acetone and water in the 10–500 Hz frequency range. The frequency dependence of the real and imaginary parts of the complex dielectric constant of the networks shows a relaxation at the low-frequency region that cannot be explained by direct current and electrical double layer. Figure 8 illustrates the dielectric spectra of the gel by changing the concentration of acetone [265].

By careful experiments, relaxation time and relaxation intensity have been evaluated. As a result, as seen in Fig. 9, an ionized gel exhibits a sudden first-order phase transition at a certain acetone concentration. Also, due to the low concentration of the carrier, a lower dielectric constant than the solution state is observed. By analyzing these results carefully, it was concluded that there are 3 to 4 water molecules per each chemical repeat unit when the gel networks are complete.

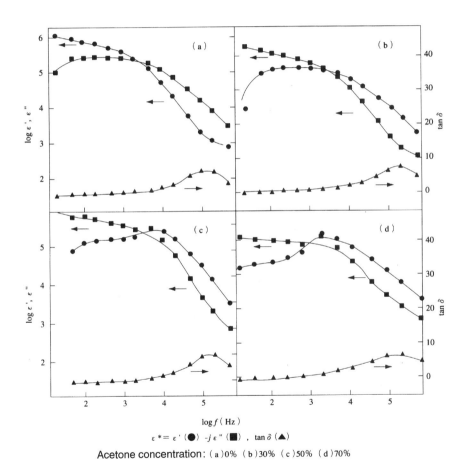

$\varepsilon * = \varepsilon'$ (●) $-j\varepsilon''$ (■) , $\tan \delta$ (▲)

Acetone concentration: (a)0% (b)30% (c)50% (d)70%

Fig. 8 Dielectric spectrum of a polyacrylamide.

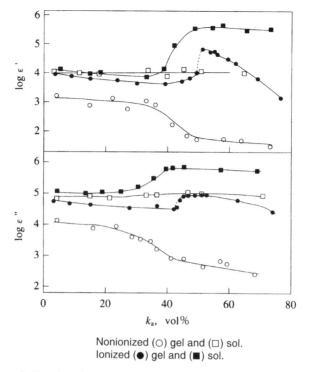

Nonionized (○) gel and (□) sol.
Ionized (●) gel and (■) sol.

Fig. 9 The relationship between complex dielectric constant and acetone concentration [265].

2.10.4 Conclusions

The dielectric measurement is a widely applicable technique that requires no extensive experience. As discussed in the foregoing, it is very useful in estimating the fairly microscopic structure within polymer gels by fitting empirical equations to the experimental spectra, obtaining various physical quantities, such as relaxation intensity or relaxation time, and quantitatively analyzing them. In order to ensure the possibility of the application of polymer gels to devices, which is currently drawing considerable attention, it is essential to clarify the fundamental principle including the electrical properties of polymer gels. From the fundamental point of view, further understanding of polymer properties as a whole can be achieved by solving the structures of the solution, solid, and gel states. Further progress in this field is strongly desired. This paper could not cover all aspects of the dielectric properties of polymer gels. Also, some of the descriptions or subjective interpretations may not be as clear as possible. For this, the author takes full responsibility.

Those readers who wish further knowledge are referred to monographs [232–243, 261] which are excellent sources.

2.11   PERMEABILITY OF GASES

HOTAKA ITO

2.11.1   Introduction

Recently, there has been a decrease in the number of young people who wear eyeglasses. However, this conflicts with the report that the number of people who need eyeglasses has increased in the past ten years. This discrepancy is merely the result of the widespread use of contact lens. Also, along with the rise in sophistication of health care, the increase in the number of older people continues to require high-quality medical devices. For example, no matter how sophisticated the technique is, cardiac surgery cannot be performed without the help of artificial heart and lung equipment. The common characteristic of such indispensable items as contact lens and artificial heart and lung equipment is the permeability of oxygen. Initially, poly(methyl methacrylate) was used for contact lenses because of its excellent optical properties. Unfortunately, discomfort due to the hardness of the material and poor oxygen permeability were shortcomings [266]. Later, a more comfortable water-containing gel, poly(2-hydroxyethyl methacrylate) (hereinafter called poly HEMA), was developed, and the oxygen permeability was improved, making it possible to wear the contacts for a prolonged period of time [267]. Similarly, an obstacle to the commercialization of artificial heart and lung equipment was material with poor oxygen permeability. Teflon, which has poor oxygen permeability, was used as the material for the membrane [268]. Furthermore, in the case of the artificial lung, in addition to the oxygen permeability, blood-type compatibility must be considered. Accordingly, hydrophilic membrane materials that have excellent compatibility with human tissue are being developed [269].

Gels, as stated in their definition, have a solvent in their structure, and because they are characterized by having a solvent within the network structure, their original properties will not be realized if measurements are made in a dry condition. For example, if water is absorbed by a polymer membrane, water acts as a plasticizer for the polymer. Also, if a large amount of water exists, gaseous molecules diffuse through the water rather than through the polymer chains. In fact, the diffusion rate of oxygen molecules is known to be two orders of magnitude faster than the rate through polymer chains [270].

The characteristics of gel membranes (hydrophilic membrane) depend strongly on the measurement environment. Therefore, in order to measure the permeability of a gas through a gel membrane, it is important to develop a technique in which measurements can be made at a swollen state in the presence of water.

2.11.2   Measurement Technique for Permeation Coefficient of Gases

When the permeability of gases of membranes is measured, the measurement techniques can be divided into a pressure gradient technique and an isopressure technique (see Fig. 1) [271]. In the pressure gradient technique two spaces separated by a membrane are prepared, a measuring gas with different partial pressures is then introduced and the pressure change or volume change is measured when the gas moves from the high-pressure side to the low-pressure side. Also, the differential method is to determine the rate of change of the gas as a function of time, whereas the integral method is to measure the total accumulated amount. The pressure technique is to evacuate the system sufficiently by a vacuum pump and the system is closed to the vacuum pump. Then, the permeating gas, such as oxygen or carbon dioxide, is introduced into the high-pressure side, and the pressure change at the low-pressure side is measured by a detector, such as the McLaud pycnometer or Baratron pycnometer. From the permeation curve that is, the pressure change of the low-pressure side as a function of time, the permeation coefficient is calculated.

On the other hand, because the pressure of both sides is the same with the isopressure technique, measurement can be made even in the presence of water. Generally, a technique called the electrode method is used to measure the permeability. This electrode method fills 0.5 N

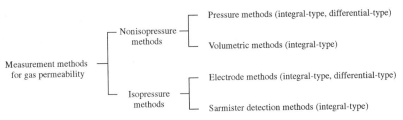

Fig. 1   Measurement methods for gas permeability.

potassium chloride in the electrode side that is in contact with water through the membrane.

It was probably Yasuda and Stone who first attempted to measure oxygen permeability using the oxygen electrode in the presence of water [272]. Later, the measurement technique was improved. Aiba *et al.* [273] developed a technique termed the direct method in which a sample is adhered directly to the electrode. On the other hand, Hwang *et al.* [274] developed the indirect method where a membrane is adhered to the interface between the gaseous and liquid phases. Initially, both methods studied mainly hydrophobic membranes. Today, however, both techniques have been improved to use both hydrophobic and hydrophilic (gel) membranes, and are established as the gaseous permeation measurement techniques in the presence of water. In the following, the measurement devices and their procedures will be discussed.

2.11.2.1   *Isopressure technique*

2.11.2.1.1   Direct method
The direct method adheres a sample membrane directly onto the oxygen electrode filled with oxygenated water. Current will flow in the electrode in proportion to the amount of oxygen that has permeated through the membrane. By determining the amount of current, the permeability of the gas can be determined in the presence of water. Minoura and Nakagawa [275] improved the method by adding a support to the oxygen electrode so that a weak membrane that contains a water-like gel can be used. The structure of the electrode is shown in Fig. 2, whereas the flow diagram of the equipment is illustrated in Fig. 3. Platinum is used as the electrode and it is fixed through the insulator to the tip of the electrode. Another improvement includes the addition of a supporter so that swollen gel can

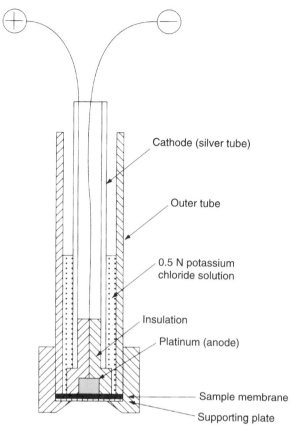

Fig. 2   Oxygen electrode (cross-sectional diagram) [275].

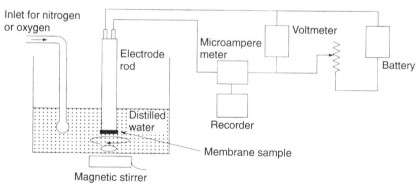

Fig. 3   A measurement device for oxygen permeability using the direct method [275].

be supported. Considering the installation and removal of the membrane, and the adhesion of the membrane and platinum, a compression scale is used. In this method, the oxygen electrode is in contact with nitrogen and the current is zero. Then, the membrane is placed in contact with the oxygen-saturated water and the amount of oxygen permeation is measured. At the anode (Pt), the following reaction is taking place depending on the amount of oxygen:

$$O_2 + 2H_2O + 4e \rightarrow 4OH^-$$

On the other hand, the cathode side has the following reaction:

$$4Ag + 4OH^- \rightarrow 2H_2O + 2Ag_2O + 4e$$

Due to the movement of electrons created by this reaction at both electrodes, current will flow. The diffusion coefficient of oxygen is slower than the oxidation–reduction reaction at the electrode. The steady-state current i, which is proportional to the diffusion coefficient of oxygen, can be expressed in microamperes by the following equation:

$$i_\infty = NFAP_s \frac{P}{l} \tag{1}$$

where N is the number of electrons that participate in the reaction ($= 4$); F is the Faraday constant ($= 96,500\,\text{coulomb/mol} = 96,500/22.4 \times 10^3$ coulomb/cm^3 (STP)); A is the area of the anode (cm^2); and P_s is the partial pressure of surface oxygen ($= 76\,\text{cmHg}$). Here, coulomb $=$ A (ampere) $\times$ s (time).

The permeability curve of oxygen obtained by this method will be a derivative type (see Fig. 4). The abscissa is the current that is proportional to time and the ordinate is the current that is proportional to the amount of oxygen. Inserting the value of this steady-state current i_∞ into Eq. (1), the permeation coefficient in the unit of cm^3(STP) $\cdot$ cm/cm^2 $\cdot$ s $\cdot$ cmHg can be obtained [273]:

$$P = \frac{i_\infty l}{NFAP_s} \tag{2}$$

However, the permeation coefficient of oxygen obtained from this measurement contains resistance of not only the sample membrane but also the fluid. If the permeation coefficient of only the membrane is desired, measurement should be repeated varying the membrane thickness. In fact, the resistance to oxygen diffusion includes the fluid, sample membrane, and the electrolyte solution at the electrode. Because each

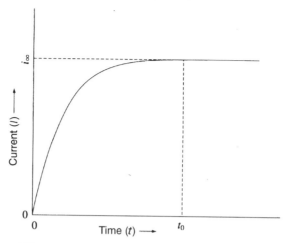

Fig. 4 Permeability curve as measured by the direct method using an oxygen electrode.

resistance is proportional to the thickness of the membrane, the resistance to the permeation of oxygen can be expressed as follows:

$$\frac{l}{P} = \frac{l_L}{P_{\mathrm{L}}} + \frac{l_m}{P_m} + \frac{l_{\mathrm{E}}}{P_{\mathrm{E}}} \tag{3}$$

where l_{L}, l_m, l_{E} are the thickness of the fluid interphase, membrane, and electrolyte interphase, respectively. Also, P_{L}, P_m, and P_{E} are the mass transfer coefficient at the fluid interphase, permeation coefficient of the sample membrane, and mass transfer coefficient of electrolyte interphase, respectively. Thus, Eq. (1) can be rewritten as follows:

$$\frac{NFAP_{\mathrm{s}}}{i_{\infty}} = \frac{l_{\mathrm{L}}}{P_{\mathrm{L}}} + \frac{l_m}{P_m} + \frac{l_{\mathrm{E}}}{P_{\mathrm{E}}} \tag{4}$$

From the slope of the straight line obtained from the plot of the left-hand side of Eq. (4) by changing the thickness of the membrane, the resistance of only the membrane can be obtained. Furthermore, from its reciprocal, the permeation coefficient of the membrane is obtained. Table 1 lists the permeability of oxygen for various polymer membranes in the presence of water measured by this technique [276]. A comparison is also made with the same polymer in the dry state. Judging from this result, a hydrophobic membrane such as polyethylene shows similar permeability of oxygen in the presence or absence of water. In contrast, as the hydrophilicity

Table 1 The oxygen permeability of various membranes using the direct method at 20 °C.

Membranes	The oxygen permeability $P \times 10^{10}$ cm^3 (STP) $\cdot$ cm^2 $\cdot$ s $\cdot$ cmHg	
	Dry state	Wet state
Low density polyethylene	4.11	4.36
Polypropylene	0.43	0.64
Nylon 6	0.051	0.247
Polycarbonate	0.53	1.40
Vinylon	0.0005	3.39
Cellophane	0.0073	5.19
Cornea	–	38.0
Poly(glutamic acid)	0.0006	84.0

increases, the difference becomes pronounced. In the case of polyglutamic acid, it has been confirmed that the difference is as large as five orders of magnitude.

2.11.2.1.2 Indirect method

Unlike the direct method, the indirect method places a fluid on both sides (see Fig. 5). Also, another configuration is to place a fluid on one side and a gas on the other side. In this section, the case where one side has a fluid and the other pure oxygen will be explained for the sake of simplicity [277]. The basic principle is the same for the case where both sides are filled with a fluid [278].

Oxygen diffuses into deaerated water through the sample membrane. In this case, the diffusion solvent, water, exists between the sample membrane and the oxygen electrode. The resistance in this case is also the sum of the resistance of the membrane itself and the fluid interphase. At this point, the fluid interphase will be briefly explained [279]. When systems having the same pressure but different concentration come in contact through a membrane, gaseous molecules move from the high-concentration phase to the low-concentration side. In this case, the concentration of oxygen in the bulk solution is different from the solution that is in contact with the membrane and a steep concentration gradient exists in the vicinity of the membrane surface. Such a region is called the interphase. The nature of the interphase is the resistance to diffusion due to the restriction of the motion of the solution that is in contact with the membrane surface and the lack of the diffusion velocity vector normal to

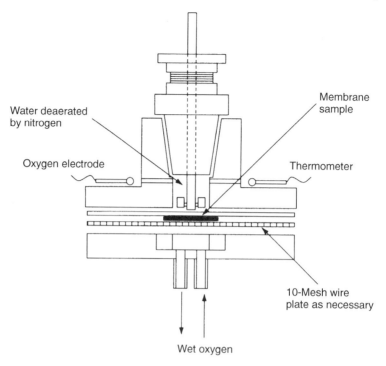

Fig. 5 Measurement device for oxygen permeability using the indirect method (gas-liquid).

the membrane surface. To evaluate the sample membrane itself without the influence of the interphase requires some methods to separate this interphase resistance from the membrane resistance.

When the foregoing statement is expressed in a equation, it becomes $1/P = 1/P_L + 1/P_m$, where P, P_L and P_m are the total mass transfer coefficient, interphase mass transfer coefficient, and permeation coefficient of the membrane, respectively. Their reciprocals are the total resistance $1/P$, interphase resistance $1/P_L$, and membrane resistance $1/P_m$. When both sides are fluid, then there will be one more interphase, and the equation becomes $1/P = 1/P_L + 1/P_m + 1/P'_L$. The unit of the permeation coefficient is $cm^3(STP) \cdot cm/cm^2 \cdot s \cdot cmHg$ when it is expressed by the partial pressure difference $(mol/cm^2 \cdot s \cdot cmHg)$, but it is cm/s if it is expressed by concentration difference (mol/cm^3).

The gas transfer rate from the gaseous phase to liquid phase is proportional to the concentration difference and gas-liquid interface area.

If the area of the membrane, the volume of the solution, the concentration of dissolved oxygen, the concentration of oxygen in the gaseous phase, and time, are expressed as A, V, C, C^* and t, the following equation holds:

$$V \frac{dC}{dt} = P \times A(C^+ - C)$$

If both sides are organized into concentration and time and further integrated, we obtain [277]

$$\log\left[\frac{(C^* - C_0)}{(C^* - C)}\right] - P \times A \times \frac{t}{2.3V}$$

If the left term is for the ordinate and time is taken as the abscissa, the slope of the graph is $P \times A/2.3V$. As both A and V are constant, after each known value is inserted, the total mass transfer coefficient P or total resistance $1/P$ can be determined. Furthermore, because the interphase resistance is a diffusion resistance, naturally it depends on the stirring condition of the solution side. In other words, the interphase resistance value is determined by the stirring speed of the solution. Hence, it is necessary to measure the total resistance $1/P$ at various stirring speeds (see Fig. 6). By plotting $1/P$ into the ordinate and $(1/n)^c$, where c is the rotation (stirring) speed, into the abscissa, Fig. 7 is obtained.

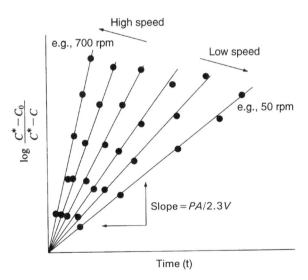

Fig. 6 Rotational speed dependence of permeability.

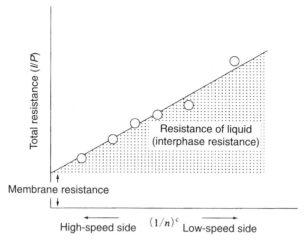

Fig. 7 Wilson plot.

This is the so-called Wilson plot [280]. Because the rotation speed c is the constant, which depends on the condition of the equipment (instrument constant), it is necessary to determine this instrument constant when new equipment is manufactured. To determine the instrument constant, the Wilson plot is made for commercially available membranes (as long as some permeability exist, any membrane can be used). Then, the instrument constant is determined in such a way as to obtain the straightest line for all Wilson plots, though it should be within the range of 0.5–0.8.

In the Wilson plot obtained, when the rotation speed is infinite (intercept at the y-axis), the point with zero interphase resistance, that is, only the membrane resistance, can be obtained (the reciprocal of the membrane resistance is the permeation coefficient of the membrane).

Katoh and Yoshida evaluated the oxygen permeability (membrane resistance) of various membranes in the presence of water using this equipment. Table 2 lists the results obtained [277]. The membranes used are mainly hydrophobic. A silicone membrane that is supported by a mesh or a composite of a porous polypropylene membrane and homogeneous membrane are the subject of the permeation measurement. For example, the permeation coefficient of a silicone membrane (Silastic #500-1) with a thickness of 125 μm has a permeation coefficient of 1.04 cm/min (the membrane resistance is 0.96 min/cm). One of the advantages of this method is that not only can the membrane resistance be measured, but

Table 2 The oxygen permeability of various membranes using indirect method at 36.5 °C.

	Membrane thickness (μm)	Membrane resistance (min/cm)	Permeability (cm/min)
Silicone membrane[1] (Manufactured by Fuji Systems)	70	0.47	2.13
Silicone membrane[1] (Manufactured by Fuji Systems with a reinforcement)	150	0.20	5.00
Silicone membrane[2] (Manufactured by Dow Corning Corp.)	125	0.96	1.04
Teflon	25	21.2	0.0047

[1] Trademark Phycon
[2] Trademark Silastic #500-1

interphase resistance can also be obtained simultaneously. In fact, it was reported that even at a rotation speed of 600 rpm, the total resistance was 2.16 min/cm, which is twice that of the membrane resistance. This clearly indicates that the majority of resistance for the transfer of oxygen is in the liquid phase side. Taking into consideration actual conditions of use, the interphase resistance will not become zero. Accordingly, it is essential to minimize this resistance at the interphase by choosing the proper material design.

On the other hand, when a hydrophilic membrane, that is, gel membrane, is studied using this equipment, it is possible to make the measurement by modifying the equipment with a support using 10 mesh wire plate. The permeation coefficient of poly HEMA membrane with a thickness of 203 mm using this device is 0.17 cm/min (membrane resistance is 5.89 min/cm) [281]. In this case, the total resistance at 600 rpm was similarly obtained and the total resistance was found to be approximately 7.3 min/cm. This value is different from the hydrophobic membrane and is similar to the value of the membrane resistance. It is interesting to note that when the hydrophilicity of the membrane surface that is in contact with the fluid increases, the interphase resistance tends to decrease.

2.11.3 Conclusions

In general, hydrophilic polymer membranes such as poly(vinyl alcohol) exhibit low gas permeability under a dry state. For example, the permeability of dry poly(vinyl alcohol) is 5.2×10^{-14} $(cm^3(STP) \cdot cm/cm^2 \cdot$

s · cmHg) and this membrane functions as a gas barrier. By contrast, the silicone membrane shown in Table 2 is known as a high permeability material with the permeability of $6.1 \times 10^{-8} (\text{cm}^3 (\text{STP}) \cdot \text{cm/cm}^2 \cdot \text{s} \cdot \text{cmHg})$. The permeability of the oxygen through the silicone membrane under a dry state is as high as a million times that of the poly(vinyl alcohol) membrane [282]. When each membrane is compared under a wet condition, the difference was reduced by a factor of approximately 100. This is obviously due to the increased diffusion of oxygen through the water in the poly(vinyl alcohol) gel. The characteristics of gel membranes appear for the first time when they are wet. It can be imagined how important the function is of water in a swollen membrane with respect to the permeation of gases. Recently, the relationships between the permeability and the structure of water in a gel, such as free water, bound water, and restricted water, has been studied [283, 284].

2.12   OPTICAL PROPERTIES

TOYOAKI MATSUURA

2.12.1   Introduction

Most information obtained by living things comes through the eyes. However, the information obtained through visible light is limited. Nonetheless, optical properties for polymeric materials and gels are one of the important properties. Also, indices of refraction and transmissivity are important among those gels that form living tissue, especially the gels that form the cornea.

2.12.2   Usefulness of Index of Refraction Measurements

Refraction is the ratio of the speed of a wave, which can be either light, sound, or electric waves, from one material to another. Assuming that the speed of light in a material is v, dielectric constant ε and permittivity μ, the index of refraction is given by the following equation:

$$n = \frac{c}{v} \sqrt{\varepsilon\mu} \tag{1}$$

From this equation, the Lorentz–Lorenz formula that correlates the index of refraction and molecular structure is obtained:

$$\frac{n^2 - 1}{n^2 + 2} = \frac{4}{3}\,\pi N \alpha \tag{2}$$

where N is the number of molecules in a unit volume and α is the polarizability.

If the material (gel) is homogeneous, when light propagates from material A with the index of refraction n_1 into material B with index of refraction n_2, it changes the direction of propagation. If the angle of incidence and refraction angle are θ_1 and θ_2, respectively, then Snell's law holds:

$$n_1 \sin\theta_1 = n_2 \sin\theta_2 \tag{3}$$

The refractive index of a homogenous gel can be measured relatively easily by the Abbé refractometer described later based on Snell's law. Also, a spherical gel that does not possess a flat plane or a large fibrous gel can also be measured by immersing the gel in liquids with various refractive indices. In general, α changes according to the kind of molecule and its bonding state. Also, N is determined by the state of aggregation of molecules, in particular, by the concentration of the gel networks. Furthermore, as both dielectric constant and permittivity are a function of frequency, the refractive index depends on the wavelength. This is important when we attempt to understand the ingenious mechanism of the adjustment of refraction by the human eye. If the refractive index varies depending upon the position, N and α also vary. Thus, a distribution of index of refraction will be observed. For more in-depth information on the

Table 1 Measurement methods of refractive indices.

	Number of gel planes necessary for measurement	Measurement precision	Characteristics
Focal point method	2	$10^{-2} - 10^{-3}$	Simple but poor precision
Minimum angular deviation method	2	$10^{-3} - 10^{-4}$	Gel formation is necessary
Interference method	2	10^{-7}	Most precise
Critical angle method	1	$10^{-4} - 10^{-5}$	Easy measurement
Differential refractivity method	0	10^{-6}	Precise

minimum angular deviation method, see the *Optical Methods Handbook* [285].

2.12.3 Summary of Refractive Index Measurements

In this section, the refractive index is limited to the visible region. Birefringence (a famous example in the human body is the cornea) will be described in a later section. These measurement methods (see Table 1) will be described briefly and the Abbé refractometer, which is a useful device for studying gels, will be discussed in detail.

2.12.3.1 Focal point method

The thickness d of a gel film with both surfaces parallel to each other under a microscope with magnification approximately 200 times is measured.

The movement of the microscope d' is determined by focusing onto the top and bottom surface and taking the difference between the two positions. The refractive index of the gel is given by the following equation:

$$n = \frac{d}{d'} \qquad (4)$$

The precision in this case is not high but, when the gel is transparent, it is a convenient method (see Reference [285] for more information).

2.12.3.2 Minimum angular deviation method

A prism is made of a gel with the azimuthal angle a. Determine the difference b between the angle of incidence and refraction. The following equation will give n:

$$n = \frac{\sin(1/2(a+b))}{\sin(a/2)}, \quad a \leq 2 \; \sin^{-1}\left(\frac{1}{n}\right) \qquad (5)$$

The requirement of preparing a gel with an optically flat surface with the azimuthal angle a is a disadvantage of this method. However, advantages are that, in principle, there is no limitation in the measurement range of n, a wide range of measuring wavelengths can be used, and a weak light can also be used. Hence, this method is suitable for measuring the dispersion of n. Measurements can be made by placing the gel in a hollow quartz cell. Thus, the measurement temperature can be changed easily.

2.12.3.3   Interference method

In principle, this method [286] uses the differential refractometer and is the most precise. Two lights passing in different directions interfere with each other. The phase lag between them will determine the refractive index difference between these two points. This method can be used for measurement with micrometer-thick gel film.

Well-known commercially available equipment (which are all principally the same) include the Mach–Zehnder interference microscope (Mizoshiri Optical Industries), the differential interference microscope, and Interphako (Carl–Zeiss). However, for the ease of use for a gel on a thin film to bulk gels, Interphako is excellent. By connecting a constant temperature bath to the sample stage, measurements in the temperature range of − 10 to 200°C is possible. Figure 1 illustrates the principle behind the Interphako microscope. The light that passes through the sample is divided into two paths by a beamsplitter. The optical axis of one beam

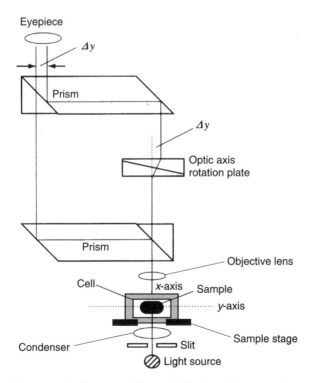

Fig. 1   Conceptual diagram of Interphako interference microscope.

deviates by Δy by a shearing device, and is recombined with the other beam to interfere. When there is a minute difference in both waves, these waves interfere and interference fringes are observed when the phase lag is odd multiples of half the wavelength. By observing the change of these interference fringes, the refractive index can be determined precisely. Moreover, the measurement of inhomogeneity within the sample is possible.

Figure 2 depicts the structure of the Mach–Zehnder type interference microscope. The parallel beam is divided into sample beam I and reference beam II, and recombined at M_1. When there is no sample in place and M_1, M_2, M_3, and M_4 are perfectly parallel, no interference patterns are observed. When the light paths I and II are slightly tilted and are allowed to cross near the sample, a parallel interference pattern can be observed.

When an interference pattern shown in Fig. 3 is obtained, the following equation holds:

$$\frac{R}{D}\,\lambda = (n - N)t \tag{6}$$

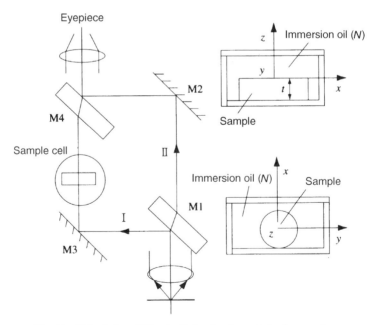

Fig. 2 Principle of Mach–Zehnder-type interferometer.

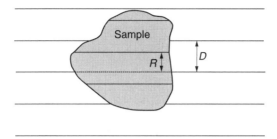

Fig. 3   The interference pattern that appears in a sample.

where N is the refractive index of the immersing liquid and γ is the wavelength of the measuring light.

2.12.3.4   Critical angle method

Among all the measurement methods described thus far, the most widely applicable method used for gels is the following [288] which uses Snell's law. It uses the Abbé refractometer (see Fig 4). This method has relatively high precision, is easy to measure, uses a small amount of sample, and does not require a monochrome source. At present, the wavelength can be changed by an additional light source attached as an accessory. The measurement temperature can be controlled by attaching a constant temperature bath to the sample stage from $-10°C$ to $140°C$. If a measurement below room temperature is desired, it is necessary to avoid moisture condensation on the sample that interferes with the measurement. Figure 5 illustrates an Atago refractometer and Fig. 6 shows index determination of a gel using the Abbé refractometer.

Next, the actual refractive index measurement of gels will be described. If the gel is relatively hard and in a bulk form, a flat surface of an approximately 10×20 mm sample will be placed in contact with the prism as shown in Fig. 6(a). The measuring beam is incident from the direction of the arrow. Also, as shown in Fig. 6(b), the totally reflected beam can be used (total reflection method).

Gel samples allow excellent contact with the prism surface. However, there are a few occasions when a contact aid fluid may be necessary. The refractive index of the contact aid fluid should be about 0.01–0.02 higher than the sample gel. In particular, when the gel surface is not smooth, the contact aid fluid is gradually changed to vary the excess

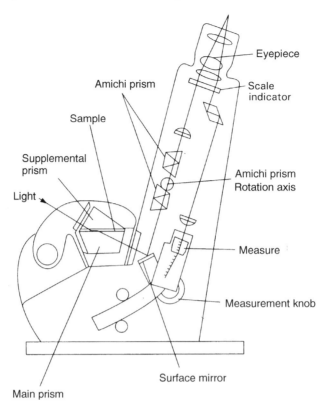

Fig. 4   Structure of Abbé refractometer.

refractive index from 0.005–0.01 to find the optimum refractive index of the fluid. This fluid must not dissolve, shrink, or swell the gel. For hydrophilic gels, methyl-α-bromonaphthalene salicylic acid or a solution with low polarity such as this will be suitable. For nonpolar polymers, potassium iodomercury is used. If a film-shaped gel is available, the measurement technique shown in Fig. 6(c) can be used by adjusting the size of the film to that of the prism. Because the temperature control is applied to two prisms, the temperature fluctuation can be minimized. Figure 6(d) is the image seen through the eyepiece.

2.12.4   Measurement of Refractive Index Distribution of Gels

The measurement of refractive index distribution [285–288] is done using mostly the method developed for measuring the internal refractive index

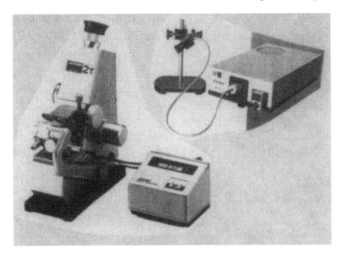

Temperature is controlled by circulating ethylene glycol using a constant temperature bath; the light source accessory can vary light wavelength using the grating spectrometric method in the visible wavelength range (450–700 nm).

Fig. 5 Atago refractometer (Model 2T).

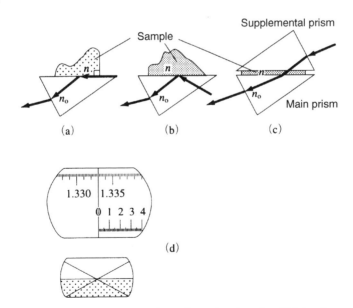

Fig. 6 Refractive index determination of a gel using the Abbé refractometer.

distribution of a GI-type optical fiber. These methods can be applied to
determine the refractive index distribution of gel samples. In practice, the
interference technique is superior to others in terms of precision and
usefulness. Readers are referred to the monographs for detailed information. In particular, these techniques are applied to the measurement of
refractive index distribution of the lens in the ophthalmologic field [289].
The lens is almost spherical and the analysis of the measurement results is
relatively simple. The refractive index distribution is said to be approximated by a second-order polynomial (see Fig. 7). Here, Z_0 is determined
by measuring the position where the incident light becomes almost a plane
wave by a Mach–Zehnder type interference microscope. Also, a is the
radius of the lens. It is more difficult to measure the refractive index
distribution of natural gels such as the lens. This is because, first of all, the
assumption of straight-line propagation of light will not hold, and then the
interference fringes become too narrow to measure easily. Another method
involves the following steps. First, the lens is cut into the direction of the
equator and the light axis, which is then further microtomed cryogenically.
The light beam is shone parallel to the sample surface using a Schlieren
device and the deviation of the light flux of each portion is measured.

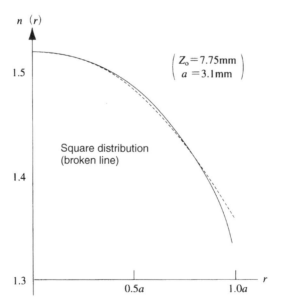

Fig. 7 Distribution of the refractive index of the lens of a rainbow trout.

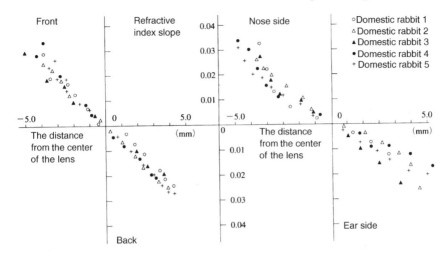

Fig. 8 Refractive index gradient of the lens of a domestic rabbit.

Similarly, a comparison is made with water as a reference and the change of the refractive index is determined [290, 291]. Figure 8 shows the refractive index distribution of the lens of a domestic rabbit. Although primitive, there is a report where the lens of a domestic rabbit is frozen and many thin sections are prepared by microtome. The refractive index of each specimen is measured and the data obtained are gathered [292] (see Fig. 9). Furthermore, in the case of monkeys, the refractive index distribution was shown to change during the focal adjustment (to focus precisely the image of a subject, the ability to change the refractive index is needed [293]). Also, there is another report where a gel showed a critical phenomenon when the gel approached the transition point, leading to the change in refractive index [294].

2.12.5   Measurement of Birefringence of Gels

Refractive index depends on polarizability as shown in Eq. (2). In particular, because gels that form the human body almost always show anisotropy with varying degrees of polarization and refractive index, birefringence is observed (for example, it has long been known that the cornea shows birefringence [295]). Even optically homogeneous gels can sometimes become optical anisotropic materials under tension, electric field, and magnetic field.

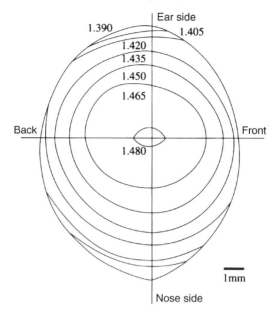

Fig. 9   Refractive index distribution of the lens of a domestic rabbit.

2.12.5.1   Abbé refractometer

Similar to the technique described before, birefringence can be readily measured by adding a polarization plate at the eyepiece of the Abbé refractometer and further using a monochromatic light such as a sodium lamp.

2.12.5.2   Measurement using Babinet's compensator

Assume that the refractive indices of a thin gel film measured by polarized light along the x and y axes are n_x and n_y. Then, the retardation δ can be given by the following equation:

$$\delta = \frac{2\pi t}{\lambda} \, (n_x - n_y) \tag{7}$$

This equation expresses the phase lag between x-axis polarization and y-axis polarization when the light passes through normal to the sample. The polarizer and analyzer are placed normal to each other. They are placed at $45°$ with respect to the x and y axes. Both components deviate by δ only when they pass through the sample. Figure 10 illustrates the principle of the measurement. Babinet's compensator is positioned between the sample

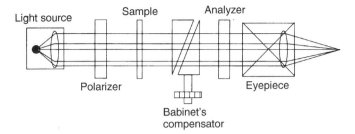

Fig. 10 Birefringence measurement device.

and the analyzer. When there is no sample, dark parallel lines will be seen by Babinet's compensator. Upon placement of an anisotropic sample, the parallel dark lines shift due to the deviation of the phase lags that are normal to each other. Compensating the shift by adjusting the micro-adjustment screw, δ will be determined and $(n_x - n_y)$ can be calculated. Figure 11 shows the changes in the cornea following swelling [295].

However, corneal retardation varies significantly depending on position and depth. This indicates that the cornea has a direct role in correcting visual aberrations.

2.12.6 Transparency of Gels

Transparency of gels can be divided qualitatively into highly transparent, translucent (semitransparent), and opaque. The light incident upon the

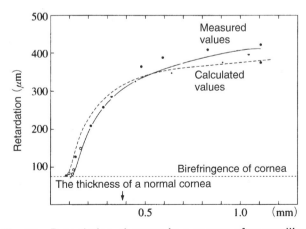

Fig. 11 Retardation changes in a cornea after swelling.

transparent material reflects partly at the surface. The other part is absorbed within the gel and the remainder will transmit. The transmitted light can be divided into parallel transmitted light that propagates in parallel and scattered transmitted light. The luminous transmittance is used as a measure of the brightness and haze (the ratio of scattered

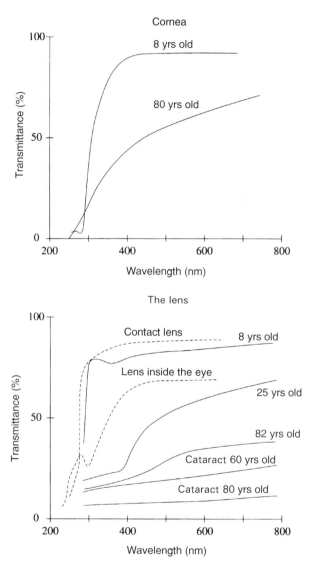

Fig. 12 Age-related changes in corneal transmittance.

transmitted light and parallel transmitted light in percentage) as the measure of clarity. Assume that the intensity of the transmitted light is I from the material with a thickness l (refractive index: nD) with the intensity of the incident light, I_0. The reflectivity R and transmittance T are expressed by the following equations:

$$R = \frac{(nD - 1)^2}{(nD + 1)^2} \tag{8}$$

$$T = \frac{I}{I_0} = \frac{(1 - R)^2 e^{-\alpha l}}{1 - R^2 e^{-2\alpha l}} \tag{9}$$

where α is the absorption coefficient (cm^{-1}) and $100\, I/I_0$ is called the light transmittance. Finally, Fig. 12 summarizes the age-dependent changes of transmittance of the cornea and lens [296]. As the aging process continues, the reduction of transmittance becomes quite apparent, in particular, the reduction in the short wavelength region is noteworthy. Also, in the strong opaqueness seen in cataracts the transmittance reduces over the entire frequency range. Hence, the transmittance distribution differs markedly after a lens is transplanted and this creates clinical problems.

REFERENCES

1   Tasumi, T. and Shimanouchi, T. (1965). *J. Chem. Phys.*, **43**: 1245.
2   Kobayashi, M., Akita, K., and Tadokoro, H. (1968). *Makromol. Chem.*, **11**: 324.
3   Kobayashi, M., Tsumura, K., and Tadokoro, H. (1968). *J. Polym. Sci. Polym. Phys. Ed.*, **6**: 1493.
4   Kobayashi, M. and Ueno, Y. *Macromolecules* (to be published).
5   Kobayashi, M., Nakaoki, T., and Ishihara, N. (1990). *Macromolecules*, **23**: 78.
6   Nakaoki, T. and Kobayashi, M. (1991). *J. Mol. Struct.*, **242**: 315.
7   Kobayashi, M. and Kozasa, T. (1993). *Appl. Spectrosc.*, **47**: 1417.
8   Kobayashi, M., Yoshioka, T., Imai, M., and Itoh, Y. (1995). *Macromolecules*, **28**: 7376; (1995). *Physica B*, **213 & 214**: 734.
9   Kobayashi, M. and Yoshioka, T. (1993) *Surface. Jpn.*, **31**: 826.
10 Nakaoki, T. and Tobayashi, M. (1991). *Rep. Progr. Polym. Phys., Jpn.*, **34**: 359.
11 Kato, T. and Takahashi, A. (1982). *Kobunshi Jikkengaku*, **11**: 364; Nemoto N. (1982). *Kobunshi Jikkengaku*, **11**: 446.
12 Pecora, R. (ed.) (1985). *Dynamic Light Scattering*, New York: Plenum, pp. 181, 347.
13 Richards, R.W. (ed.) (1995). *Scattering Method in Polymer Science*, New York: Ellis Horwood, pp. 6, 26.
14 Brown, W. (ed.) (1993). *Dynamic Light Scattering*, Oxford: Oxford Sci. Pub., p. 512.
15 Ramer, O. (ed.) (1986). *Biological and Synthetic Polymer Networks*, New York: Elsevier Appl. Sci.

16   Cohen Addad, J.P. (ed.) (1996). *Physical Properties of Polymer Gels*, New York:
17   DeRossi, D., Kajiwara, K., Osada, Y., and Yamauchi, A. (eds.) (1989). *Polymer Gels*, New York: Plenum.
18   Burchard, W. (1983). *Adv. Polym. Sci.*, **48**: 1.
19   Chu, B. (ed) (1990). *Quasielastic Light Scattering by Macromolecular, Supramolecular and Fluid Systems*, Bellingham: SPIE Opt. Eng. Press, pp. 321, 635.
20   Burchard, W. and Ross-Murphy, S.B. (eds.) (1990). *Physical Networks Polymers and Gels*, New York: Elsevier Applied Science.
21   Schmidt, P.W. (1991). *J. Appl. Cryst.*, **24**: 414.
22   Adam, M., Delsanti, M., Munch, J.P., and Durand, D. (1987). *J. Physique*, **48**: 1809.
23   de Gennes, P.D. (1979). *Scaling Concepts in Polymer Physics*, Ithaca, New York: Cornell University Press, pp. 62, 128.
24   Pezron, I., Djabourov, M., and Leblod, J. (1991). *Polymer*, **17**: 3201.
25   Burchard, W. and Ross-Murphy, S.B. (eds.) (1990). *Physical Networks Polymers and Gels*, New York: Elsevier Appl. Sci., p. 253.
26   Ramer, O. (ed.) (1986). *Biological and Synthetic Polymer Networks*, New York: Elsevier Appl. Sci., p. 383.
27   Burchard, W., Kajiwara, K., and Neger, D. (1982). *J. Polym. Sci. Polym. Phys. Ed.*, **20**: 157.
28   Asnagli, D., Giglio, M., Bossi, A., and Righetti, P.G. (1995) *J. Chem. Phys.*, **102**: 9736.
29   Bale, H.D. and Schmidt, P.W. (1984). *Phys. Rev. Lett.*, **53**: 596.
30   Korenaga, T., Oikawa, H., and Nakanishi, H. (1995). *Proc. 4th Symposium on Polymer Gels*, p. 81.
31   Kanaya, T., Ohkura, M., Takeshita, H., Kaji, K., Furusaka, M., Yamaoka, H., and Wignall, G.D. (1995). *Macromolecules*, **28**: 3168.
32   Munch, J., Candau, S., Herx, J., and Hild, G. (1977). *J. Physique*, **38**: 971.
33   Oikawa, H. and Murakami, K. (1991). *Macromolecules*, **24**: 1117.
34   Oikawa, H. (1992). *Polymer*, **33**: 1116.
35   Oikawa, H. and Nakanishi, H. (1993). *Polymer*, **34**: 3358.
36   Brown W. (ed.) (1993). *Dynamic Light Scattering*, Oxford: Oxford Sci. Pub., p. 299.
37   Koike, A., Nemoto, N., Inoue, T., and Osaki, K. (1995). *Macromolecules*, **28**: 2339.
38   Martin, J.E. and Wilcoxon, J.P. (1988). *Phys. Rev. Lett.*, **61**: 373.
39   Oikawa, H., Harasawa, M., and Ono, K. (1993). *J. Int. Mat. Syst. Struct.*, **4**: 146.
40   Fang, L. and Brown, W. (1992). *Macromolecules*, **25**: 6897.
41   Moussaid, A., Candau, S.J. and Joosten, J.G.H. (1994). *Macromolecules*, **27**: 2102.
42   Glatter, O. and Kratchy, O. (eds.) (1982). *Small-Angle X-ray Scattering*, London: Academic Press.
43   Kajiwara, K. and Hiragi, Y. (1996). In *Application of Synchrotron Radiation to Materials Analysis*, H. Saisho and Y. Goshi, eds., Amsterdam: Elsevier.
44   Debye, P. and Bueche, A.M. (1949). *J. Appl. Phys.*, **20**: 518.
45   Flory, P.J. (1956). *Principles of Polymer Chemistry*, Ithaca, New York: Cornell University Press.
46   Debye P. (1964). In *Light Scattering from Dilute Polymer Solutions*, D. McIntyre and F. Gornick, eds., London: Gordon & Breech.
47   Kajiwara, K., Kohjiya, S., Shibayama, M., and Urakawa, H. (1991). In *Polymer Gels: Fundamentals and Biomedical Applications*, D. DeRossi, D. Kajiwara, Y. Osada and A. Yamauchi, eds., New York: Plenum.
48   de Gennes, P.G. (1979). *Scaling Concepts in Polymer Physics*, Ithaca, New York: Cornell University Press.

49 Panyukov. S. and Rabin, Y. (1996). *Macromolecules*, **29**: 7960.

50 Guinier, A. and Fournet, G. (1955). *Small-Angle Scattering of X-rays*, New York: Wiley.

51 Horkay, F., Hecht, A.M., and Geissler, E. (1994). *Macromolecules*, **27**: 1795.

52 Shimode, M., Mimura, M., Urakawa, H., Yamanaka, S., and Kajiwara, K. (1996). *Sen'I Gakkaishi*, **52**: 301.

53 Yuguchi, Y., Urakawa, H., and Kajiwara, K. *Macromol. Sym.* (in press).

54 Chandrasekaran, R., Puigjaner, L.C., Joyce, K.L., and Arnott, S. (1988). *Carbohydr. Res.*, **181**: 23.

55 Suzuki, A., Yamazaki, M., and Kobiki, Y. (1996) *J. Chem. Phys.*, **104**: 1751.

56 Wilson, T. (ed.) (1990). *Confocal Microscopy*, London: Academic Press.

57 Corle, T.R. and Kino, G.S. (1996). *Confocal Scanning Optical Microscopy and Related Imaging Systems*, San Diego: Academic Press.

58 Pawley, J.B. (ed.) (1995). *Handbook of Biological Confocal Microscopy*, 2nd edition, New York: Plenum.

59 Jinnai, H., Nishikawa, Y., Koga, T., and Hashimoto, T. (1995). *Macromolecules*, **28**: 4782.

60 Rathjen, C.M., Park, C.-H., and Goodrich, P.R. (1995). *Polym. Gels Networks*, **3**: 101.

61 Hirokawa, Y., Jinnai, H., Nishikawa, Y., Kanazawa, Y., and Hashimoto, T. (1995). *Proc. 4th Symp. on Polym. Gels*, 67.

62 Miyazawa, S. and Aihara, K., eds. *Electron Microscopy and the Related Devices*, Chapter 3, Igaku Shuppan Center.

63 Tanaka, K. (1989). *Challenge to the Ultramicro World*, Iwanami Shuppan.

64 Shimakura, S. (1980). *Microbes*, **5**: 64.

65 Danilatos, G.D. (1988). *Adv. Electron. Electron Phys.*, **71**: 109.

66 Fujikawa, S. (1988). *Electron Microsc. Rev.*, **1**: 113.

67 Suzuki, M., Tateishi, T., Matsuzawa, M., and Saito, M. (1996). *Macromol. Symp.*, **109**: 55.

68 Binnig, G., Rohrer, H., Gerber, Ch., and Weibel, E. (1982). *Phys. Rev. Lett.*, **49**: 57.

69 Spong, J.K., Mizes, H.A., LaComb, L.J., Jr., and Dovek, M.M. (1989). *Nature*, **338**: 137.

70 Mizutani, W., Shigeno, M., Kajimura, K., and Ono, M. (1992). *Ultramicroscopy*, **42–44**: 236.

71 Mintmire, J.W., Harrison, J.A., Colton, R.J., and White, C.T. (1992). *J. Vac. Sci. Technol.*, **A10**: 603.

72 Sakurai, T., Hashizume, T., and Sakai, A. (1992). *Frontier of Physics*, vol. 29, Japan: Kyoritsu Publ.

73 Chen, C.J. (1993). *Introduction to Scanning Tunneling Microscopy*, New York: Academic Press.

74 Nishi, T. and Tanaka, H. (1993). In *Formation and Properties of Polymer Aggregates*, Society of Polymer Science, ed., Kyoritsu Publ., pp. 376–384.

75 Mikoshiba, N., Morita, S., Ono, M., and Kajimura, K. (1993). *Scanning Tunneling Microscopy*, Society of Electronic Information and Communication, ed., Corona Publ.

76 Hayashi, T., Yamamura, H., Nish, T., and Kakimoto, M. (1992). *Polymer*, **33**: 3751.

77 Clemmer, C.R. and Beebe, T.P., Jr. (1991). *Science*, **251**: 640.

78 Nakajima, K., Ikehara, T., and Nishi, T. (1996). *Carbohydr. Polym.*, **30**: 77.

79 Chandrasekaran, R., Puigjaner, L.C., Loyce, K.L., and Arnott, S. (1988). *Carbohydr. Res.*, **181**: 23.

80 Binnig, G., Quate, C.F., and Gerber, Ch. (1986). *Phys. Rev. Lett.*, **56**: 930.
81 Ueyama, H., Ohta, M., Sugawara, Y., and Morita, S. (1995). *J. Appl. Phys. Jpn.*, **B34**: L1086.
82 Nakajima, K., Kageshima, M., Ara, N., Yoshimura, M., and Kawazu, A. (1993). *Appl. Phys. Lett.*, **62**: 1892.
83 Kobiki, H., Yamazaki, M., and Suzuki, J. (1995). *Polymer Preprints, Jpn.*, **44**: 2822.
84 Nakajima, K., Yamaguchi, H., Lee, J.-C., Kageshima, M., Ikehara, T., and Nishi, T. *J. Appl. Phys., Jpn.* (in press).
85 Landau, L. and Lipshitz, E. (1992). *Theory of Elasticity*, Tokyo: Tokyo Tosho, pp. 37–44.
86 Rohrer, H. (1992). *Ultramicroscopy*, **42**: 1.
87 (1995). *New Experimental Methods in Polymers-Structure of Polymers: Nuclear Magnetic Resonance*, Soc. Polymer Science, Jpn., Ed.,Kyoritsu Publ.
88 Ando, K. (ed.) (1994). *Solid State NMR of Polymers*, Kodansha Scientific.
89 Chujo, R. (ed.) (1992). *NMR of Polymers and Biomolecules*, Tokyo: Tokyo Kagaku Dojin.
90 Bloembergen, N., Purcell, E.M., and Pound, R.V. (1948). *Phys. Rev.*, **73**: 679.
91 Ohta, H., Ando, I., Fujishige, S., and Kubota, K. (1991). *J. Polym. Sci.*, **B29**: 963.
92 Tanaka, N., Matsukawa, S., Kurosu, H., and Ando, I. (to be submitted).
93 von Meerwall, E.D. (1983). *Adv. Polym. Sci.*, **54**: 1.
94 Stilbs, P. (1987). *Prog. Nucl. Magn. Reson. Spectrosc.*, **19**: 1.
95 Kargel, J., Pfeifer, H. and Heink, W. (1988). *Adv. Magn. Res.*, **12**: 1.
96 Abragam, A. (1964). *Magnetization of Nuclei*, Kyoto: Yoshioka Shoten, pp. 67–70.
97 Hahn, E.L. (1950). *Phys. Rev.*, **80**: 580.
98 Farrar, T.C. and Becker, E.D. (1971). *Pulse and Fourier Transform NMR: Introduction to Theory and Methods*, Kyoto: Yoshioka Shoten, p. 32. (Translated from book of the same title published by Academic Press.)
99 Stejskal, E.O. and Tanner, J.E. (1965). *J. Chem. Phys.*, **42**: 288.
100 Yasunaga, H. and Ando, I. (1993). *Polym. Gels Networks*, **1**: 83.
101 Matsuoka, S. and Ando, I. (1996). *Macromolecules*, **29**: 7136.
102 Yasunaga, H., Kobayashi, M., Matsukawa, S., Kurosu, H., and Ando, I. (1997). In *Annual Reports on NMR Spectroscopy*, vol. 34, London: Academic Press, pp. 39–104.
103 Fujito, T., Deguchi, K., Ohuchi, M., Imanari, M., and Albright, M.J. (1981). *Proc. 20th Meeting on NMR*, Tokyo, 68; Yasunaga, H. and Ando, I. (1993). *J. Mol. Struct.*, **301**: 129.
104 Kobayashi, M., Ando, I., Ishii, T., and Amiya, S. (1995). *Macromolecules*, **28**: 6677.
105 Terao, T., Maeda, S., and Saika, A. (1983). *Macromolecules*, **16**: 1535.
106 Kobayashi, M. and Ando, I. *J. Mol. Struct.* (in press).
107 Torchia, D.A. (1978). *J. Magn. Reson.*, **30**: 613.
108 Kamei, Y. (1987). *Techniques in Magnetic Resonance*, Tokyo: Kogyu Chosakai.
109 (1995). *New Experimental Methods in Polymers 5, Polymer Structures: (1) Nuclear Resonance Spectroscopy*, Tokyo: Kyoritsu Publ., pp. 331–382.
110 Hahn, E.L. (1950). *Phys. Rev.*, **80**: 580.
111 Kumer, A., Welti, D., and Ernst, R.R. (1975). *J. Magn. Reson.*, **18**: 69.
112 Edelstein, W.A., Hutchison, J.M.S., Johnson, G., and Redpath, T. (1980). *Phys. Med. Biol.*, **25**: 751.
113 Yasunaga, H. and Ando, I. (1993). *J. Mol. Struct.*, **301**: 129.
114 Cory, D.C. (1992). *Ann. Rep. NMR Spectrosc.*, **24**: 87.
115 Yasunaga, H. and Ando, I. (1993). *Polymer Gels Networks*, **1**: 267.

116 Yasunaga, H., Kurosu, H., and Ando, I. (1992). *Macromolecules*, **25**: 6505.
117 Carr, H.Y. and Purcell, E.M. (1954). *Phys. Rev.*, **94**: 630.
118 Meiboom, S. and Gill, D. (1958). *Rev. Sci. Instrum.*, **29**: 688.
119 Tanaka, T., Nishi, I., Sun, S.-T., and Ueno-Nishio, S. (1982). *Science*, **218**: 467.
120 Shibuya, T., Yasunaga, H., Kurosuy, H., and Ando, I. (1995). *Macromolecules*, **28**: 4377.
121 Kurosu, H., Shibuya, T., Yasunaga, H., and Ando, I. (1996). *Polym. J.*, **28**: 80.
122 Clark, A.H. and Ross-Murphy, S.B. (1987). *Adv. Polym. Sci.*, **83**: 57.
123 Saito, H., Ohki, T., and Sasaki, T. (1977). *Biochemistry*, **16**: 908.
124 Saito, H., Miyata, E., and Sasaki, T. (1978). *Macromolecules*, **11**: 1244.
125 Yokota, K., Abe, A., Hosaka, S., Sakai, I., and Saito, H. (1978). *Macromolecules*, **11**: 95.
126 Saito, H. (1979). *Realm of Chem., Jpn.*, **33**: 609.
127 Saito, H. (1979). *ACS Symp. Ser. 150*, Washington, DC: American Chemical Society, pp. 609–618.
128 Saito, H. (1992). *ACS Symp. Ser. 489*, Washington, DC: American Chemical Society, pp. 296–310.
129 Saito, H. (1995). *Ann. Rev. NMR Spectrosc.*, **31**: 157.
130 Saito, H., Shimizu, H., Sakagami, T., Tuzi, S., and Naito, A. (1995). In *Magnetic Resonance in Food Science*, P.S. Belton, I. Delgadillo, A.M. Gill, and G.A. Webb, eds., London Royal Soc. Chem., pp. 257–271.
131 Saito, H. (1992). *J. Org. Synthesis Chem., Jpn.*, **50**: 488.
132 Saito, H. (1986). *Magn. Reson. Chem.*, **24**: 835.
133 Saito, H. and Ando, I. (1989). *Ann. Rep. NMR Spectrosc.*, **21**: 209.
134 Saito, H., Yokoi, M., and Yoshioka, Y. (1989). *Macromolecules*, **22**: 3892.
135 Saito, H., Yoshioka, Y., Yokoi, M., and Yamada, J. (1990). *Biopolymers*, **29**: 1689.
136 Saito, H., Ohki, T., and Sasaki, T. (1979). *Carbohydr. Res.*, **74**: 227.
137 Saito, H., Yoshioka, Y., Uehara, N., Aketagawa, J., Tanaka, S., and Shibata, Y. (1991). *Carbohydr. Res.*, **217**: 181.
138 Yoshioka, Y., Uehara, N., and Saito, H. (1992). *Chem. Pharm. Bull.*, **40**: 1221.
139 Foord, S.A. and Atkins, E.D.T. (1989). *Biopolymers*, **28**: 1345.
140 Saito, H., Yokoi, M., and Yamada, J. (1990). *Carbohydr. Res.*, **199**: 1.
141 Saito, H., Yamada, J., Yoshioka, Y., Shibata, Y., and Erata, T. (1991). *Biopolymers*, **31**: 933.
142 Saito, H., Yamada, J., Yukumoto, T., Yajima, H., and Endo, R. (1991). *Bull. Chem. Soc., Jpn.*, **64**: 3528.
143 Saito, H., Tabeta, R., Shoji, A., Ozaki, T., Ando, I., and Miyata, T. (1984). *Biopolymers*, **23**: 2279.
144 Saito, H. and Yokoi, M. (1992). *J. Biochem.*, **111**: 376.
145 Naito, A., Tuzi, S., and Saito, H. (1994). *Eur. J. Biochem.*, **224**: 729.
146 Sipe, J.D. (1992). *Ann. Rev. Biochem.*, **61**: 947.
147 Gil, A.M., Masui, K., Naito, A., Tatham, A.S., Belton, P.S., and Saito, H. (1997). *Biopolymers*, **19**, 289.
148 Lansbury, P.T., Jr., Costa, P.R., Griffiths, J.M., Simon, E.J., Auger, M., Halverson, K.J., Cocisko, D.A., Hendsch, Z.S., Ashburn, T., Spencer, R.G.S., Tidor, B., and Griffin, R.G. (1995). *Nature Struct. Biol.*, **2**: 990.
149 Naito, A. (1996). *Kobunshi*, **45**: 340.
150 Shimomura, T. (1995). *Preprints of the Workshop on Super Absorbing Polymers*, pp. 42–50.

151 (1992). *Report on the Standardization of Reliability Evaluation Method for New Organic and Composite Materials*, Polymer Materials Center, pp. 129–136.

152 (1992). *Report on the Standardization of Reliability Evaluation Method for New Organic and Composite Materials*, Polymer Materials Center, pp. 180–211.

153 (1992). *Report on the Standardization of Reliability Evaluation Method for New Organic and Composite Materials*, Polymer Materials Center, pp. 197–211.

154 Masuda, F. (1987). *Superabsorbent Polymer*, Kyoritsu Publ., pp. 51–80.

155 (1990). *Chemistry Review No.8: Organic Polymer Gels*, Soc. Chemistry, Jpn., ed. Gakkai Publ. Center.

156 Yamauchi, A. and Hirokawa, N. (1990). *Functional Gel*, Kyoritsu Publ.

157 Irie, M., Nagata, Y., and Katayama, S. (1989). *Mechanochemistry*, Maruzen Publ.

158 Bard, A.J. and Faulkner, L.R. (1980). *Electrochemical Methods*, New York: Wiley.

159 Watanabe, M., Tedenuma, Y., Ban, M., Sanui, K., and Ogata, N. (1993). *J. Intelligent Mat. Systems Struct.*, **4**: 216.

160 Watanabe, M. (1995). *Proc. 11th Polymer Gel Study Group, Soc. Polym. Sci., Jpn.* pp. 1–4.

161 Watanabe, M. (1995). *Hyomen Gijitsu*, **46**: 324.

162 Aoki, K. (1993). *Denki Kagaku*, **61**: 618.

163 Nagaoka, T. (1993). *Denki Kagaku*, **61**: 621.

164 Nishihara, H. (1993). *Denki Kagaku*, **61**: 624.

165 Nakatani, S., Uchida, T., and Kitamura, N. (1993). *Denki Kagaku*, **61**: 628.

166 Wightman, R.M. and Wipf, D.O. (1989). *Electroanalytical Chemistry*, Vol. 15, A.J. Bard, ed., Paris: Marcel Dekker, pp. 267–353.

167 Fleishmann, M., Pons, S., Rolison, D.R., and Schmidt, P.P. (eds.) (1987). *Ultramicroelectrodes*, Morganton, North Carolina.

168 Geng, L. and Murray, R.W. (1986). *Inorg. Chem.*, **25**: 311.

169 McDevitt, J.T., Ching, S., Sullivan, M., and Murray, R.W. (1989). *J. Am. Chem. Soc.*, **111**: 4528.

170 Watanabe, M. and Ogata, N. (1991). *Kino Zairyo*, **11**: 24.

171 Watanabe, M. (1993). *Kobunshi*, **42**: 702.

172 Watanabe, M. (1995). *Kagaku Kogyo*, **46**: 369.

173 Shoup, D. and Szabo, A. (1982). *J. Electroanal. Chem.*, **140**: 237.

174 Aoki, K., Akimoto, K., Tokuda, K., Matsuda, H., and Osteryoung, J. (1984). *J. Electroanal. Chem.*, **171**: 219.

175 Nakayama, D., Mukaizawa, T., Sasaki, K., and Watanabe, M., (1994). *Polym. Prepr. Jpn.*, **43**: 2607.

176 Hermans, P.H. (1949). In *Gels in Colloid Science*, vol. II, H.R. Kruyt, ed., New York: Elsevier Publishing, pp. 483–651.

177 Philippoff, W. (1965). In *Relaxation in Polymer Solutions and Gels in Physical Acounstics*, Vol. II, Part B, W.P. Mason, ed., New York: Academic Press.

178 Djabourov, M., Leblond J., and Papan, P. (1988). *J. Phys. (France)*, **49**: 333.

179 Tschoegel, W. (1989). *The Phenomenological Theory of Linear Viscoelastic Behavior*, Berlin: Springer, pp. 314–364.

180 Nakamura, K. (1996). *Foods and Food Ingredients J., Jpn.*, **167**: 4.

181 Ferry, J.D. (1970). *Viscoelastic Properties of Polymers*, New York: Wiley, pp. 292–327.

182 Nakamura, K. and Nakagawa, T. (1975). *J. Polym. Sci. Polym. Phys. Ed.*, **13**: 2299.

183 Nakamura, K. (1975). *J. Poly. Sci. Polym. Phys. Ed.*, **13**: 137.

184 Nakamura, K., Ishii, T., and Kamidaira, T. (1988). *J. Biorheology Soc., Jpn.*, **2**: 207.

185 Winter, H.H. and Chambon, F. (1986). *J. Rheology*, **30**: 367.

186 Jonscher, A.K. (1975). *Colloid Polym. Sci.*, **253**: 231.
187 Tokita, M. and Hikichi, K. (1987). *Phys. Rev. A*, **35**: 4329.
188 Gauthier-Manuel, B., Guyon, E., Roux, S. Gits, S., and Lefausheux, F. (1987). *J. Phys.*, **48**: 869.
189 Hodgson, D.F. and Amis, E.J. (1990). *Macromolecules*, **23**: 2512.
190 Nakamura, K. (1991). In *Interactions between Solvent Molecules and Networks in Biogels in Polymer Gels*, D. DeRossi, K. Kajiwara, Y. Osada, and A. Yamauchi, eds., New York: Plenum Press, pp. 57–75.
191 Richardson, R.K. and Ross-Murphy, S.B. (1987). *Int. J. Biol. Macromol.*, **9**: 257.
192 Tokita, M., Fujiya, Y., and Hikichi, K. (1984). *Biorheology*, **21**: 751.
193 Clark, A.H., and Ross-Murphy, S.B. (1987). *Adv. Polym. Sci.*, **83**: 57.
194 Shimizu, N. (1989). *Vibrational Analysis by Personal Computer*, Kyoritsu Publ. pp. 126–139.
195 Yamaguchi, M. (1990). *Gosei Jushi*, 12.
196 Seto, K. and Yamada, K. (1982). *Defense University Science and Engineering Research Report*, 202, pp. 174–186.
197 (1968). *Chemistry and Technology of Silicones, 258*, Florida: Academic Press.
198 Ward, I.M. (1990). *Mechanical Properties of Solid Polymers*, 2nd ed., New York: Wiley.
199 Soc. Polym. Sci., Jpn., ed. (n.d.) *Rheology Handbook*, Maruzen Publ.
200 Yamazaki, H., *et al.* (1992). *Kikoron*, **920**: C, 240.
201 Hagino, I., Nagata, Y., Fushimi, T., and Yamauchi, A. (1991). *Gels*, Sangyo Tosho, p. 90.
202 Shimizu, N. (1995). Lecture Note.
203 Oldham, K.B. and Spanier, J. (1974). *The Fractal Calculus*, New York: Academic Press, pp. 45–60.
204 Miller, K.S. and Ross, B. (1993). *An Introduction to the Fractional Calculus and Fractional Differential Equations*, New York: Wiley.
205 Shimizu, N. and Yamazaki, H. (1993). *Kiron*, **59**: C, 3717.
206 Sekine, T., Shimizu, N., Yamazaki, H., Zako, M., and Matsumoto, K. (1994). *Kiron*, **60**: C, 3497.
207 Yamashita, S., Sedo, K., and Shimizu, N. (1991). *Kiron*, **57**: C, 91.
208 Sedo, K. *et al.* (1993). *Kikoron*, **930**: 392.
209 Shimizu, N., Yamazaki, H., Honda, H., Sekine, H., and Zako, M. (1993). *Kikoron*, **930**: 7.
210 Sekine, H. and Shimizu, N. (1993). *Kikoron*, **930**: 368.
211 Matsumoto, K., Kanda, T., and Zako, M. (1995). *Kikoron*, Dynamics & Design Conference, Aug. 23, 1995.
212 Soc. Thermal Analysis (1994). *Fundamental and Applications of Thermal Analysis*, Soc. Thermal Analysis, Jpn., Realize-sha.
213 Kanbe, H. and Ozawa, F. (1992). *Thermal Analysis*, New edition, Kodansha Scientific.
214 Kodama, M. (1995). *30th Thermal Analysis Lecture Series: Fundamentals and Applications of Thermal Analysis Textbook for Beginners*, Soc. Thermal Analysis, Jpn., p. 10.
215 Watase, M. and Kishinari, K. (1987). *Makromol. Chem.*, **188**: 1177.
216 Yoshida, H. and Takahashi, M. (1993). *Food Hydrocolloids*, 7: 387.
217 Nishinari, K. (1993). *Sen-I Gakkaishi*, **49**: 84.
218 Eldridge, J.E. and Ferry, J.D. (1954). *J. Phys. Chem.*, **58**: 992.
219 Nishinari, K. (1991). *Hyomen*, **29**: 546.

220 Williams, P.A., Clegg, S.M., Day, D.H., Phillips, G.O., and Nishinari, K. (1991). In *Food Polymers Gels and Colloids*, RSC Publication, p. 39.

221 Quinn, F.X., Hatakeyama, T., Takahashi, M., and Hatakeyama, H. (1994). *Polymer*, **35**: 1248.

222 Tanaka, F. (1996). *Hyomen*, **31**: 483.

223 Kaji, K. (1989). *Microcrystals in Gels, Polymer Gel Annual Review*, Soc. Polym. Sci. Polym. Gel Study Group, Jpn., p. 7.

224 Kobayashi, M. and Yoshioka, T. (1993). *Hyomen*, **31**: 826.

225 Yoshida, H., Hatakeyama, T., and Hatakeyama, H. (1992). *ACS Sym. Ser.* **489**: 217.

226 Yoshida, H. (1995). *Kobunshi Kako*, **44**: 345.

227 Nakamura, K. and Hatakeyama, R. (1985). *Sen-I Gakkaishi*, **41**: 369.

228 Yoshida, H. (1995). *Hyomen*, **33**: 723.

229 Yoshida, H., Hatakeyama, T., and Hatakeyama, H. (1993). *J. Intelligent Matr. Systems Struc.*, **4**: 543.

230 Aoki, H., Takahashi, H., Hatta, I., and Kodama, M. (1996). *Proc. 32th Thermal Analysis Symposium*, p. 60.

231 Todoki, M. (1996). *Proc. 11th ICTAC Symposium*, Philadelphia, p. 249.

232 Oka, S. and Nakata, O. (1960). *Dielectricity of Solids*, Iwanami Publ.

233 Soc. of Electric Science, Jpn., (ed.) (1973). *Phenomenology of Dielectricity.*

234 Chelekowski, A. (1980). *Dielectric Physics*, Amsterdam: Elsevier Science.

235 Furukawa, T. (1989). *Applied Material Characterization*, Chapter 4, Soc. of Appl. Phys., ed., Ohm Publ.

236 de Gennes, P.G. (1984). *Physics of Polymers*, Chapter 5, Takano, H., and Nakanishi, H. Transl., Yoshioka Shoten.

237 Flory, P.J. (1975). *Polymer Chemistry*, Chapter 9, S. Oka and Kanamaru, Transl., Maruzen Publ.

238 Yamauchi, A. and Hirokawa, N. (1990). *Functional Gels*, Society of Polymer Science, Kyoritsu. Publ.

239 (1990). *Organic Polymer Gels*, Soc. of Chemistry, Jpn., Gakkai Publ Center.

240 Yamauchi, A., Hatakeyama, R., Nishinari, K., Kato, T., Watase, M. and Katayama, S. (1993). *Formation and Properties of Polymer Aggregates*, Chapter 2, Soc. of Polymer Science, Jpn., ed., Kyoritsu Publ.

241 Tanaka, F. (1994). *Physic of Polymers*, Chapter 4, Shokabo.

242 Stoufer, (1988). *Fundamentals of Permeability*, T. Oda, Transl., Yoshioka Shoten.

243 Fedder, J. (1991). *Fractal*, T. Matsushita, Y. Hayakawa and S. Sato, Transl., Keigaku Publ.

244 Soc. of Electric Science, Jpn., (ed.) (1952). *Electric Circuitry.*

245 Kinoshita, S. (1986). *Tutorial on Electric Circuitry*, Kyoritsu Publ.

246 Dobrowski, S., Davies, G., McIntyre, J.E. and Ward, I. (1991). *Polymer*, **32**: 2887.

247 Sheppard, F., Tucker, C. and Samira, S. (1992). *Polym. Mater. Sci. Eng.*, **66**: 226.

248 Colomban, P. and Novak, A. (1992). *Chem. Solid State Mater.*, **2**: 272.

249 Ilavsky, M., Liptak, J. and Nedbal, J. (1993). *Polym. Networks Blends*, **3**: 21.

250 Aliotta, F., Fontanella, M., Galli, G., Lanza, M., Magliardo, P. and Salvato, G. (1993). *J. Phys. Chem.*, **97**: 733.

251 Yoshino, K. (1993). *Nobel Symp.*, **81ST**: 121.

252 Voice, A., Southall, J., Rogers, V., Matthews, K., Davies, G., McIntyre, J.E., and Ward, I. (1994). *Polymer*, **35**: 3363.

253 Liptak, J., Ilavsky, M., and Nedbal, J. (1995). *Polym. Networks Blends*, **5**: 55.

254 Liao, Y. and Levon, K. (1995). *Polym. Adv. Technol.*, **6**: 47.

255   Chen, P., Adachi, K., and Kodaka, T. (1993). *Polym. J.*, **25**: 473.

256   Takahashi, M. (1987). *Kobunshi*, **36**: 786.

257   Ress, D.A. (1981). *Pure & Appl. Chem.*, **53**: 1; Morris, E., Rees, D.A., and Robinson, G. (1980). *J. Mol. Biol.*, **138**: 349.

258   Paoletti, S., Smidsrod, O., and Grasdalen, H. (1984). *Biopolymers*, **23**: 1171.

259   Furukawa, H., Kimura, Y., Ito, K., and Hayakawa, R. (1993). *R. P. P. J.*, **36**: 55.

260   Furukawa, H., Kimura, Y., Ito, K., and Hayakawa, R. (1995). *Proc. 7th Polymer Gel Symp.*, p. 17.

261   Ito, K., and Hayakawa, R. (1993). *Fundamentals of Polymer Properties*, Chapter 1, Soc. Polym. Sci., Jpn., ed., Kyoritsu Publ.

262   Hashimoto, Y., Inamori, I., Chiba, A., and Tajitsu, Y. (1995). *Proc. 7th Polymer Gel Symp.*, p. 23.

263   Tajitsu, Y., *J. Mat. Sci.* (in press).

264   Aliotta, F., Fontanella, M.E., Galli, G., Lanza, M., Migliardo, P., and Salvato, G. (1993). *J. Phys. Chem.*, **97**: 773.

265   Ilavsky, M., Liptak, J., and Nedbal, J. (1993). *Polym. Networks Blends*, **3**: 21.

266   Manabe, R. (1986). *Practical Contact Lenses*, Igaku Shoin.

267   Wichterle, O., Bartl, P., and Rosenberg, M. (1960). *Nature*, **185**: 498.

268   Effier, D.B., Kolff, W.J., Groves, L.K., and Sones, F.M. (1956). *J. Thoracic Surg*, **32**: 620.

269   Ito, H., *et al.* (1997). *Jinko Zoki*, **26**: 519.

270   Himmelblau, D.M. (1964). *Chem. Rev.*, **64**: 542.

271   Nakagawa, T. (1981). *Evaluation Methods for Artificial Membranes*, Japan Artificial Membrane Soc., ed., Kitami Shobo, p. 59.

272   Yasuda, H., and Stone, W. Jr. (1966). *J. Polym. Sci., A-1*, **4**: 1314.

273   Aiba, S., Ohashi, M., and Huang, S.Y. (1968). *Ind Eng. Chem. Fundam.*, **7**: 497.

274   Hwang, S.T., Tang, T.E., and Kammermyer, K. (1971). *J. Macromol. Sci., Phys.*, **B5**: 1.

275   Minoura, T., and Nakagawa, T. (1976). *Nippon Kagaku Kaishi*, **8**: 1271.

276   Nakagawa, T., and Miroura, T. (1972). *Proc. 5th Packaging Symp.*, Tokyo.

277   Katoh, S., and Yoshida, F. (1971). *Chem. Eng. J.*, **3**: 276.

278   Smith, K.A., Colton, C.K., Merrill, E.W., and Evans, L.B. (1968). *Chem. Eng. Progr. Symp. Series*, **84**: 45.

279   Yoshida, F., and Sakai K. (1993). *Chemical Engineering and Artificial Organs*, Kyoritsu Publ.

280   Wilson, E.E. (1915). *Trans. Am. Soc. Mech. Engrs.*, **37**: 47.

281   Ito, H. (1997). PhD. Dissertation, Tokyo Science University, Tokyo.

282   Nakagawa, T. (1987). *Separation Membranes-From Fundamentals to Applications*, Sangyo Tosho.

283   Wisniewski, S., and Kim, S.W. (1980). *J. Membrane Sci.*, **6**: 299.

284   Peppas, N.A., and Reinhart, C.T. (1983). *J. Membrane Sci.*, **15**: 275.

285   Kubota, H., Ukita, Y., and Aida, G., (eds.) (1967). *Optical Methods Handbook*, Asakura Shoten, p. 270.

286   Soc. Polym. Sci., Jpn., (ed.) (1990). *Introductory Polymer Characterization*, Kyoritsu Publ.

287   Soc. Polym. Sci., Jpn., (ed.) (1985). *Testing and Evaluation Methods of Polymer Materials*, Baifukan Publ., pp. 263–295.

288   Soc. Polym. Sci. Experimental Polym. Sci. Study Group (ed.) (1984). *Thermodynamic, Electrical and Optical Properties*, Kyoritsu Publ., pp. 346–369.

289   Iwata, K., and Nagata, R. (1975). *Ringan*, **29**: 837.

290   Nakajima, A., Hirano, A., and Saito, K. (1960). *Ringan*, **14**: 1666.
291   Nakao, S., and Fujimoto, S. (1968). *J. Opt. Soc. Am.*, **58**: 1125.
292   Nakao, S., Nagata, R., and Sakai, M. (1966). *J. Appl. Phys., Jpn.*, **5**: 447.
293   Nakao, S., Fujimoto, T., Higuchi, S.M., Ito, Y., Iwata, K., and Nagata, R. (1968). *Ringan*, **22**: 1251.
294   Matsuura, T. (1994). *Shikaku no Kagaku*, **15**: 2.
295   Furukawa, Y. (1968). *Nippon Ganka Kiyo*, **19**: 1212.
296   Nishinobu, M. (1990). *Fundamentals of Opthalmic Optics*, Kimbara Publ. pp. 139–140.

INDEX

413